AF595074

Integration of Renewable and Distributed Energy Resources in Power Systems

Integration of Renewable and Distributed Energy Resources in Power Systems

Editors

Tomás Gómez San Román
José Pablo Chaves-Ávila

MDPI • Basel • Beijing • Wuhan • Barcelona • Belgrade • Manchester • Tokyo • Cluj • Tianjin

Editors

Tomás Gómez San Román
Comillas Pontifical University
Spain

José Pablo Chaves-Ávila
Comillas Pontifical University
Spain

Editorial Office
MDPI
St. Alban-Anlage 66
4052 Basel, Switzerland

This is a reprint of articles from the Special Issue published online in the open access journal *Energies* (ISSN 1996-1073) (available at: https://www.mdpi.com/journal/energies/special_issues/integration_energy_resources).

For citation purposes, cite each article independently as indicated on the article page online and as indicated below:

LastName, A.A.; LastName, B.B.; LastName, C.C. Article Title. *Journal Name* **Year**, *Article Number*, Page Range.

ISBN 978-3-03943-487-9 (Hbk)
ISBN 978-3-03943-488-6 (PDF)

Contents

About the Editors

Tomás Gómez San Román is Professor of Electrical Engineering at the Engineering School of Universidad Pontificia Comillas in Madrid, Spain. Currently, he is the Director of Instituto de Investigación Tecnológica (IIT) at Comillas. He obtained the Degree of Doctor Ingeniero Industrial from Universidad Politécnica, Madrid, in 1989, and the Degree of Ingeniero Industrial in Electrical Engineering from Comillas in 1982. He joined IIT in 1984. From 1994 to 2000, he was also Director of IIT, and from 2000 to 2002, the Vice-Rector of Research, Development, and Innovation of Comillas. Prof. Gómez has a wealth of experience in industry joint research projects in the field of Electric Energy Systems in collaboration with Spanish, Latin American, and European institutions. He has served as project manager and/or principal investigator in more than 80 research projects. His areas of interest are in operation and planning of transmission and distribution systems, power quality assessment and regulation, and economic and regulatory issues in the electrical power sector. He has published more than 100 articles in different specialized journals, such as *IEEE PES Transactions*, in addition to conference proceedings and has co-authored the book "Electricity Economics: Regulation and Deregulation" in Wiley-IEEE Press. He is a Senior Member of IEEE. He is President of the Council of the Power System Computation Conference and serves or has served on the technical committees of other conferences such as Probabilistic Methods Applied to Power Systems and IEEE Power Tech. He has been Visiting Researcher at the Energy Analysis Department of the Lawrence Berkeley National Laboratory in California. From May 2011 to October 2013, he served as commissioner at the Spanish Energy Commission (CNE).

José Pablo Chaves-Ávila is Research Professor at the Institute for Research in Technology (IIT) at the Engineering School (ICAI) of the Comillas Pontifical University. Since October 2016, he has been leading and operating agent of the Academy of the International Smart Grids Action Network. From August 2020, he was appointed as member of the Expert Advisory List of the Comisión Regional de Interconexión Eléctrica (CRIE). In 2008, he obtained his bachelor's in Economics from University of Costa Rica (Costa Rica). As part of the Erasmus Mundus Master in Economics and Management of Network Industries, he holds a master's in the Electric Power Industry from ICAI and master's in Digital Economics and Network Industries from Paris Sud-11 University (Paris, France). In 2014, he obtained the Erasmus Mundus Joint Doctorate on Sustainable Energy Technologies and Strategies Degree at Delft University of Technology (The Netherlands), a joint program with Comillas Pontifical University and the Royal Institute of Technology (KTH), Sweden. His areas of interests are energy economics, integration of renewable resources and distributed energy resources in the electricity sector, smart grids, and regulation of the electricity and gas sectors. He has more than 30 national and international publications in journals and conference proceedings on these topics. José Pablo has been Visiting Scholar at the European University Institute (Italy), the Lawrence Berkeley National Laboratory (USA), and the Massachusetts Institute of Technology (MIT), USA. During his postdoc, he participated in the Utility of the Future project, a joint project between IIT and MIT.

Preface to "Integration of Renewable and Distributed Energy Resources in Power Systems "

Decarbonization, digitalization, and decentralization are transforming our power systems towards a more sustainable energy system. The integration of large amounts of renewable generation is required to achieve an almost 100% decarbonized power system. Many of the new renewable generators, wind and solar, are connected to distribution grids. Digitalization allows changing the traditional paradigm of passive grid operation and planning to smart grids with higher levels of automation and self-control, maintaining the security of supply and economic efficiency. Communications and cybersecurity are required to interconnect millions of sensors and equipment connected not only on the network but also on the customer side. In this Special Issue, we have the opportunity to delve deeply into some of the advances that are required to achieve this transformation. Microgrids are a new concept to operate, in a clever manner, thousands of new control equipment located on customer premises, such as flexible demand, renewable generation sources, and storage installations. In addition, microgrids is an alternative to create more resilient systems against extreme weather conditions or massive cyberattacks. New planning techniques for urban districts and grid development tackling uncertainty regarding load growth and flexibility are also needed. Finally, new operational and planning criteria are explored to integrate a massive penetration of distributed generation, electric vehicles, and controllable loads.

Tomás Gómez San Román, José Pablo Chaves-Ávila
Editors

Article

Photovoltaic Generation Impact Analysis in Low Voltage Distribution Grids

Gregorio Fernández [1,*], Noemi Galan [1,*], Daniel Marquina [1,*], Diego Martínez [1,*], Alberto Sanchez [2], Pablo López [2,*], Hans Bludszuweit [1,*] and Jorge Rueda [2,*]

1 Fundacion CIRCE, Parque Empresarial Dinamiza, Avenida Ranillas 3-D, 1st Floor, 50018 Zaragoza, Spain
2 Grupo Cuerva, C/Santa Lucia, 1 K. Churriana de la Vega, 18194 Granada, Spain; asanchez@grupocuerva.com
* Correspondence: gfernandez@fcirce.es (G.F.); ngalan@fcirce.es (N.G.); dmarquina@fcirce.es (D.M.); dmartinez@fcirce.es (D.M.); plopez@grupocuerva.com (P.L.); hbludszuweit@fcirce.es (H.B.); jruedaq@grupocuerva.com (J.R.)

Received: 20 July 2020; Accepted: 14 August 2020; Published: 22 August 2020

Abstract: Due to a greater social and environmental awareness of citizens, advantageous regulations and a favourable economic return on investment, the presence of photovoltaic (PV) installations in distribution grids is increasing. In the future, not only a significant increase in photovoltaic generation is expected, but also in other of the so-called distributed energy resources (DER), such as wind generation, storage, electric vehicle charging points or manageable demands. Despite the benefits posed by these technologies, an uncontrolled spread could create important challenges for the power system, such as increase of energy losses or voltages out-of-limits along the grid, for example. These issues are expected to be more pronounced in low voltage (LV) distribution networks. This article has two main objectives: proposing a method to calculate the LV distributed photovoltaic generation hosting capacity (HC) that minimizes system losses and evaluating different management techniques for solar PV inverters and their effect on the hosting capacity. The HC calculation is based on a mixture of deterministic methods using time series data and statistical ones: using real smart meters data from customers and generating different combinations of solar PV facilities placements and power to evaluate its effect on the grid operation.

Keywords: photovoltaics; distributed energy resources (DERs); grid impact; power quality; low-voltage distribution network; inverter regulation

1. Introduction

The interest in renewable energies has grown considerably all over the world in recent time. Among renewable energies, solar photovoltaics is by far the most widespread technology adopted for distributed power generation. The global photovoltaic energy generation capacity is increasing rapidly due to efficiency improvements and cost reductions in recent years. It is expected that by 2022, solar PV will be the largest renewable energy source worldwide [1].

Not too many years ago, photovoltaic systems were considered suitable only for isolated systems in remote locations, because solar PV was costly and there were serious concerns regarding negative impacts on the grid [2]. Large cost reductions, together with improved inverter technology and new standards and regulations have made solar PV also attractive for grid connected applications, covering a wide range from utility-scale facilities to small, distributed installations.

PV systems can be connected to the grid at any voltage level and can be classified into three main categories according to their size (power generation) [1]:

- Large-scale (or utility-scale) systems have an installed power higher than 100 MWp and are typically connected to high-voltage transport grids.

- Medium power systems, from 1 MWp to 100 MWp, are connected to medium voltage distribution grids.
- Small systems, with a generation power below from 1 MWp, are distributed along low-voltage grids and very small systems (below 10 kWp) are typically connected to single-phase residential systems. The number of small systems in self-consumption applications has been increasing rapidly in recent years as a way to reduce electric energy bills for residential and commercial customers. In fact, a new term has emerged as a consequence of this phenomenon of self-consumption: "prosumers" [3], which indicated the important evolution of the end user from a passive consumer to a more active producer and consumer. This paper is focused on small and very small PV systems power level, connected always in prosumers' facilities, and the related impacts on LV grids.

The electric power system (transmission and distribution) was designed for unidirectional power flows, from generation to consumption points. This classical conception of the electric system limits system flexibility and adaptability to new technologies, as distributed generation or storage in LV grids. Consequently, new requirements and technologies become necessary to deploy in a safe manner a massive integration of renewable energy sources in distribution grids, both medium and low voltage [4]. These new requirements and technologies are incorporated in the concept of the "smart grid".

As explained in the previous paragraphs, PV generation integrated in AC LV grids is a popular and accessible source of renewable energy for many potential consumers. Solar resource availability in most areas of the world makes this option for electric generation widely adoptable. Furthermore, it can provide benefits to the grid in terms of reduced losses or overload compensation, as well as improve voltage control and management.

Despite the potential benefits of distributed PV, solar radiation variability is an important issue which may lead to instability at the common connection point, with respect to voltage and frequency variations [5]. A high penetration of distributed generators (prosumers) could create different failure possibilities, not only in the LV grid, but also in the medium voltage (MV) grid. Intermittent renewable energy integration presents significant technical and economic challenges. So that, an optimal strategy to operate and control the distribution grid in presence of high photovoltaic generation levels is essential. Main issues of solar PV distributed generation are well known and are listed here below [6,7]:

- *Voltage deviations*. Energy injection of distributed resources along LV lines can change voltage profiles, generating overvoltage and undervoltage whose value can change along the day.
- *Phase imbalances*. In most scenarios, LV photovoltaic facilities are single-phase, generating imbalances respect to the other phases, this phenomenon causes an increase in neutral current and, therefore, losses increase.
- *System losses*. Total line losses decrease as photovoltaic penetration increases and energy generated is consumed locally. However, if the active power injected by the PV systems to the connection point exceeds the load demand, losses increase as not all the energy can be consumed and it is exported to the grid.
- *Reverse power flow*. Electric grids have been designed to withstand unidirectional power flows, so a power flow in the opposite direction could generate problems, affecting grid components as transformers, especially if protection devices are not designed to deal with reverse power flows.
- *Harmonic injection*. Harmonic currents generate voltage harmonics in LV grids that distort the waveform affecting connected loads. Some PV inverters can introduce harmonics into the grid, which means losses increase in transformers that cause heating in its windings and also heating in protection systems, leading them to malfunctions.
- *Short-circuit currents*. High photovoltaic penetration levels cause greater short-circuit currents, which could lead to greater damage to the grid equipment.

To evaluate the effect of renewable generation systems, the penetration level and its limit in the shape of hosting capacity are used. For the first term, the adopted definition in this article is the relation between the total installed peak power and the total power contracted by consumers, kWp/kW in a given LV grid [8]. For the second one, hosting capacity, there are many definitions as shown in [9] or [10], but the approach used in this paper is the maximum penetration level of distributed solar PV generation that could be installed without exceeding grid safe operating limits (thermal and voltages) and reducing energy losses.

As it can be seen, the calculation of the hosting capacity of a grid is not trivial and is influenced by many parameters as the grid components and its characteristics, the consumption profiles of the customers connected, the generation resources and their location in the electrical system [9,10]. In [8] a deep analysis of hosting capacity tools and methods is made and the next main HC assessment techniques classification is proposed:

- Deterministic methods
    - Constant generation methods, that assume that the output of DERs is fixed during the evaluated period. These techniques require small amounts of input data, are simple and fast but the results are approximated and the scenarios tested are reduced.
    - Time series methods, in these techniques the constant value of DERs, the main drawback of previous methods, is avoided using generation profiles. This approach is accurate but need big data sets.
- Stochastic methods introduce the variability or ignorance of many factors, as DERs power or location [11], in the calculation process. Stochastic techniques do not need large amounts of data and can provide accurate results but can be slow and complex
- Optimization based methods. DERs are placed and sized as a result of an optimization process. These methods are exact for the optimal case, not for others.
- Streamlined methods. In these methods the scenarios evaluated are reduced to an amount that represent the possible grid states. These techniques have approximated results.

There are many works calculating hosting capacities for an electric grid and others that propose and evaluate different techniques to increase this penetration level without jeopardizing the system or increasing losses. For example, Reference [12] lists and describes current and new solutions to enable a large-scale penetration of solar PV systems in distribution grids. The authors classify the solutions according to the provider: DSO, prosumer and interactive solutions. Proposed solutions are: network reinforcement, STATCOM use, prosumer storage, increase of self-consumption using tariff incentives, curtailment of power feed-in to the grid, active power control of PV inverter depending on the grid voltage (P(U)), reactive power control of PV inverter depending on the grid voltage or generated active power (Q(U) and Q(P)) or demand response managed by local price signals are some examples.

In References [13–15], as in other many sources, the use of distributed storage is proposed to increase the penetration of solar PV generation reducing its impact on the grid. To reach a widespread adoption of this solution, storage costs need to be further reduced. In several countries, government support is given to accelerate mass adoption with the objective of the desired cost reduction.

On the other hand, Reference [16] proposes and analyses the use of Set Voltage Regulators (SVR) to reduce voltage fluctuations in distribution systems with high solar PV penetration levels. This solution, although interesting, involves the renovation of expensive assets.

In Reference [17] a deep analysis of the negative effects in electric distribution grids with high solar PV penetrations is carried out. To mitigate these negative effects, the connection of PV generators as three-phase systems is proposed. This solution is proved to reduce losses and voltage issues, but it is difficult to be applied in real grids as many consumers have single-phase grid connections and a mixture of single phase and three phase solar PV generation systems is more than expected.

Reference [8] proposes the installation of single-phase solar PV generators in specific phases to reduce grid issues and increase PV penetration. To choose the proper phase, a deep analysis of load along time has to be carried out. This solution is also promising, but the installation of solar PV facilities depends mainly on the consumers and some country-specific normative make the limitation of the installation of PV generators in public grids difficult.

The main objective of [18] is to evaluate how prosumers can support the distribution system operator in the management of the grid and focuses on the changes and demand management, that the customer can offer to benefit both, prosumer and DSO. It also shows the importance of photovoltaic installations near consumption points in order to obtain a better balance between generation and consumption. A big effort is being made through different projects, as in [19] or in [20], and platforms to design, develop, test and foster these local flexibility markets.

Reference [21] shows a different approach as it includes economic and regulatory aspects in the analysis of the maximum PV penetration. The effect of self-consumption policies (economics) and regulations in PV facilities size is analysed to evaluate the PV penetration level in the electric grid and its effect on grid operation. This paper proposes proper regulation, technical and economical, as an alternative to grid reinforcement in high PV penetration grids. Reference [22] continues with the economical point of view to propose solutions to improve solar PV penetration without hampering the electric system in Latvia.

In Reference [23], a reactive power control is implemented in PV inverters managing reactive power according to voltage in the grid connection point. It also reduces the active power generation if the reactive power consumption does not produce the desired decrease in terminal voltage. Reference [24] proposes three reactive power controls in order to reduce overvoltage, two of them based on reactive power management depending on active power generated, $Q(P)$ and the other to grid voltage, $Q(U)$.

Following these ideas, the objective of this article is twofold: on the one hand, to propose a method to calculate the photovoltaic hosting capacity that minimizes system losses, and on the other, to evaluate different management techniques for solar PV inverters and their effect on the hosting capacity aforementioned. As it is shown in the next section, the proposed HC calculation method is a mixture of the deterministic methods using time series data and the statistical ones. This union aims to obtain a simple and fast-running method that provides accurate results and covers multiple possible scenarios for the deployment of distributed generation. Other works as [11] or [25] propose similar methods, but in the research shown in this document real data from users is used, solar PV facilities are installed always in consumers facilities (single or three-phase), to evaluate the effect of self-consumption, and its power is variable and related to the contracted power of the possible prosumer.

This paper is organised as follows: Section 2 describes the grid used to test different PV inverter control and the methodology used to calculate the optimal hosting capacity and the effect on it of the solar PV inverters management. Section 3 shows the main results of the tests carried out. All calculations shown in this document have been carried out on a real low-voltage distribution grid in Granada, southern Spain. Finally, Section 4 discusses the main conclusions derived from the simulations and proposes new working lines.

2. Materials and Methods

The case study presented here, has been carried out using PowerFactory DIgSILENT software (Stuttgart, Germany), with data from a real distribution network in southern Spain and consumption data form customers smart meters. A methodology has been developed in order to derive hosting capacity of distributed PV generation that minimizes grid losses and ensures a safe operation of the grid. The framework has been also employed to evaluate several control strategies of the solar PV inverters and its effect on the aforementioned hosting capacity.

For the simulations, the following inputs have been available:

- Network model in PowerFactory DIgSILENT.
- A typical consumer profiles of the set day (active and reactive power)
- Voltage profile in the transformer LV side of the selected day.
- Irradiation for PVs generation.

Next, the methodology to obtain the optimal distributed photovoltaic penetration is detailed. This analysis consists in increasing the photovoltaic penetration with steps of 10 kW, to reach the maximum level of PV penetration that could be installed by the clients randomly. In the Figure 1, it is shown a flow chart of the methodology to apply. The first step of the flowchart is to define different variables:

(a) maximum number of scenarios for each level of penetration (Max_num_scenarios). For this study 30 scenarios have been chosen.
(b) maximum penetration level (Max_Penetration). For this study 400 kW,
(c) the minimum level of penetration (Install_level). For this study it starts with 10 kW.

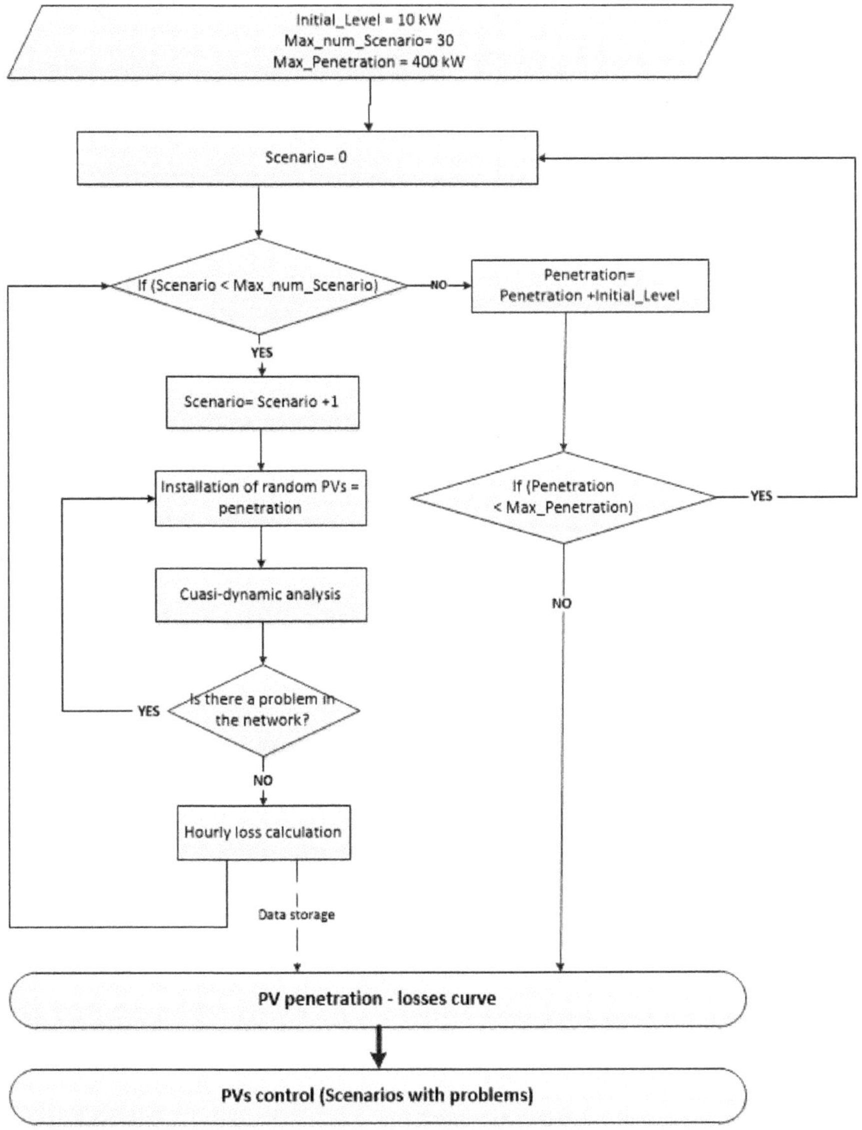

Figure 1. Methodology to obtain the optimal PV penetration level.

After defining the model parameters, the scenario counter is initialized (Scenario = 0). This variable is a counter increase (scenario = scenario + 1) until reaching 30 scenarios at each level of penetration. These 30 scenarios locate PV facilities uniform distribution in the network at existing customer connections, applying a Monte Carlo approach. The total power installed with PVs is set as the level of penetration. The size of each PV system is assumed to be equal to the customer's contracted power and the phase to which it is connected. The total PV power installed is defined by the level of penetration. After completing all scenarios, penetration level is increased by 10 kW. This process is repeated until maximum penetration level is reached.

Networks should not have congestion problems or voltage deviations. Limits are set by each facility. In this case it is Grupo Cuerva that stablishes that voltage limits are 1.07 p.u. and 0.93 p.u., while the congestion limit of lines and transformer is 100%. In order to verify if there are any problems of congestion or voltage deviations, a quasi-dynamic study is executed each scenario, which consists in a power flow for each hour of a day.

If no problems are detected, total losses are calculated and stored for each of the 24 h. Once there are 30 valid scenarios, the penetration of PVs is increased (Penetration = Penetration + 10 kW) until the maximum penetration is reached (Max_Penetration). After that, the curve "PV penetration-losses" is obtained. Finally, the optimal PV penetration level is derived from the minimum of the PV penetration-loss curve. According to the model definition, network losses decrease with increasing PV penetration until a certain level, where losses start to increase again.

After obtaining the optimal penetration value that minimizes systems losses, ten random uses cases are created with the optimal renewable penetration (240 kW), this choice is based on the study made to obtain the minimum losses in the system. The ten use cases are created with different PVs location (more information about the network topology in Appendix B). These ten use cases include congestion or voltage deviations. The control strategies are then evaluated in terms of ability to solve voltage problems caused by distributed PV generation. The controls tested are:

- *Constant power reactive control (Q = constant)*: PVs consume or generate reactive power regardless of the active power installed. An amount of reactive power to be consumed or generated must be set. In order to evaluate efficiently this control strategy, six different values for Q are studied:
    - Consumption of 10% active power PVs (Q = 10% P)
    - Consumption of 20% active power PVs (Q = 20% P)
    - Consumption of 30% active power PVs (Q = 30% P)
    - Generation of 10% active power PVs (Q = 10% P)
    - Generation of 10% active power PVs (Q = 20% P)
    - Generation of 10% active power PVs (Q = 30% P)
- *Reactive power control with constant power factor (PF = constant)*: The power factor of PVs is set as a constant. To evaluate this control strategy, six PF values are considered indicating if it has inductive or capacitive characteristic (to consume or to generate, respectively):
    - PF = 0.95 inductive
    - PF = 0.95 capacitive
    - PF = 0.9 inductive
    - PF = 0.9 capacitive
    - PF = 0.85 inductive
    - PF = 0.85 capacitive
- *Power factor control depending on the active power generated (PF = f(P))* [26]: PVs consume reactive power as a function of the active power generated at each moment. In this case the power factor depends on the active power generation. For this control strategy, two parameters are defined: (1) the minimum active power, that indicates which is the value where reactive power

consumption starts; (2) the minimum value of the power factor. Figure 2 shows the *PF* variation curve used to set control strategies such as:

- $P = 0.5$ p.u. and $PF = 0.9$
- $P = 0.5$ p.u. and $PF = 0.95$
- $P = 0.65$ p.u. and $PF = 0.9$
- $P = 0.65$ p.u. and $PF = 0.95$

- *Reactive power control depending on terminal voltage ($Q = f(U)$)*: PVs consume or generate enough reactive power to ensure that the grid terminals are within the permitted range. If the voltage is below a marked limit (i.e., 0.98 p.u.), the PV generates reactive power; and if the voltage is above a marked limit (i.e., 1.02 p.u.), the PV consumes reactive power. The maximum reactive power generated is reached when the voltage is 0.95 p.u. and the maximum reactive power consumed is reached when the voltage is 1.05 p.u. The three strategies implemented in this control are:

  - $PF = 0.95$
  - $PF = 0.9$
  - $PF = 0.85$

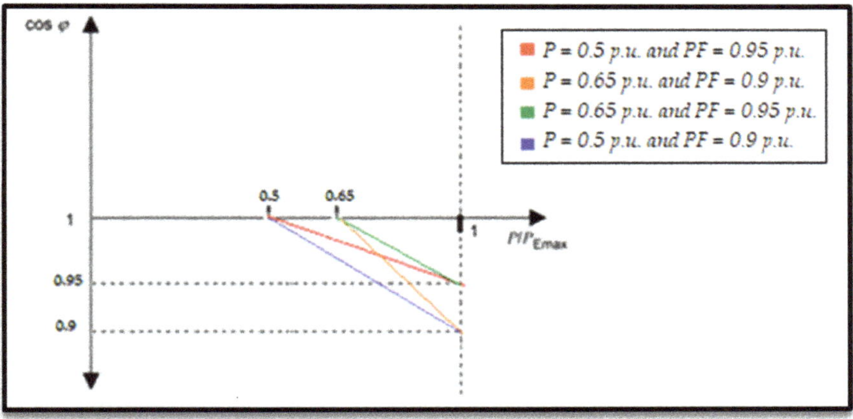

Figure 2. *PF* variation curve depending on the active power generated.

In order to test this methodology, a network model has been developed in PowerFactory DIgSILENT. A real model provided by Grupo Cuerva has been chosen, this real model is a district of a rural town. This model contains 266 consumers, from which 45 have a three-phase connection and 221 single-phase. The total installed power of all customer is 1, 320 MW approximately. This network is divided in 11 feeders (Figure A1 in Appendix A, each of the feeders is indicated in different colours) which have different lengths and different numbers of consumers. Figure 3 shows the LV network topology (More information about the network topology in Appendix A).

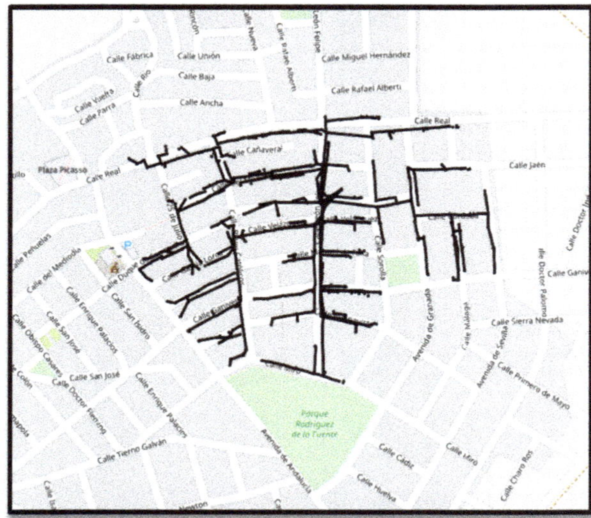

Figure 3. LV network topology of the case study.

3. Results

The first part of this section provides the results of the study of optimal PV penetration and in the second part results from the different use cases are presented for each control strategy.

3.1. Optimal PV Penetration

As explained in Section 2, a method has been developed to obtain the optimum value (minimum losses in the LV network) of distributed PV penetration within a LV network, where PV systems are associated by Monte Carlo method to consumers within the LV grid.

For this purpose, in each level of penetration (10 kW, 20 kW ...) until the maximum is reached (Max_Penetration), 30 scenarios are carried out in which Monte Carlo distributed generation is installed, that is, for each level of penetration 30 different scenarios will be obtained. For each of these scenarios a quasi-dynamic study (load flow for each hour) is carried out, allowing to obtain the value of the losses.

Each blue point in the Figure 4 represents the losses of scenario (total losses in the grid) in each PV penetration level, In other words, each level of penetration (PV generation) in Figure 4 has 30 blue points that represent the losses of each of the 30 scenarios (30 scenarios are defined as Max_num_scenarios in Figure 1, i.e., the 30 scenarios for each level of penetration in which losses are calculated). The tendency line that occurs between the different scenarios is the line that occurs when connecting the average points of losses of each level of penetration (PV generation).

Figure 4 confirms the expected effect, that distributed PV reduces total losses until certain penetration level and further increasing of PV installations increases also grid losses. The Monte Carlo method also reveals important differences depending on the actual distribution of the PV system within the grid.

As mentioned before, the increase of distributed PV generation may create problems in the network, such as voltage deviations ($U > 1.07$ p.u. or $U < 0.93$ p.u.) and congestions. One of the main issues is overvoltage as PV production may produce inverse power flows which increase voltage.

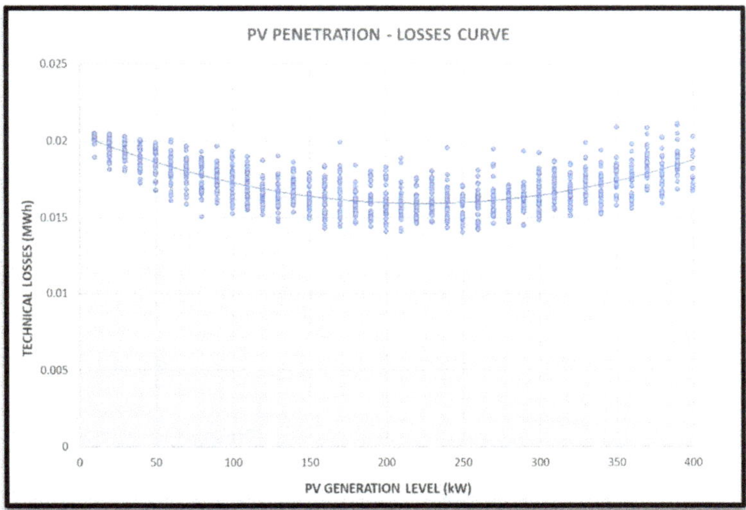

Figure 4. Total grid losses from all scenarios and PV penetration levels.

Figure 5 shows, in orange, the scenarios with problems in the network (congestion and/or voltage deviations) that have been obtained by Monte Carlo method, while trying to reach the 30 scenarios (30 scenarios are defined as Max_num_scenarios in Figure 1, i.e., the 30 scenarios for each level of penetration in which losses are calculated), without problems (blue).

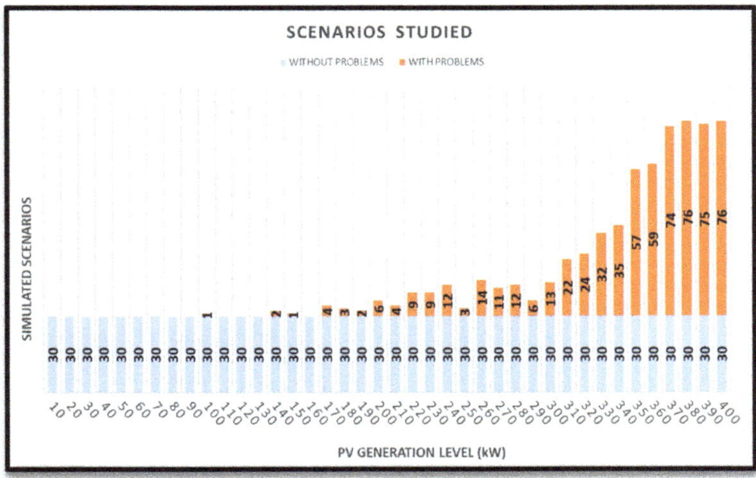

Figure 5. Comparison of simulated scenarios with and without problems.

As it can be seen in Figure 5, as distributed PV generation increases, more scenarios were found with problems of voltage deviations, which complicated the finding of the 30 scenarios without voltage deviations. For example, in the scenario with a penetration of 240 kW of distributed PV, 12 scenarios with grid problems have been obtained, this means that to find 30 scenarios without problems, 42 scenarios had to be analysed.

3.2. PV Control

In the following paragraphs, the control strategies explained in Section 2 are compared identifying the advantages and disadvantages of each control. To highlight the advantages and disadvantages of each option, three indicators have been used:

- Losses
- Voltage deviations (undervoltage and overvoltage)
- Congestions

To test each of the controls, ten use cases are created with different PV locations (see Appendix A for the locations of the PVs in each use case) but all of them with the same penetration 240 kW (total installed power for the PVs). In each of the use cases, a quasi-dynamic study (hourly load flow) will be performed over the period of one day in which the three indicators mentioned above will be calculated.

3.2.1. Constant Power Reactive Control

As explained in Section 2, the constant reactive power control consumes or generates the reactive power according to the value of the active power. Six values for reactive power have been considered (10, 20 and 30% of active power, capacitive and inductive)

The results obtained for these six control strategies in terms of losses (first indicator) can be seen in Figure 6. Figure 6 shows that the lowest values of losses are obtained when reactive power is being generated, since in all use cases, the lowest losses are obtained when the reactive power generated is 20% of the active power. This is because the reactive power is consumed by the loads and the reactive power transported by the network is minimized. While it is consumed, the highest losses are obtained in all use cases (grey points).

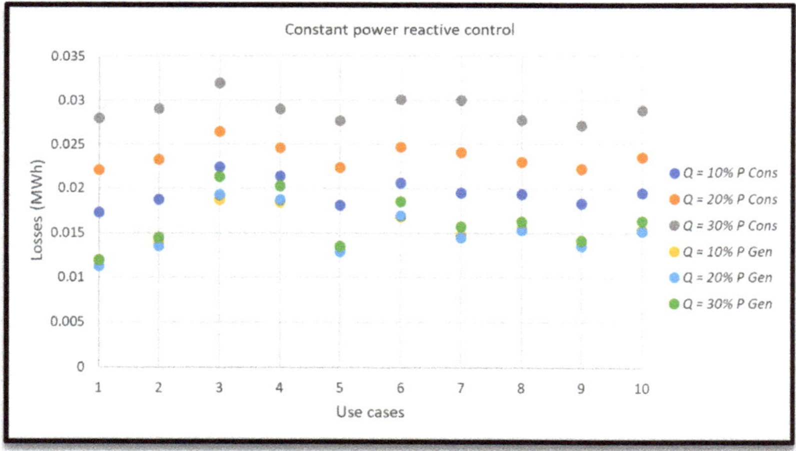

Figure 6. Constant power reactive control: Losses (More information see Table A3 in Appendix C).

The second key indicator in this analysis is voltage deviation. In this case, no undervoltage event happened but overvoltage events occur. These overvoltage events are visualized in Figure 7.

When generating more reactive power (green and blue lines in Figure 7) there are more overvoltage events than when reactive power is consumed (grey line in Figure 7). For example, in use case 4, when reactive power is generated at 30% of active power, there are 25 overvoltage events in the network. However, if the reactive power is consumed at 30% of the active power, only one overvoltage event has been observed. Finally, the key congestion indicator has been analysed and the shows no congestion in any use case.

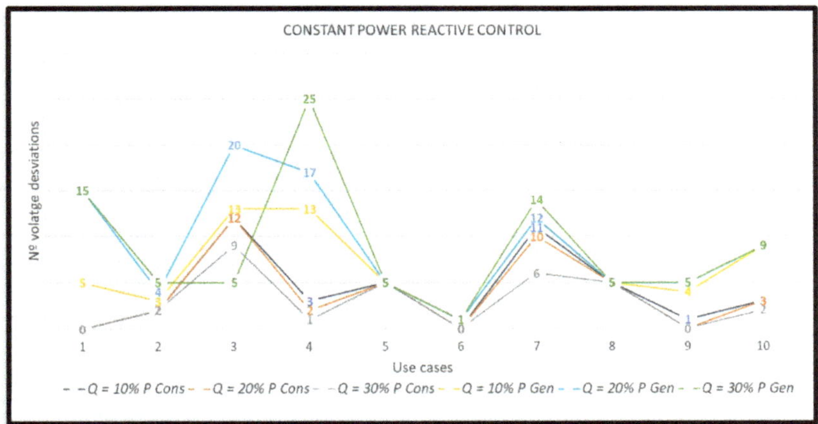

Figure 7. Constant reactive power control: Voltage deviations. (More information see Table A4 in Appendix C).

3.2.2. Reactive Power Control with Constant Power Factor

This control strategy considers a constant power factor, fixing reactive power proportional to active power generated. As mentioned in Section 2, six different *PF* levels have been considered (0.85, 0.9 and 0.95 inductive and capacitive).

The first indicator analysed is the losses, the results obtained for each use case can be seen in Figure 8. General conclusions are very similar to the first control strategy, where capacitive power factors are leading to lowest losses.

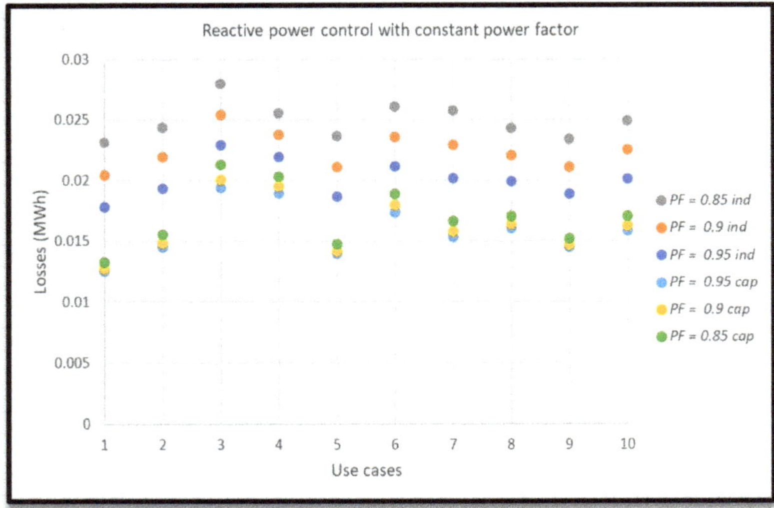

Figure 8. Reactive power control with constant power factor: Losses (More information see Table A5 in Appendix C).

As can be seen in Figure 9, results regarding overvoltages are also very similar to the first control strategy with fixed reactive power control. In comparison, a slightly higher number of overvoltage events are observed for fixed *PF* control. This effect can be explained due to the fact that reactive power is highest when active power peaks, which are the hours where reversed power flows can be observed.

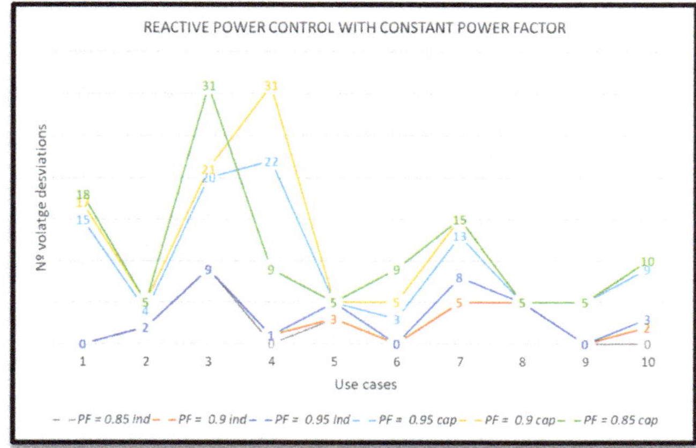

Figure 9. Reactive power control with constant power factor: Voltage deviations (More information see Table A6 in Appendix C).

Finally, the result of this congestion indicator shows zero as there is no congestion in any of the use cases.

3.2.3. Power Factor Control Depending on Generated Active Power

The PV inverter consumes reactive power as a function of generated active power, the control parameters analysed are:

- $P = 0.5$ p.u. and $PF = 0.9$
- $P = 0.5$ p.u. and $PF = 0.95$
- $P = 0.65$ p.u. and $PF = 0.9$
- $P = 0.65$ p.u. and $PF = 0.95$

The results obtained evaluating the losses for this control strategy can be seen in Figure 10. As it can be seen in Figure 10, the loss values obtained applying the different control parameters are very close to each other. Thus, the effect on network losses is very limited.

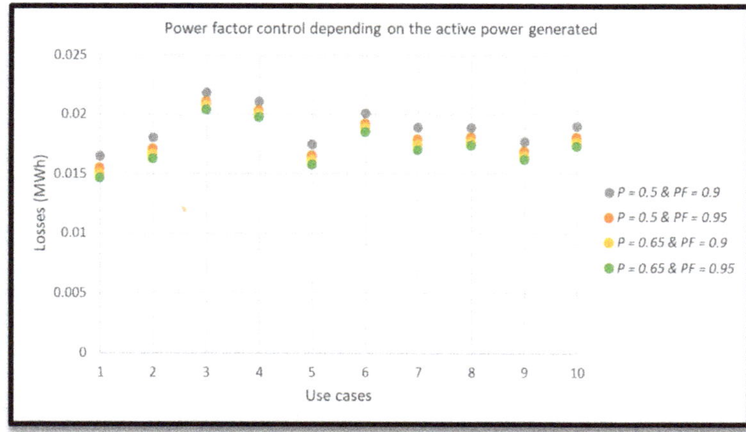

Figure 10. Reactive power control with constant power factor: Losses (More information see Table A7 in Appendix C).

Overvoltage events are shown in Figure 11. Similar to the other two control strategies, no under-voltage is observed. The number of overvoltage events in this case is considerably lower, which is mainly due to the fact that no capacitive reactive power has been considered.

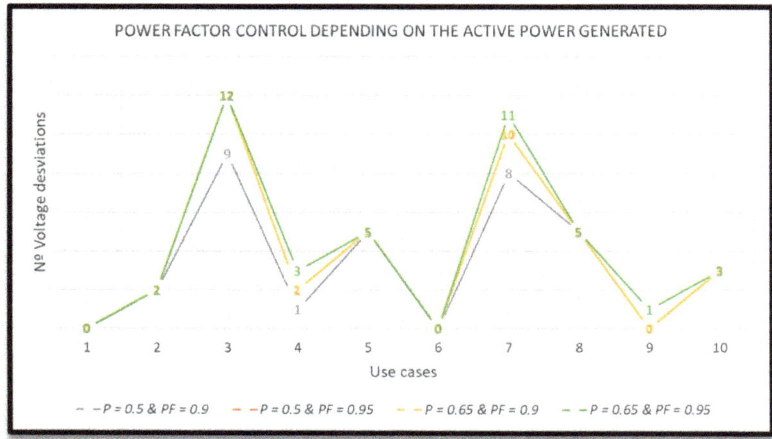

Figure 11. Power factor control depending on the active power generated: Voltage deviations (More information see Table A8 in Appendix C).

In Figure 11, it can be seen that the control strategy of reactive power generation with 0.5 p.u. of active power and *PF* = 0.9 p.u., is the one where less overvoltage events are obtained. Therefore, if only the overvoltage events are considered, this would be the optimal strategy.

3.2.4. Reactive Power Control Depending on Terminal Voltage

With this strategy, PV facilities aim keeping terminal voltages within desired limits, adjusting reactive power accordingly withing a predefined range. In this case study, reactive power limits have been defined in terms of *PF*, considering 3 cases.

- *PF* = 0.95 p.u.
- *PF* = 0.9 p.u.
- *PF* = 0.85 p.u.

The first key indicator to analyse is losses. Figure 12 shows network losses for this control strategy of *PF* = 0.95 p.u. (grey dots), *PF* = 0.9 p.u. (yellow dots) and *PF* = 0.85 p.u. (green dots).

Figure 12 shows that the lowest losses are obtained when the power factor is 0.95, i.e., when less reactive power is delivered per voltage level.

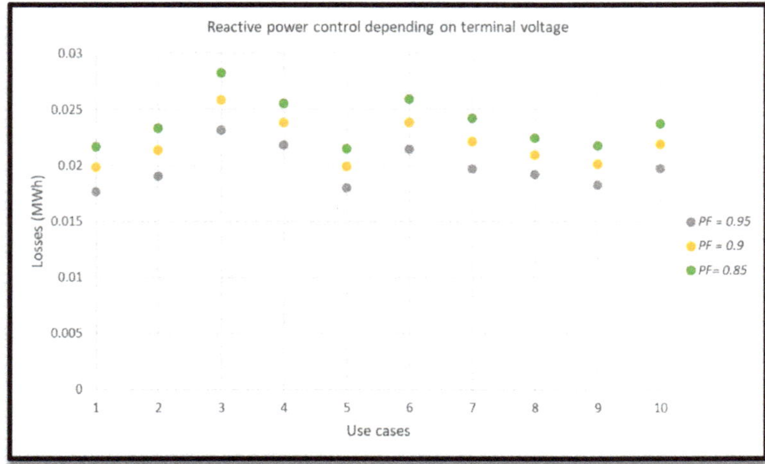

Figure 12. Reactive power control depending on terminal voltage: Losses (More information see Table A9 in Appendix C).

The number of overvoltage events is shown in Figure 13, where is becomes clear that with the lowest *PF* (0.85) limit, overvoltages can be reduced most. This is not surprising, as the more reactive power is allowed to be consumed, the higher is the effect on terminal voltages. Nevertheless, it should be highlighted here, that this reduction of overvoltages comes with the downside of increased network losses.

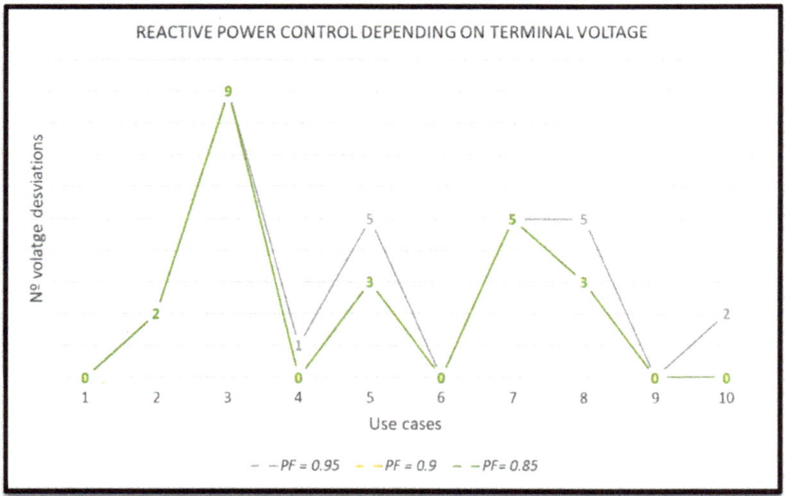

Figure 13. Reactive power control depending on terminal voltage: Voltage deviations (More information see Table A10 in Appendix C).

As happened in the previous control, in the use case there are no congestions and therefore the last key indicator, congestions, is zero.

3.2.5. Comparison of Control Strategies

In this section, all evaluated control strategies are compared regarding losses and overvoltages. Figure 14 shows a comparison of network losses. Each control has several points according to the different control strategies: 6 dots for Q constant represented by a blue dot in Figure 14, 6 for PF constant represented by red dots in Figure 14, 4 for $PF = f(P)$ represented by green dots in Figure 14 and 3 for $Q = f(U)$) represented by yellow dots in Figure 14.

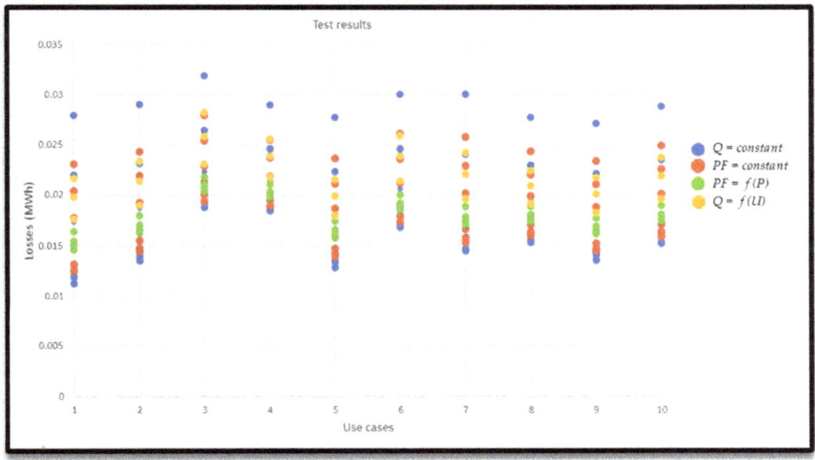

Figure 14. Results all controls: Losses.

In the Figure 14, identifying the optimal control is difficult because there is no clear tendency in the controls. For example, in the constant value control of Q, a very small value of losses is observed if the reactive power is low and a very high value of losses if the reactive power consumed is high. The same situation occurs when the constant PF control is applied. However, the three dots (green dots in Figure 14) that define the $PF = f(P)$ are close and the losses are small enough in order to select this control as optimal control.

Figure 15 shows the number of overvoltage events in each use case for each control strategy. In Figure 15 it can be seen the effectiveness of each control. It is represented by the number of terminals in overvoltage. The results show that $Q = f(U)$ (yellow line Figure 15) is the most effective control as in all use cases has the lowest number of overvoltage events.

For the selection of the optimal control, it is necessary to take into account both key indicators (losses, voltage deviations) and to reach an agreement on which level of losses and overvoltage events are considered as valid. As a conclusion, the optimal controls would be $Q = f(u)$ and $PF = f(p)$ since they are the controls that cause the least overvoltage events and the loss values are sufficiently small.

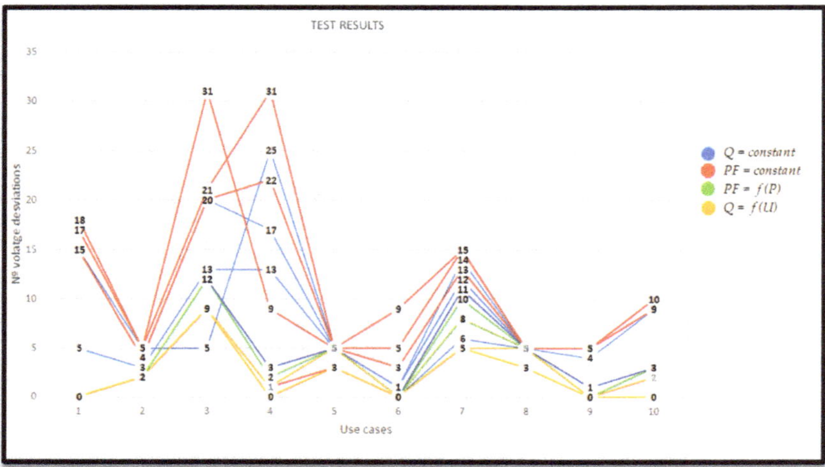

Figure 15. Results all controls: Voltage deviations.

4. Conclusions

In this paper, two main topics have been analysed: on the one hand, to propose a method to calculate the photovoltaic hosting capacity that minimizes system losses, and on the other, to evaluate different management techniques for solar PV inverters and their effect on the hosting capacity aforementioned

In a first step, optimal PV penetration levels have been derived using a Monte Carlos approach, considering optimal PV penetration when network losses are minimised, that is to say, the HC has been obtained in each node where self-consumption can be introduced, which minimizes the losses of the system.. This approach is very promising for planning purposes, if a detailed network model is available. Similar methods are proposed in the works of [11] or [25], however in this article the scope has been extended using real data from the smart meters, the installed PVs can be installed in single phase or three-phase (depending on the installation of the consumer), the size of the installed PV will depend on the contracted power of the consumer, with this it is possible to evaluate the effect of PV self-consumption in a distribution network. Some improvements can be proposed for future studies: (i) it has been used a standard day and it is proposed to use a whole year of smart meter data to evaluate the seasonal effect (autumn, spring, winter and summer), (ii) it has been assumed that PV facilities power is equal to the contracted power and it could be interesting modelling the variability of the installed PV facilities and for the last (iii) it will be evaluated other methods to expand the penetration of renewables such as distributes storage technologies and / or demand response policies, either through direct signals or through a local market flexibility.

The second topic was dedicated to the evaluation of 4 different control strategies of solar PV inverters, considering the possibility of reactive power control to mitigate excessive voltage deviations. In addition, the effect on total network losses have been documented. Congestion problems were also expected, but were not observed in this case study, given the restrictive condition which limits installed solar PV power at consumer premises to their respective contracted power. The effectiveness of this limit could be demonstrated hereby.

In the tested grid, 240 kW of installed PV power has been found as the optimal penetration level in terms of technical losses. The 4 control strategies have been tested for this penetration level, and ten use cases were generated in which overvoltage occurred. Control strategies with constant reactive power or PF setpoints were outperformed by more dynamic approaches considering generated power to modify PF and terminal voltages to modify reactive power.

Author Contributions: Conceptualization, N.G., G.F. and A.S.; methodology, N.G. and G.F.; software, D.M. (Daniel Marquina) and D.M. (Diego Martinez); validation, P.L.; formal analysis, G.F. and N.G.; investigation, D.M. (Diego Martinez) and P.L.; resources, D.M. (Daniel Marquina) and G.F.; writing—original draft preparation, G.F. and N.G.; writing—review and editing, D.M. (Daniel Marquina), D.M. (Diego Martinez), A.S., P.L., J.R. and H.B.; visualization, G.F., N.G., D.M. (Daniel Marquina) and D.M. (Diego Martinez); supervision, N.G.; project administration, G.F. All authors have read and agreed to the published version of the manuscript.

Funding: This research has received funding from the European Union's Horizon 2020 Framework Programme for Research and Innovation under grant agreement no. 864319, project PARITY.

Acknowledgments: We thank CUERVA Group for supporting the research carried out and to provide information about their distribution grids.

Conflicts of Interest: The authors declare no conflict of interest.

Appendix A

In this section the main characteristics of the network model used for the developments are detailed. Figure A1 shows a single-line diagram of the network, in which each of the feeders that conform the network can be seen in different colours.

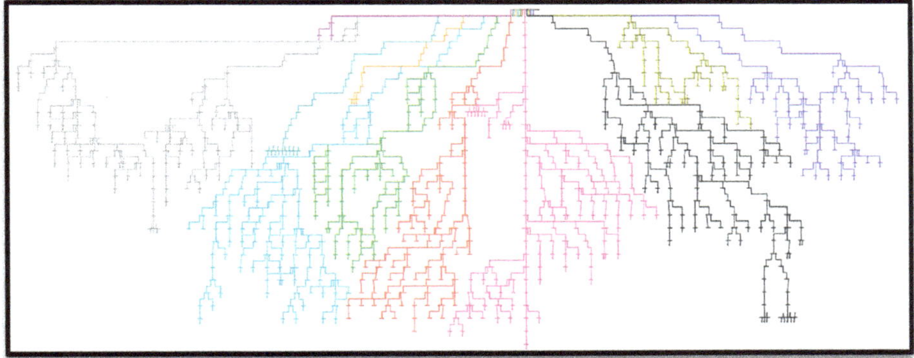

Figure A1. Network Topology.

The cables used to model the network are listed in Table A1, this table shows the characteristics of the type of cable which are: voltage of the conductor (Voltage), maximum allowed current (Current), the cable is type, resistance (R), reactance (X) and inductance (L).

Table A1. Type of conductors.

Type of Conductor	Voltage (kV)	Current (kA)	Phases	R (Ohm/Km)	X (Ohm/Km)	L (mH/Km)
RV 2X25 AL	0.4	0.074	1	1.44	0.078	0.2482817
RV 2X50 AL	0.4	0.105	1	0.77	0.0777	0.2473268
RV 3X150 AL	0.4	0.2	3	0.249	0.072	0.2291831
RV 3X25 AL	0.4	0.074	3	1.44	0.078	0.2482817
RV 3X50 AL	0.4	0.105	3	0.77	0.0777	0.2473268
RV 4X50 AL	0.4	0.105	3	0.77	0.0777	0.2473268
RV 4X95 AL	0.4	0.155	3	0.39	0.0733	0.2333211
RZ 3X16 CU	0.4	0.075	1	1.45	0.0813	0.2587859
RZ 3X6 CU	0.4	0.053	3	3.95	0.0901	0.2867972
RZ 4X10AL	0.4	0.054	3	3.61	0.086	0.2737465
RZ 4X16 AL	0.4	0.058	3	2.27	0.0813	0.2587859
RZ 4X25 AL	0.4	0.074	3	1.44	0.078	0.2482817

To characterize the network, real measurements of the clients obtained from the smart meters have been used, in Figure A2 an example of the consumption of two customers profiles is shown.

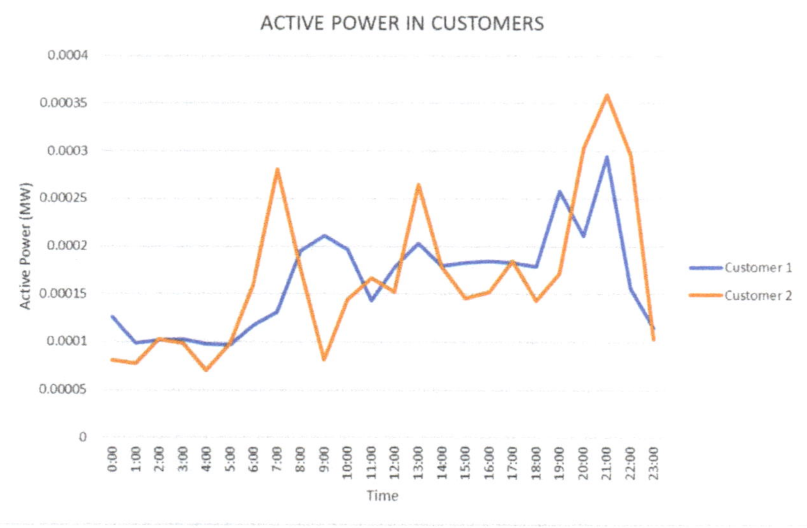

Figure A2. Active power profiles of different customers.

The voltages of the secondary substation are shown in Figure A3, In Figure A3 the voltage of phase 1 (phase R) can be seen in blue, the voltage of phase 2 (phase S) can be seen in orange and the voltage of phase 3 (phase T) can be seen in grey.

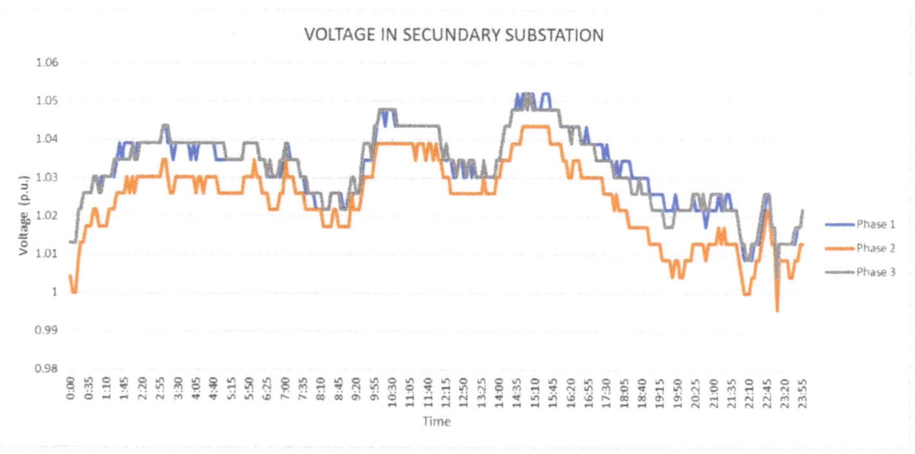

Figure A3. Voltage in secondary substation.

The characteristics of the secondary substation are described in the Table A2.

Table A2. Characteristics of the secondary substation.

Name of Transformer	HV-Side (kV)	LV-Side (kV)	Rated Power (MVA)
Trafo	20	0.4	0.65

Appendix B

Use case 1:

Figure A4 shows the location of the photovoltaic installations in case of use 1, the locations of the photovoltaic installations are shaded in blue.

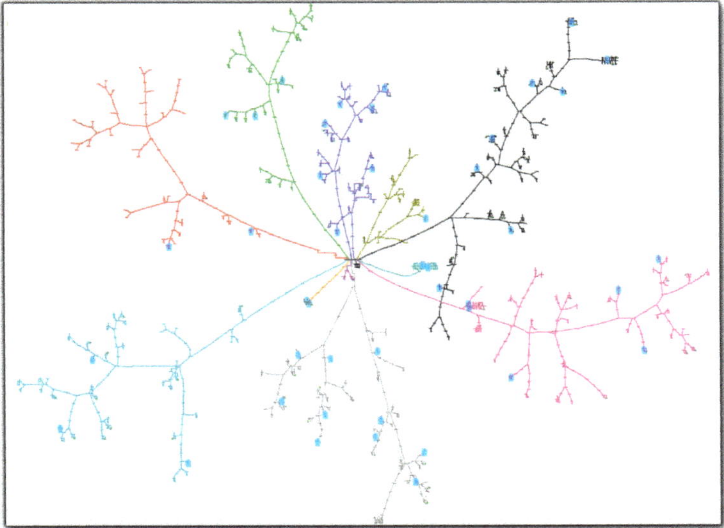

Figure A4. Use case 1—Location of PVs.

Use case 2:

Figure A5 shows the location of the photovoltaic installations in case of use 2, the locations of the photovoltaic installations are shaded in blue.

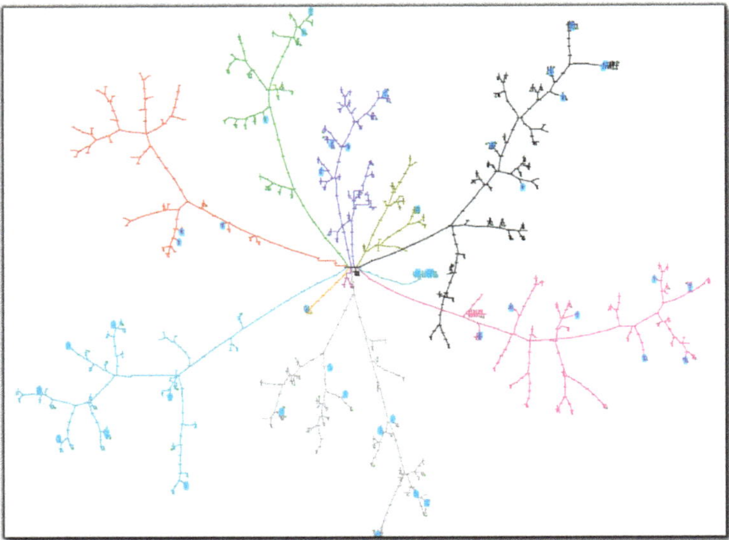

Figure A5. Use case 2—Location of PVs.

Use case 3:

Figure A6 shows the location of the photovoltaic installations in case of use 3, the locations of the photovoltaic installations are shaded in blue.

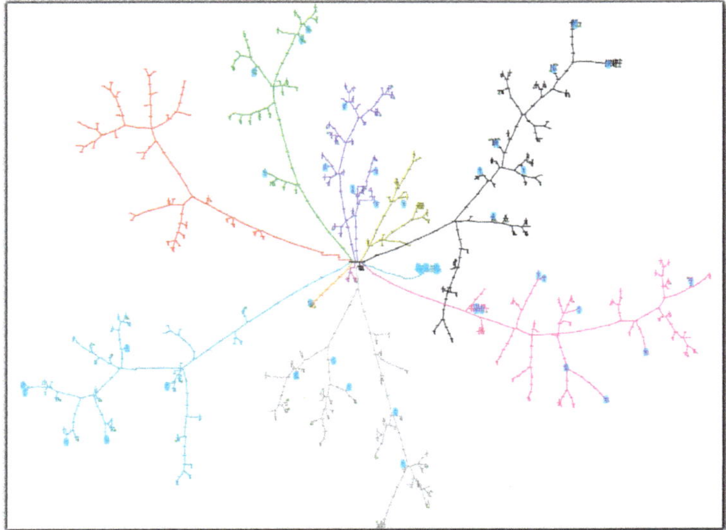

Figure A6. Use case 3—Location of PVs.

Use case 4:

Figure A7 shows the location of the photovoltaic installations in case of use 4, the locations of the photovoltaic installations are shaded in blue.

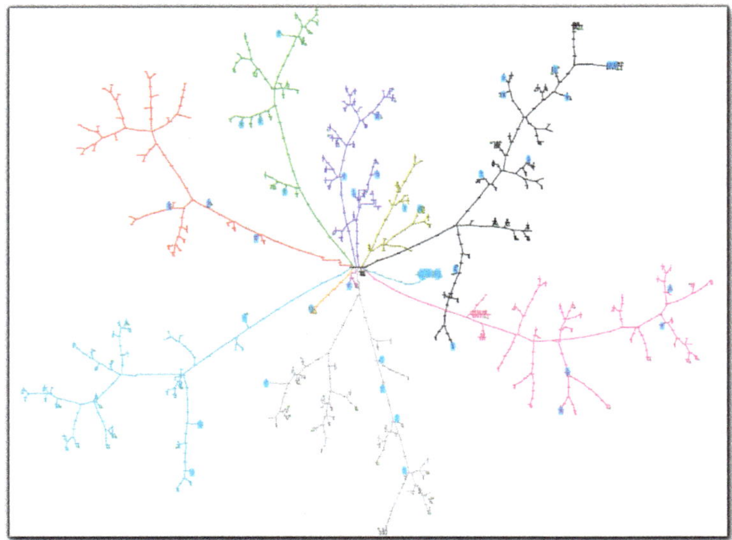

Figure A7. Use case 4—Location of PVs.

Use case 5:

Figure A8 shows the location of the photovoltaic installations in case of use 5, the locations of the photovoltaic installations are shaded in blue.

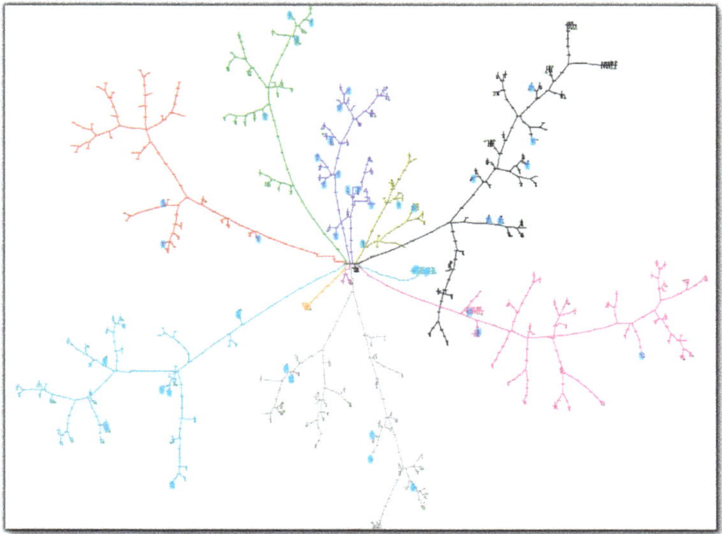

Figure A8. Use case 5—Location of PVs.

Use case 6:

Figure A9 shows the location of the photovoltaic installations in case of use 6, the locations of the photovoltaic installations are shaded in blue.

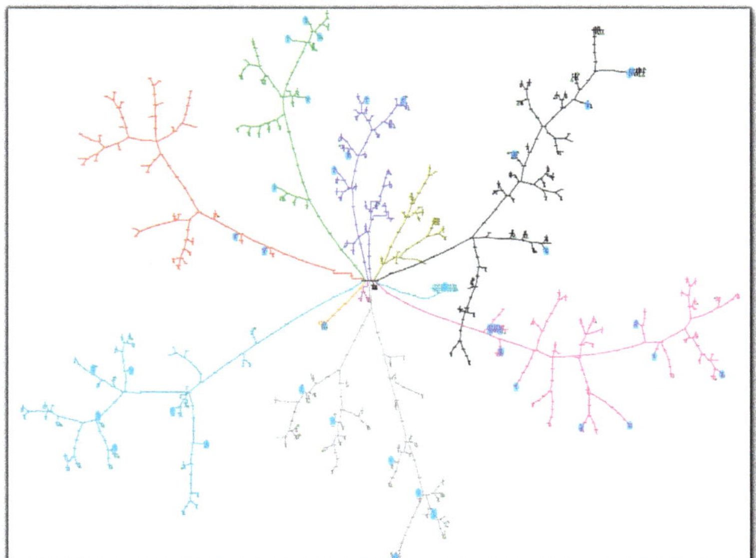

Figure A9. Use case 6—Location of PVs.

Use case 7:

Figure A10 shows the location of the photovoltaic installations in case of use 7, the locations of the photovoltaic installations are shaded in blue.

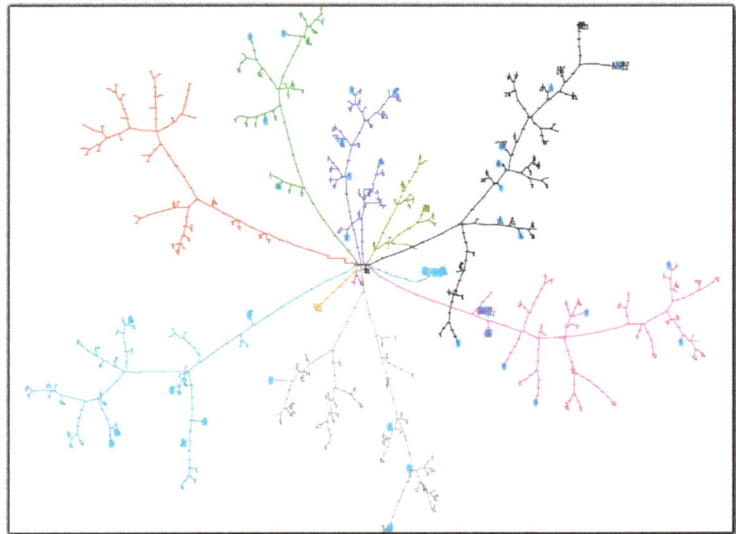

Figure A10. Use case 7—Location of PVs.

Use case 8:

Figure A11 shows the location of the photovoltaic installations in case of use 8, the locations of the photovoltaic installations are shaded in blue.

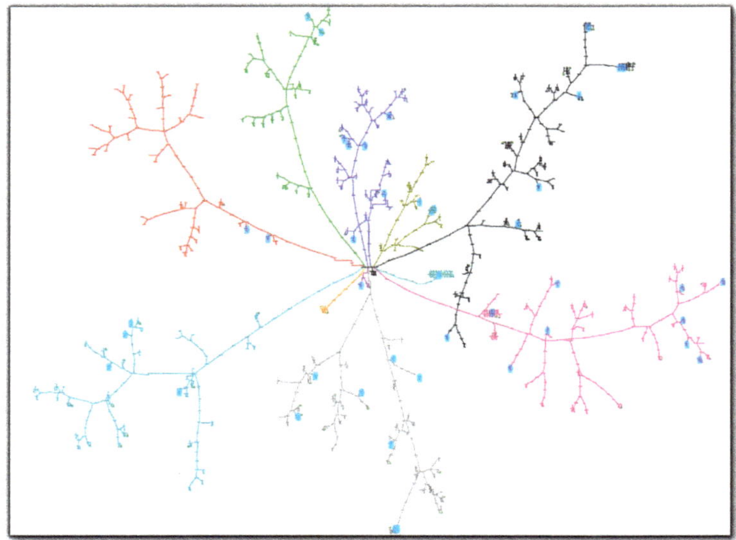

Figure A11. Use case 8—Location of PVs.

Use case 9:

Figure A12 shows the location of the photovoltaic installations in case of use 9, the locations of the photovoltaic installations are shaded in blue.

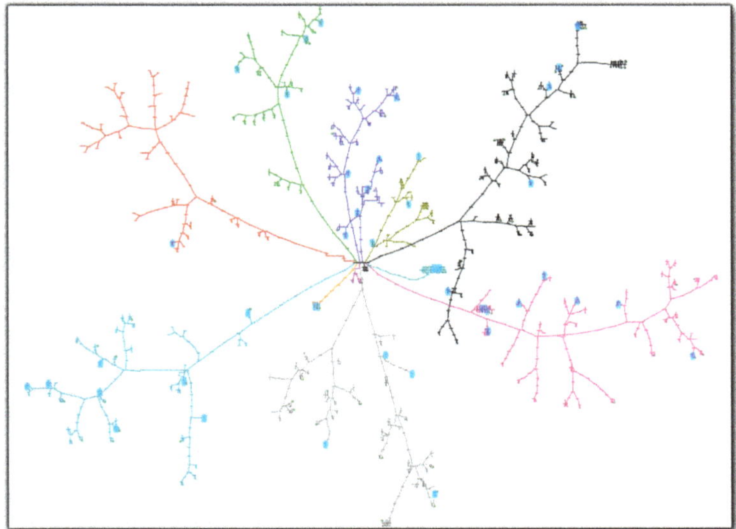

Figure A12. Use case 9—Location of PVs.

Use Case 10:

Figure A13 the location of the photovoltaic installations in case of use 10, the locations of the photovoltaic installations are shaded in blue.

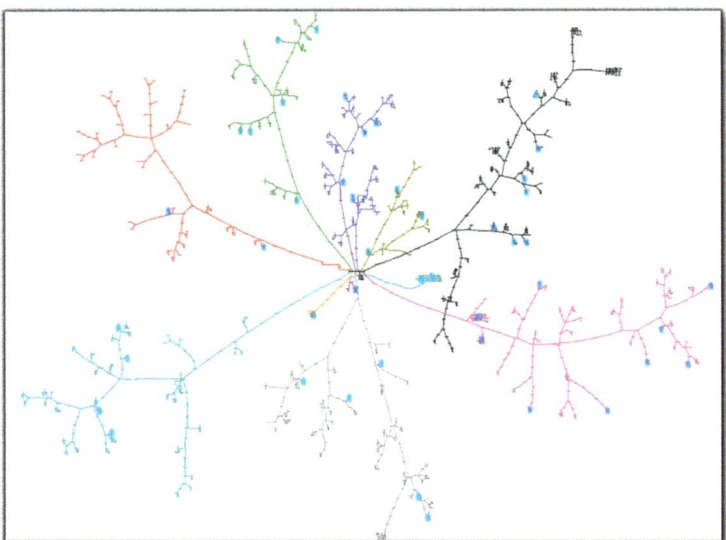

Figure A13. Use case 10—Location of PVs.

Appendix C

This section details the results of each use case for each of the control strategies. The tables show the results of voltage deviations and losses.

Constant power reactive control

Table A3. Constant power reactive control: Losses.

Use Cases	Constant Power Reactive Control (Q = *Constant*)					
	Q = 10% P Cons	Q = 20% P Cons	Q = 30% P Cons	Q = 10% P Gen	Q = 20% P Gen	Q = 30% P Gen
Use Case 1	0.0173	0.0220	0.0279	0.0119	0.0112	0.0119
Use Case 2	0.0187	0.0232	0.0290	0.0139	0.0135	0.0145
Use Case 3	0.0223	0.0264	0.0319	0.0187	0.0192	0.0213
Use Case 4	0.0213	0.0246	0.0289	0.0184	0.0187	0.0202
Use Case 5	0.0181	0.0223	0.0277	0.0134	0.0129	0.0135
Use Case 6	0.0206	0.0246	0.0300	0.0168	0.0170	0.0185
Use Case 7	0.0194	0.0240	0.0300	0.0147	0.0145	0.0157
Use Case 8	0.0193	0.0229	0.0277	0.0155	0.0153	0.0163
Use Case 9	0.0183	0.0221	0.0271	0.0140	0.0135	0.0141
Use Case 10	0.0195	0.0235	0.0288	0.0153	0.0152	0.0163

Table A4. Constant reactive power control: Voltage deviations.

Use Cases	Constant Power Reactive Control (Q = *Constant*)					
	Q = 10% P Cons	Q = 20% P Cons	Q = 30% P Cons	Q = 10% P Gen	Q = 20% P Gen	Q = 30% P Gen
Use Case 1	0	0	0	5	15	15
Use Case 2	3	2	2	3	4	5
Use Case 3	12	12	9	13	20	5
Use Case 4	3	2	1	13	17	25
Use Case 5	5	5	5	5	5	5
Use Case 6	0	0	0	1	1	1
Use Case 7	11	10	6	12	12	14
Use Case 8	5	5	5	5	5	5
Use Case 9	1	0	0	4	5	5
Use Case 10	3	3	2	9	9	9

Reactive power control with constant power factor

Table A5. Reactive power control with constant power factor: Losses.

Use Cases	Reactive Power Control with Constant Power Factor (PF = *Constant*):					
	PF = 0.85 ind	PF = 0.9 ind	PF = 0.95 ind	PF = 0.95 cap	PF = 0.9 cap	PF = 0.85 cap
Use Case 1	0.0231	0.0204	0.0178	0.0125	0.0127	0.0132
Use Case 2	0.0244	0.0219	0.0193	0.0145	0.0148	0.0155
Use Case 3	0.0280	0.0254	0.0229	0.0194	0.0201	0.0213
Use Case 4	0.0255	0.0238	0.0219	0.0189	0.0195	0.0203
Use Case 5	0.0236	0.0211	0.0187	0.0140	0.0142	0.0147
Use Case 6	0.0261	0.0236	0.0212	0.0174	0.0179	0.0189
Use Case 7	0.0258	0.0229	0.0202	0.0153	0.0158	0.0166
Use Case 8	0.0243	0.0220	0.0199	0.0160	0.0164	0.0171
Use Case 9	0.0234	0.0211	0.0188	0.0145	0.0147	0.0152
Use Case 10	0.0249	0.0225	0.0201	0.0159	0.0163	0.0171

Table A6. Reactive power control with constant power factor: Voltage deviations.

Use Cases	Reactive Power Control with Constant Power Factor (*PF* = *Constant*):					
	PF = 0.85 ind	*PF* = 0.9 ind	*PF* = 0.95 ind	*PF* = 0.95 cap	*PF* = 0.9 cap	*PF* = 0.85 cap
Use Case 1	0	0	0	15	17	18
Use Case 2	2	2	2	4	5	5
Use Case 3	9	9	9	20	21	31
Use Case 4	0	1	1	22	31	9
Use Case 5	3	3	5	5	5	5
Use Case 6	0	0	0	3	5	9
Use Case 7	5	5	8	13	15	15
Use Case 8	5	5	5	5	5	5
Use Case 9	0	0	0	5	5	5
Use Case 10	0	2	3	9	10	10

Power factor control depending on the active power generated

Table A7. Power factor control depending on the active power generated: Losses.

Use Cases	Power Factor Control Depending on the Active Power Generated (*PF* = *f*(*P*)):			
	P = 0.5 & *PF* = 0.9	*P* = 0.5 & *PF* = 0.95	*P* = 0.65 & *PF* = 0.9	*P* = 0.65 & *PF* = 0.95
Use Case 1	0.0164	0.0155	0.0150	0.0146
Use Case 2	0.0180	0.0171	0.0166	0.0163
Use Case 3	0.0218	0.0211	0.0207	0.0204
Use Case 4	0.0210	0.0203	0.0201	0.0198
Use Case 5	0.0174	0.0165	0.0161	0.0158
Use Case 6	0.0201	0.0192	0.0189	0.0185
Use Case 7	0.0189	0.0179	0.0174	0.0170
Use Case 8	0.0189	0.0181	0.0177	0.0174
Use Case 9	0.0177	0.0169	0.0165	0.0162
Use Case 10	0.0190	0.0181	0.0177	0.0173

Table A8. Power factor control depending on the active power generated: Voltage deviations.

Use Cases	Power Factor Control Depending on the Active Power Generated (*PF* = *f*(*P*)):			
	P = 0.5 & *PF* = 0.9	*P* = 0.5 & *PF* = 0.95	*P* = 0.65 & *PF* = 0.9	*P* = 0.65 & *PF* = 0.95
Use Case 1	0	0	0	0
Use Case 2	2	2	2	2
Use Case 3	9	12	12	12
Use Case 4	1	2	2	3
Use Case 5	5	5	5	5
Use Case 6	0	0	0	0
Use Case 7	8	10	10	11
Use Case 8	5	5	5	5
Use Case 9	0	0	0	1
Use Case 10	3	3	3	3

Reactive power control depending on terminal voltage

Table A9. Reactive power control depending on terminal voltage: Losses.

Use Cases	Reactive Power Control Depending on Terminal Voltage ($Q = f(U)$)		
	$PF = 0.95$	$PF = 0.9$	$PF = 0.85$
Use Case 1	0.0176	0.0199	0.0217
Use Case 2	0.0190	0.0214	0.0234
Use Case 3	0.0231	0.0259	0.0282
Use Case 4	0.0218	0.0238	0.0256
Use Case 5	0.0180	0.0199	0.0215
Use Case 6	0.0214	0.0238	0.0259
Use Case 7	0.0197	0.0221	0.0242
Use Case 8	0.0191	0.0209	0.0224
Use Case 9	0.0182	0.0201	0.0217
Use Case 10	0.0196	0.0218	0.0237

Table A10. Reactive power control depending on terminal voltage: Voltage deviations.

Use Cases	Reactive Power Control Depending on Terminal Voltage ($Q = f(U)$)		
	$PF = 0.95$	$PF = 0.9$	$PF = 0.85$
Use Case 1	0	0	0
Use Case 2	2	2	2
Use Case 3	9	9	9
Use Case 4	1	0	0
Use Case 5	5	3	3
Use Case 6	0	0	0
Use Case 7	5	5	5
Use Case 8	5	3	3
Use Case 9	0	0	0
Use Case 10	2	0	0

References

1. Mansouri, N.; Lashab, A.; Sera, D.; Guerrero, J.M.; Cherif, A. Large photovoltaic power plants integration: A review of challenges and solutions. *Energies* **2019**, *12*, 3798. [CrossRef]
2. Munkhchuluun, E.; Meegahapola, L. Impact of the Solar Photovoltaic (PV) Generation on Long-Term Voltage Stability of a Power Network. In Proceedings of the 2017 IEEE Innovative Smart Grid Technologies—Asia (ISGT–Asia), Auckland, 2017; pp. 1–6. Available online: https://ieeexplore.ieee.org/abstract/document/8378456/authors#authors (accessed on 20 April 2020).
3. Eguía, P.; Torres, E.; Etxegarai, A. Impact of Photovoltaic Self-consumption on Power Losses and Voltage Levels of MV Distribution Networks. 2017. Available online: https://ieeexplore.ieee.org/abstract/document/8260161 (accessed on 20 April 2020).
4. Tavakkoli, M.; Pouresmaeil, E.; Godina, R.; Vechiu, I.; Catalão, J.P. Optimal management of an energy storage unit in a PV-based microgrid integrating uncertainty and risk. *Appl. Sci.* **2019**, *9*, 169. [CrossRef]
5. Aziz, T.; Ketjoy, N. PV Penetration Limits in Low Voltage Networks and Voltage Variations. *IEEE Access* **2017**, *5*, 16784–16792. [CrossRef]
6. Patil, A.; Girgaonkar, R.; Musunuri, S.K. Impacts of Increasing Photovoltaic Penetration on Distribution Grid—Voltage Rise Case Study. In Proceedings of the 2014 International Conference on Advances in Green Energy, (ICAGE), Thiruvananthapuram, India, 17–18 December 2014; pp. 100–105. Available online: https://ieeexplore.ieee.org/abstract/document/7050150 (accessed on 20 April 2020).
7. Gandhi, O.; Kumar, D.S.; Rodriguez-Gallegos, C.; Srinivasan, D. Review of power system impacts at high PV penetration Part I: Factors limiting PV penetration. *Solar Energy* **2020**, 1–21, In Press, Corrected Proof. Available online: https://www.sciencedirect.com/science/article/abs/pii/S0038092X20307118 (accessed on 20 April 2020).

8. Wajahat, M.; Khalid, H.A.; Bhutto, G.M.; Leth Bak, C. A Comparative Study into Enhancing the PV Penetration Limit of a LV CIGRE Residential Network with Distributed Grid-Tied Single-Phase PV Systems. *Energies* **2019**, *12*, 2964. [CrossRef]
9. Ul Abideen, M.Z.; Ellabban, O.; Al-Fagih, L. A review of the tools and methods for distribution networks' hosting capacity calculation. *Energies* **2020**, *13*, 2758. [CrossRef]
10. Bollen, M.H.; Rönnberg, S.K. Hosting capacity of the power grid for renewable electricity production and new large consumption equipment. *Energies* **2017**, *10*, 1325. [CrossRef]
11. Klonari, V.; Toubeau, J.; Lobry, J.; Vallée, F. Photovoltaic integration in smart city power distribution: A probabilistic photovoltaic hosting capacity assessment based on smart metering data. In Proceedings of the 2016 5th International Conference on Smart Cities and Green ICT Systems (SMARTGREENS), Rome, Italy, 23–25 April 2016; pp. 1–13.
12. Mateo, C.; Frías, P.; Cossent, R.; Sonvilla, P.; Barth, B. Overcoming the barriers that hamper a large-scale integration of solar photovoltaic power generation in European distribution grids. *Sol. Energy* **2017**, *153*, 574–583. [CrossRef]
13. Yuan, J.; Weng, Y.; Tan, C.W. Quantifying Hosting Capacity for Rooftop PV System in LV Distribution Grids. 2019, pp. 1–10. Available online: https://arxiv.org/abs/1909.00864 (accessed on 20 April 2020).
14. Hong, B.; Li, Q.; Miao, W.; Yan, H.; Liu, J. Energy Storage Application in Improving Distribution Network's Solar Photovoltaic (PV) Adoption Capability. *Energy Procedia* **2018**, *145*, 472–477. [CrossRef]
15. Denholm, P.; Margolis, R.M. Evaluating the limits of solar photovoltaics (PV) in electric power systems utilizing energy storage and other enabling technologies. *Energy Policy* **2007**, *35*, 4424–4433. [CrossRef]
16. Jeong, Y.C.; Lee, E.B.; Alleman, D. Reducing voltage volatility with step voltage regulators: A life-cycle cost analysis of Korean solar photovoltaic distributed generation. *Energies* **2019**, *12*, 652. [CrossRef]
17. Torres, I.C.; Negreiros, G.F.; Tiba, C. Theoretical and experimental study to determine voltage violation, reverse electric current and losses in prosumers connected to low-voltage power grid. *Energies* **2019**, *12*, 4568. [CrossRef]
18. Bremdal, B.A.; Mathisen, G.; Degefa, M.Z. Flexibility offered to the distribution grid from households with a photovoltaic panel on their roof. In Proceedings of the 2018 IEEE International Energy Conference (ENERGYCON), Limassol, Cyprus, 3–7 June 2018; pp. 1–6. Available online: https://ieeexplore.ieee.org/abstract/document/8398848 (accessed on 20 April 2020).
19. PARITY Project. Available online: https://parity-h2020.eu/ (accessed on 20 April 2020).
20. FLEXCoop Project. Available online: http://www.flexcoop.eu/ (accessed on 20 April 2020).
21. Mateo, C.; Cossent, R.; Gómez, T.; Prettico, G.; Frías, P.; Fulli, G.; Meletiou, A.; Postigo, F. Impact of solar PV self-consumption policies on distribution networks and regulatory implications. *Sol. Energy* **2018**, *176*, 62–72. [CrossRef]
22. Sauhats, A.; Zemite, L.; Petrichenko, L.; Moshkin, I.; Jasevics, A. Estimating the economic impacts of net metering schemes for residential PV systems with profiling of power demand, generation, and market prices. *Energies* **2018**, *11*, 3222. [CrossRef]
23. Ren, B.; Zhang, M.; An, S.; Sun, X. Local Control Strategy of PV Inverters for Overvoltage Prevention in Low Voltage Feeder. In Proceedings of the 2016 IEEE 11th Conference on Industrial Electronics and Applications (ICIEA), Hefei, China, 5–7 June 2016; pp. 2071–2075.
24. Zhang, Y.; Wang, X. PV reactive voltage regional autonomy control strategy for Q (U) improvement in distribution network. In Proceedings of the 2019 4th International Conference on Intelligent Green Building and Smart Grid (IGBSG), Yi-chang, China, 6–9 September 2019; pp. 497–501.
25. Duwadi, K.; Ingalalli, A.; Hansen, T.M. Monte Carlo Analysis of High Penetration Residential Solar Voltage Impacts using High Performance Computing. In Proceedings of the 2019 IEEE International Conference on Electro Information Technology (EIT), Brookings, SD, USA, 20–22 May 2019; pp. 1–6.
26. German Standard VDE-AR-N 4105. Available online: https://www.vde.com/en/fnn/topics/technical-connection-rules/power-generating-plants (accessed on 20 April 2020).

Article

Modeling and Detection of Future Cyber-Enabled DSM Data Attacks

Kostas Hatalis [1,*], Chengbo Zhao [1], Parv Venkitasubramaniam [2], Larry Snyder [3], Shalinee Kishore [2] and Rick S. Blum [2]

1 Department of Electrical and Computer Engineering, Lehigh University, Bethlehem, PA 18015, USA; chz317@lehigh.edu
2 Faculty of Electrical and Computer Engineering, Lehigh University, Bethlehem, PA 18015, USA; pav309@lehigh.edu (P.V.); shk2@lehigh.edu (S.K.); rblum@lehigh.edu (R.S.B.)
3 Faculty of Industrial and Systems Engineering, Lehigh University, Bethlehem, PA 18015, USA; larry.snyder@lehigh.edu
* Correspondence: kmh511@lehigh.edu

Received: 16 July2020; Accepted: 17 August 2020; Published: 21 August 2020

Abstract: Demand-Side Management (DSM) is an essential tool to ensure power system reliability and stability. In future smart grids, certain portions of a customer's load usage could be under the automatic control of a cyber-enabled DSM program, which selectively schedules loads as a function of electricity prices to improve power balance and grid stability. In this scenario, the security of DSM cyberinfrastructure will be critical as advanced metering infrastructure and communication systems are susceptible to cyber-attacks. Such attacks, in the form of false data injections, can manipulate customer load profiles and cause metering chaos and energy losses in the grid. The feedback mechanism between load management on the consumer side and dynamic price schemes employed by independent system operators can further exacerbate attacks. To study how this feedback mechanism may worsen attacks in future cyber-enabled DSM programs, we propose a novel mathematical framework for (i) modeling the nonlinear relationship between load management and real-time pricing, (ii) simulating residential load data and prices, (iii) creating cyber-attacks, and (iv) detecting said attacks. In this framework, we first develop time-series forecasts to model load demand and use them as inputs to an elasticity model for the price-demand relationship in the DSM loop. This work then investigates the behavior of such a feedback loop under intentional cyber-attacks. We simulate and examine load-price data under different DSM-participation levels with three types of random additive attacks: ramp, sudden, and point attacks. We conduct two investigations for the detection of DSM attacks. The first studies a supervised learning approach, with various classification models, and the second studies the performance of parametric and nonparametric change point detectors. Results conclude that higher amounts of DSM participation can exacerbate ramp and sudden attacks leading to better detection of such attacks, especially with supervised learning classifiers. We also find that nonparametric detection outperforms parametric for smaller user pools, and random point attacks are the hardest to detect with any method.

Keywords: demand side management; demand response; cyber-physical systems; dynamic pricing; load forecasting; attack detection

1. Introduction

Demand-Side Management (DSM) is an essential component in smart grids for planning, monitoring, and modifying consumer load levels. Furthermore, cyber-enabled DSM will allow smart grids even higher levels of automated decision-making capabilities to selectively schedule loads on local grids to improve power balance and grid stability. Such a cyber approach relies heavily on real-time, two-way communication capabilities between a central controller and various adaptable loads. Research into the reliability of the cyberinfrastructure that enables DSM is therefore vital. The main concerns in ensuring DSM security lay in the feedback mechanism of real-time electricity pricing and distributed DSM controllable loads. For instance, in residential grids, a single home consumes only a small amount of electric power, and if it becomes compromised, then it might not have a noticeable impact on the power grid. However, a carefully planned or even random cyber attack might impact other loads not under attack by taking advantage of the feedback mechanism of load management.

Two-way communication capabilities of Advanced Metering Infrastructure (AMI) enable a utility or Independent System Operator (ISO), in the retail power markets, to collect high-resolution energy usage from consumers and enable dynamic pricing to adapt to consumer demand [1]. AMIs provide an efficient way for ISOs to schedule prices and then communicate them to consumers for automatic DSM control of certain portions of a consumer's load. AMIs can also provide practical ways for ISOs to set DSM goals, such as reducing peak or decreasing aggregate load levels through price influences. However, there are several vulnerabilities in AMIs that present noteworthy security issues since they are directly accessible by users [2]. Due to the large scale deployments of AMIs, ISOs encourage the utilization of marginally cheaper hardware, resulting in constrained computational resources to allow for robust security capacities, such as intrusion monitoring.

DSM programs also utilize demand response, a specific tariff or program to motivate customers to respond to changes in price or electricity availability over time by altering their regular electricity use habits. We take this a step further and look at future cyber-enabled DSM programs [3] that will autonomously control household loads such as water heaters and HVAC units based on Real Time Pricing (RTP). As part of the reliable implementation of this future cyber-DSM, it is crucial to understand the dependency between dynamic pricing and automatic demand response as well as the risks. We hypothesize that cyber-DSM programs can be particularly vulnerable to cyber attacks such as false pricing information or direct load manipulation, especially when the participation rate in DSM is high.

Our work is thus motivated to study these vulnerabilities in DSM. We provide a mathematical formulation of the feedback between utilities and DSM systems, and then simulate, analyze, and test different detection methods for attacks on such feedback. This relationship between load and price is shown visually in Figure 1. As prices go up, demand naturally responds by decreasing. However, if AMIs are hijacked, and false lower prices are reported to DSM systems, there will be an inappropriate increase in demand. A similar effect happens if an attacker directly controls user loads. Then, a higher load usage by the attacker may inadvertently lead to higher prices for the rest of the grid. We present a standard of how attackers can exploit such a dependency.

We propose a mathematical framework of the feedback between price setting and DSM systems to study how attacks can be structured and how to detect them. Similar frameworks have been considered in previous work. In [4], the author applied a profitable attacking strategy to a price-load dependency model. Another control-theoretic approach to derive the fundamental conditions of Real Rime Pricing (RTP) stability under integrity attacks is described in [5]. The authors in that work use an elasticity model between price and load and apply an autoregressive model to obtain the forecast load and price. The main contributions of our approach can be summarized as follows:

1. We formalize the demand–price relationship using a nonlinear elastic demand model to achieve DSM goals. We simulate load and pricing data in a DSM system, in particular under a strategic conservation scheme.
2. We propose two modes of attacks on DSM systems: false pricing data injection and direct load manipulation. We prove their equivalence and highlight three types of attacks that could be undertaken by each mode. We then empirically show how an increased use of DSM can exacerbate attacks.

3. We simulate these attacks and compare sequential change-point and machine learning methods, as well as a deeper dive in parametric vs. nonparametric methods, for detecting DSM attacks. In our main results, we demonstrate the impact of DSM on detection performance and identify what kind of detectors are effective for different attacks.

In Section 2, we provide a literature review on DSM and important DSM strategies on real-time pricing and load forecasting. In Section 3, we present a block bootstrap technique for simulating the non-DSM load distribution of a micro-grid of N homes from template residential load time series. We then propose dependency models for the feedback nature of load and prices in Section 4, where we also showcase simulations of residential load and electricity prices when an automatic DSM program controls certain portions of consumer demand as a function of price. In Section 5, we present two modes of cyber attacks, direct load manipulation attacks, and price data injection attacks that can have a significant influence on the feedback of load and price. We prove these two attacks are equivalent. In Section 6, we review several sequential detection and supervised learning methods. In Section 7, we present our experimental results that compare performance under different DSM participation rates when they are under three types of attacks: ramp, sudden, and point attacks. We find that point attacks are hardest to detect, and supervised learning methods can provide better results than sequential detection methods. To provide a deeper analysis, we also conduct a comparison of parametric and nonparametric detectors. We conclude in Section 8 with possible directions of future work.

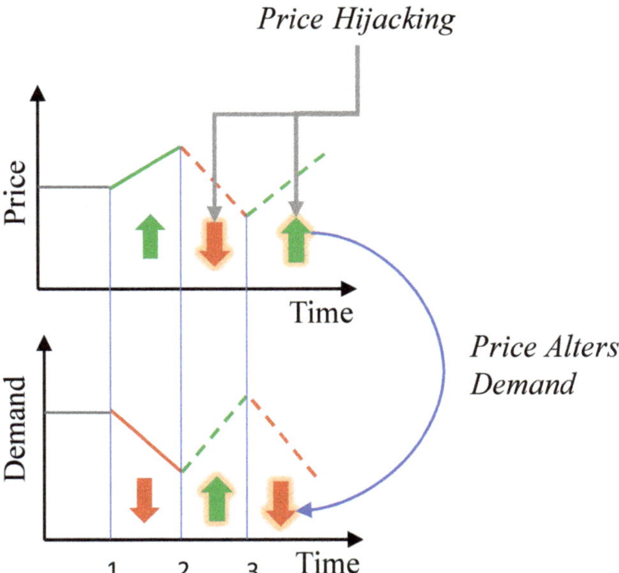

Figure 1. Feedback effect between price and DSM demand. As prices go up, demand decreases. However, if prices are hijacked and false prices are fed to DSM systems, then a false low price increases demand, and a false high price can decrease demand. The same is true if demand was altered by an attack. If load usage is increased by an attacker, then prices would increase and vice versa.

2. Background

2.1. Demand Side Management

DSM is an active and voluntary approach for reducing electricity use through activities or programs that promote electric energy efficiency, conservation, or more efficient management of electric energy loads [6]. Very often, ISO's utilize financial incentives and educational programs to modify consumer demand. The main goals of DSM are peak clipping, valley filling, load shifting, strategic load growth, flexible load shaping, and strategic conservation. We summarize these goals in Figure 2. In these goals, consumers are encouraged to use less energy during peak hours, or to move the time of energy use to off-peak times such as nighttime, or reduce overall consumption. Other applications for DSM are to aid grid operators in balancing intermittent generation from wind and solar farms due to their volatile nature, which may not coincide with energy demand at different times.

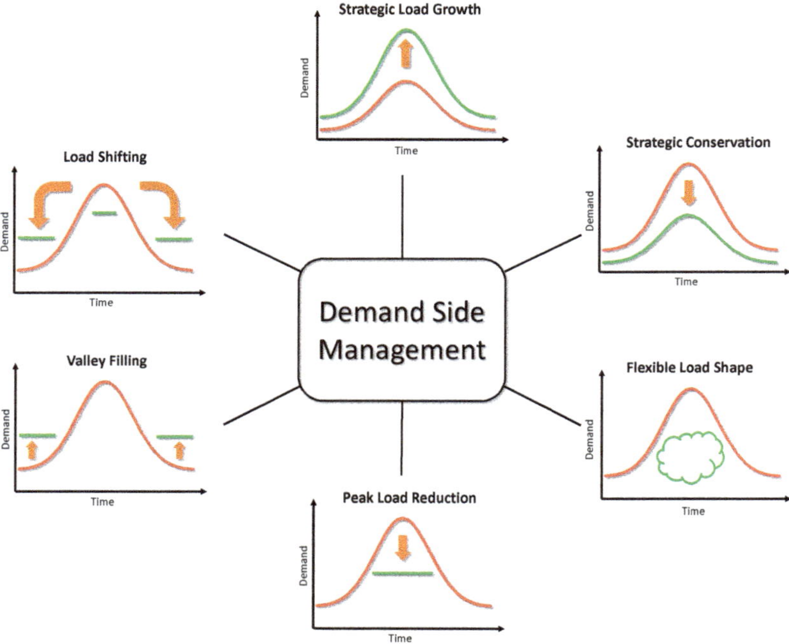

Figure 2. Various demand side management goals.

Our study focuses on modeling and simulating the DSM goal of strategic conservation due to its simplicity and essential use in the smart grid. This goal also makes it easier to study attacks on DSM by modeling strategic conservation as a general reduction in load. Attacks then could stand out more visibly than under other goal schemes like flexible load shaping. More specifically, this DSM goal aims at reducing aggregate load demand through directed reduction of electricity consumption. The successful implementation of strategic conservation programs usually requires some combination of financial incentives to customers, energy-efficient building standards, and appliance efficiency improvement. We envision that AMI and smart appliances in residential DSM programs will automatically control specific portions of consumer load as a function of real-time electricity prices to achieve the goal of strategic conservation.

Most DSM programs are formulated as an optimization problem as follows:

$$\min_{P_t} \sum_{t=1}^{T} (L_t - L_t')^2,$$

where P_t is the RTP at time t, L_t is actual load, and L_t' is the target load level the ISO is interested in achieving via DSM. The aim is to choose a price P_t for each time step such that the actual load would reach as close as possible to the target level. In [7], a DSM strategy is proposed based on a heuristic optimization strategy to shape the load curve close to the desired shape. A heuristic-based evolutionary algorithm was used to solve the above minimization problem. A multi-agent game-theoretical DSM approach is proposed in [8]. The authors use game theory and formulate an energy consumption scheduling game, where the players are the users, and their strategies are the daily schedules of their household appliances and loads. In [9], the minimization problem is solved by utilizing a feedforward neural network to map the nonlinear relationship between price and load. Recently, the DSM problem is addressed in [10] as a multi-objective optimization problem that also seeks to balance other merit functions such as energy production cost, customer preferences, and other constraints.

2.2. Load Forecasting

Load Forecasting (LF) techniques are essential for RTP and other ISO operations by predicting future energy requirements of a system from previous data and weather conditions. It is recognized as the initial building block of utility planning efforts and ensures the balance between supply and demand of energy; LF thus plays a vital role in our DSM formulations. For a given system and requirements, LF provides predictions for specific periods. These periods are divided into short-, medium-, and long-term forecasts. Short-term LF is used to predict the load on an hourly basis for up to one week when considering daily operations and cost minimization. Medium-term LF typically predicts load on a weekly, monthly, or yearly basis for efficient operational preparations. Long-term LF is used to predict the decades' load to facilitate grid and generation expansion planning. In this work, we look at short-term LF to predict loads from one hour to a week.

LF models can be divided into two approaches [3]: the first being statistical based modeling and the second being machine learning. The statistical approach can be further broken down into regression and time series models. Multiple linear regression can be used with the weighted least squares estimation technique to form a relationship between different independent covariates when load depends on such things as weather conditions. Regression models have been applied in LF in different works, such as in [11]. Time series models are also prevalent to apply to LF. The most common model is the AutoRegressive Moving Average (ARMA) model and its variants that include components such as integration (I), Fractional Integration (FI), multiVariate series (V), Seasonality (S), eXogenous (X) data, Conditional Heteroskedasticity (CH), and Nonlinearity (N). For our formulation, we employ SARIMA for LF.

Hyperparameters of ARMA models can be configured using Box–Jenkins decomposition or grid-search with the Akaike information criterion. Numerous studies have looked at all the different ARMA models for LF [12]. Other time series methods for LF include simple exponential smoothing [13] and the Holt–Winters seasonal method [14]. Time series analysis and regression analysis share many models and ideas, but they are theoretically different. Time series analysis first deals with time-indexed stationary data and accounts for the autocorrelation between time events. In regression, we assume there is no autocorrelation, and that all observations are independent and identically distributed. Furthermore, we also assume that in regression, the data are homeostatic and do not exhibit multicollinearity.

Most recently, machine learning methods have seen a huge spike in LF research. Machine learning models are data-driven, typically providing a nonlinear fit to input covariate data to predict load. The advantages of this approach include not needing preconditions for data such as stationarity (a requirement for most time series methods), excels at modeling nonlinear dependencies, and can fit large data sets. Disadvantages for most machine learning models are that most hyperparameters are continuous (difficult to tune), they require extensive feature engineering, and may get stuck in local minimums. Models for LF include support vector machines [15], feedforward neural networks [16], recurrent neural networks [17], random forests [18], and ensemble learning [19].

2.3. Real-Time Pricing

Every consumer of electricity is charged a certain amount per kiloWatt-hour (kWh) of energy. Such a charge covers the costs associated with the generation, transmission, and distribution of electricity. The two main types of costs are operational and fixed costs. During the 20th century, tariffs were used to recover costs. Lately, clever pricing schemes have been developed to meet the requirements of modern power systems [20], such as RTP, where consumers are charged with a price closest to the real price of generation at a specific interval in time. RTP plays an integral part in time-based DSM programs that make consumers choose the time of consumption of power as a response to prices [21]. Cyber-enabled DSM programs are an automated form of time-based DSM programs.

There are two types of RTP schemes, hourly pricing, and day-ahead pricing. In the first type, the price of electricity is released on an hourly basis for the next hour. In day-ahead pricing, prices are announced for the next 24 h based on predicting the load demand and the cost of generation. RTP signals combined with DSM automation at the consumer level provide benefits to both consumers and utility. An adequately designed RTP scheme increases the grid's reliability, reduces associated costs with the generation, and lowers electricity bills of consumers. Further review of RTP and other dynamic pricing schemes can be found in [3].

3. Microgrid Simulation

Before modeling the load-price dependency of a DSM system, we first need to obtain some ground truth data of what load data from a residential micro-grid look like without the presence of DSM, where we assume the elasticity of demand to price is very low. To do this, we use the power time series from several homes as templates for our grid and generate artificial N household datasets. There are several ways to generate artificially residential load data, such as using power grid simulators like MATPOWER [22] or GridLAB-D [23]. Our object is to create a model-free real-data driven alternative to grid simulators. We can generate unlimited but plausible univariate load data to serve as the base demand for sample households before applying a DSM system.

3.1. Data Source

We use the UMass Smart* [24] dataset, 2017 release, for the simulation of micro-grid load time series. The Smart* project built a data collection infrastructure that records data from a variety of sensors deployed in real homes. Their infrastructure supports both pulling data by querying individual sensors and pushing data from sensors to a gateway server, which ran on their software tools. The 2017 Home dataset release is comprised of electrical power readings from seven homes from 2014 to 2016 at a minute resolution. It includes readings from individual appliance sensors as well as total power usage of each home. We chose to use these seven datasets as template homes in simulating the power usage of a micro-grid due to the breadth of the data collected. For DSM attack research, these datasets can help model an attacker compromising individual appliances. For each home, the power consumption is given in kW for every minute. We convert this time series to kWh with a resolution of one hour, which is typical for smart meter readings and RTP modeling [3]. We do this by obtaining the average power consumption within an hour and multiplying it by the time period as such

$$E_{(kWh)} = t_{(hr)} \times \frac{1}{60}\sum_{i=1}^{60} P^{i}_{(kW)} \tag{1}$$

3.2. Block Bootstrap Simulation

In the generation of new time series from sample data, several approaches can be applied depending on the series' statistical properties. Data that are stationary can be modeled and generated using an ARMA process [25]. An ARMA model is fitted to the data, and then future data are sampled from the ARMA distribution. If there is no serial correlation, then the distribution of some sample data can be modeled using Markov Chain Monte Carlo [26], and new data can be sampled from this estimated distribution. However, in the case that data exhibited autocorrelation and non-stationarity in the presence of a periodic seasonal pattern, a natural choice is to use the block bootstrap method [27].

The bootstrap method comes from simulation statistics for estimating the distribution of a statistic such as the mean or variance. Bootstrap is particularly useful when there is no analytical form to estimate some underlying distribution. A bootstrap analysis is conducted by using the Monte Carlo algorithms with replacement. Data are sampled with replacement until a new set is formed, and then statistics are calculated from that new set. The process can be repeated to get a more precise estimate of the Bootstrap distribution and form confidence intervals for those statistics. The block bootstrap is used when the data, or the errors in a model, are correlated. The block bootstrap attempts to replicate the serial correlation by resampling blocks of data instead of individual observations. This is why the block bootstrap is used primarily with correlated time series. In block bootstrap, blocks sampled can overlap or be non-overlapping. For load time series simulation, we use block bootstrap with non-overlapping blocks to preserve the daily seasonal pattern of power consumption. The process of block bootstrap simulation of a new home power usage from a template home is shown in Figure 3.

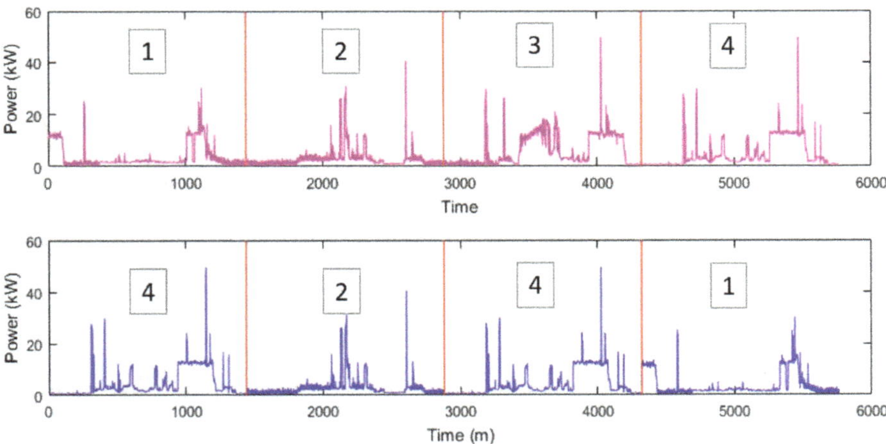

Figure 3. Process of block bootstrap simulation of a new home power usage (**bottom**) from a template home (**top**). Example simulation samples are taken from four days from the template series with replacement.

4. Dependency Model

4.1. Modeling Elastic Demand

For analyzing the feedback dependency between load and price in a DSM setting, we first need to define the supply and demand relationship of electricity. To do so, we utilize the well-known measure in economics, the Price Elasticity of Demand (PED) [28], which can be given by

$$\epsilon_d = \frac{dL}{dP} \cdot \frac{P}{L}. \tag{2}$$

PED shows the responsiveness of the load L demanded of electricity to a change in Price (P). Recent work [9] in demand response, particularly in the context of cybersecurity, has utilized this approach. Our procedure differs in that we use real demand data to determine the "unmodulated" pattern of load data. An absolute value of PED = 1 shows unitary elasticity. For instance, when $\epsilon_d = -1$, then a 1% change in the price will have a 1% change in the load demanded. As prices increase, the load will decrease. When absolute PED falls between 0 and 1, this signifies that the load demand is inelastic, while a value greater than 1 says that the demand is elastic. When $|\epsilon_d| = 0$, the demand is perfectly inelastic. A change in price does not affect the load, while $\epsilon_d = \infty$ represents perfect elasticity. If ϵ_d is constant, the whole demand curve is then

$$L = a \cdot P^{\epsilon_d} \tag{3}$$

where a is a scaling constant. An example demand curve estimated from Equation (3) can be seen in Figure 4. The figure also showcases the nonlinear relationship between load and price where as the price, the independent variable, increases, the load demanded, the dependent variable, decreases.

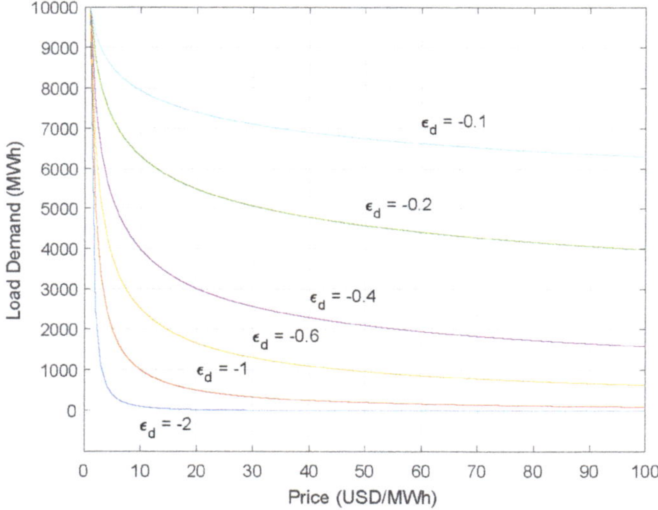

Figure 4. Load as a function of price with arbitrary price range \$1–\$100, $\alpha = 10{,}000$, and $\epsilon_d = -0.1, -0.2, -0.4, -0.6, -1, -2$.

4.2. Modeling Consumer DSM

In modeling, the relationship between load and price under a DSM program, it is important to define each customer's individual loads in determining the aggregate load. For a customer who does not participate in a DSM program, their load is determined by the stochastic demand process of users' actions such as watching TV or using the AC. Demand is impacted by multiple factors such as user preferences, weather, and time of day. In this process, electricity prices have a small influence on demand; individual customer demand is fairly inelastic to price. Following the derivation in Equation (3), we define the load usage of an individual customer i for time t as

$$\phi_{t,i} = \theta_{t,i}(P_t + P_c)^{\epsilon^d_{t,i}},$$

where P_t is the RTP for time t, P_c is constant of the retailer's market costs which does not vary with RTP, $\theta_{t,i} \geq 0$ is a scaling factor representing the stochastic process that determines the user load, and $\epsilon^d_{t,i}$ is the elasticity coefficient for the individual customers sensitivity to price changes. It can vary over time but, without DSM incentives, most users have a fairly inelastic PED.

For experimental purposes, in modeling individual user loads, we set $\phi_{t,i}$ equal to simulated bootstrapped user load profiles defined in Section 3.2. We assume that prices, user preferences $\theta_{t,i}$, and $\epsilon^{d}_{t,i}$ have been absorbed in the calculation of the simulated load series. Thus, we use $\phi_{t,i}$ as a reference point to show how much electricity a user wants to consume without the influence of a DSM program. We model the task of the demand response administrator as modifying $\phi_{t,i}$ to some desired load levels.

Realistically, only a certain portion $\kappa_{t,i} \in [0,1)$ of customer i's power usage will be under control of a cyber DSM program. There will always be some stochastic and, more importantly, an inelastic component of power usages, such as using a microwave oven or electric hairdryer. We model the DSM modulated load (including the inelastic portion) as follows:

$$l_{t,i} = (\kappa_{t,i}\phi_{t,i})P_t^{\epsilon^{dsm}_{t,i}} + (1 - \kappa_{t,i})\phi_{t,i}, \tag{4}$$

where the first part $(\kappa_{t,i}\phi_{t,i})P_t^{\epsilon^{dsm}_{t,i}}$ is the load level customer i that allows the DSM program to determine as a function of price P_t and the DSM's elasticity to price $\epsilon^{dsm}_{t,i}$. Price elasticity of DSM $\epsilon^{dsm}_{t,i}$ may vary over time and customer and affects how much power usage should be affected by price. The second portion $(1 - \kappa_{t,i})\phi_{t,i}$ of a customers load is the stochastic component. The DSM portion parameter $\kappa_{t,i}$ is defined as a function of time where users can add or remove house loads under DSM control over time. The term $\kappa_{t,i}$ can be modeled as a random variable (e.g., Uniform or Gaussian) or as a fixed constant for users. Once each customer's load is determined, total load (modified by a cyber-DSM program) for time t for N customers is calculated by

$$L_t = \sum_{i=1}^{N} l_{t,i}. \tag{5}$$

We also define the aggregate base load as

$$\Phi_t = \sum_{i=1}^{N} \phi_{t,i}, \tag{6}$$

which represents the total demand had there not been a DSM program for a time period t (IE $\kappa_{t,i} = 0, \forall_i$).

4.3. Modeling Strategic Load Conservation

We propose the following approach to model the real-time pricing from a strategic load conservation perspective, wherein, on an hourly basis, the price is a function of the aggregate load to achieve the desired load level L'_t. The approach takes as input a forecast $\hat{\Phi}_t$ of the base aggregate load to calculate price P_t. This prediction can be defined as

$$\hat{\Phi}_t = f_{pred}\left(\Phi_{t-k:t-1}, P_{t-k:t-1}, X\right), \tag{7}$$

where inputs to the prediction model are past base load Φ and price P values from time $t-k$ to $t-1$, and other predictor variables X such as time-of-day and weather information. Various types of prediction models can be used for f_{pred} such as neural networks as reviewed in Section 2.2. We now define RTP based on the formulation in Equation (3) as such

$$P_t = \left(\frac{L^*}{\hat{\Phi}_t}\right)^{1/\hat{\epsilon}^{dsm}} \qquad (8)$$
$$L^* = a \cdot P_t^{\hat{\epsilon}^{dsm}}$$

where we can have two goal scenarios for L^*

$$\begin{aligned} &\text{Goal 1: } L^* = L'_t \\ &\text{Goal 2: } L^* = L'_t + (L'_{t-1} - L_{t-1}) \end{aligned} \qquad (9)$$

The component L^* adjusts the RTP based on two goals the ISO may have. The first goal is to adjust the price to push power usage directly to the target level L'_t with the assumption that there is near 100% DSM participation by all customers. The expectation is that, if demand for time t is $\hat{\Phi}_t$, then a price point is set to push load usage to L'_t. Of course, if participation is less than 100%, which is more likely, the target level L'_t will not be reached by P_t. To push the aggregate load from all users, as close as possible to the target load level with an unknown amount of participation, a penalty would need to be added to P_t. We model this as Goal 2, where the idea is to affect the power usage of those under cyber-DSM control even more than goal 1 to compensate users who are not participating in DSM.

Some users will not be participating, or only have a small portion of their power usage under control by the cyber-DSM program. We model their remaining power usage as inelastic to RTP. Thus, to push aggregate load to a target level, taking into account that some load usage is inelastic, we need to push RTP much higher or lower to have a bigger effect in pushing DSM controlled load closer to the target load. This is what Goal 2 attempts to do, with the component $L^* = L'_t + (L'_{t-1} - L_{t-1})$ taking the target load level for time t and adding the difference from the previous target load L'_{t-1} and realized load L_{t-1} as a penalty to adjust RTP to compensate for the difference. If $L^* < 0$, then we set $L^* = 10$ or to some arbitrary small target value. By subtracting the difference between the previous load and target level, we make up for users not participating in DSM by forcing DSM users to have a higher price to push their load even lower.

The term $\hat{\epsilon}^{dsm}$ in Equation (8) is an estimate of the price elasticity of DSM of the whole grid; if individual user coefficients $\epsilon_{t,i}^{dsm}$ are unknown, then $\hat{\epsilon}^{dsm}$ can be estimated from observing past values of price and load under different levels of DSM control. Alternatively, the ISO can define ϵ^{dsm} for all household cyber DSM programs. The formulation in Equation (8) sets prices by comparing the adjusted target load for time t to the forecasted base demand $\hat{\Phi}_t$ for the same time. This demand $\hat{\Phi}_t$ would be the level if no load were under DSM influence, thus to influence and alter it, Φ_t needs to be estimated as accurately as possible.

In our approach, if the aggregate load is above the target load, RTP is set higher to decrease demand. If the aggregate load is lower than the target load, the price decreases to increase demand. Thus, as also can be observed in Equation (8), there is direct feedback between price and load Equation (4). The block diagram in Figure 5 also outlines this feedback that showcases the relationship between the utility and the grid. Generation sets the target load based on the price and supply of power, and the controller sets the price signal and the elasticity of demand coefficient for DSM systems. The price is then communicated into the grid into DSM systems, which adjust load usage appropriately. The bold red lines in Figure 5 highlight the feedback relationship between price and demand. The scope of work is in the mathematical modeling of the controller and DSM system relationship. With such a model, we present in the next section how attackers can exploit it.

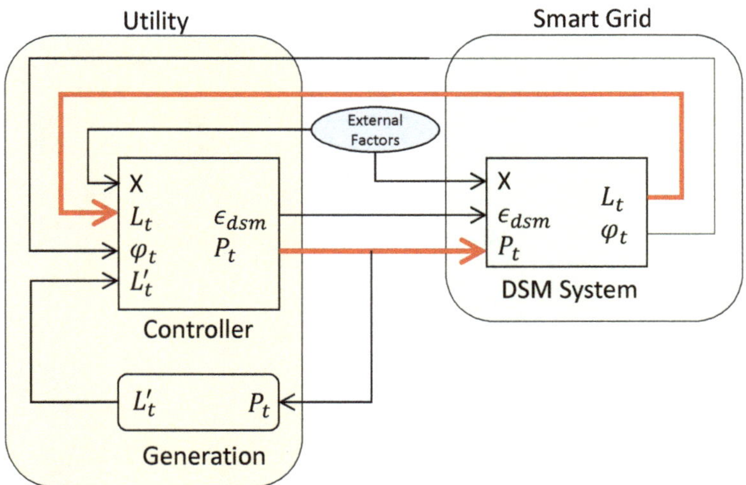

Figure 5. Block diagram highlighting the feedback between the utility and grid.

4.4. Simulation Parameters and Assumptions

For experimental purposes, in modeling individual user loads, we set $\phi_{t,i}$ equal to simulated bootstrapped user load profiles defined in Section 3.2. We assume that prices, user preferences $\theta_{t,i}$, and $\epsilon^d_{t,i}$ have been absorbed in the calculation of the simulated load series. Thus, we use $\phi_{t,i}$ as a reference point to how much electricity a user wants to consume without the influence of a DSM program. We model the task of the ISO as modifying $\phi_{t,i}$ to some desired load levels. Furthermore, we include the following assumptions in our modeling and simulation. We assume that the ISO can define ϵ^{dsm} for each household and we set it as a constant for all customers and time. For additional simplicity, we model κ_i also invariant in time, but it may vary per customer. Lastly, through AMI, we assume the ISO can obtain an estimated reading of Φ_{t-1} but not of individual user $\phi_{t-1,i}$ to preserve privacy. This way the ISO has a time series of estimated non-DSM load demand in order to provide predictions $\hat{\Phi}_t$.

For all case studies in the rest of our paper, we use the UMass Smart* dataset (2017 release) to bootstrap simulate residential load as described in Section 3. We simulate a micro-grid of $N = 200$ residential homes for a time period of T to obtain a *phi* distribution that defines a base load profile for each home i at each time t. We set the DSM demand elasticity for each user to $\epsilon^{dsm} = - - 1, \forall_i$ to allow the DSM component of customer load to be sensitive enough to price changes. For all case studies, we set a simple flat target load of $L_{target} = 200$ kWh. We note, however, that, with our pricing formulation in Equation (8), we can model any DSM goal from peak load reduction to flexible load shaping. These simulation parameters are summarized in Table 1. For forecasting $\hat{\Phi}_t$, we use the naive persistence prediction method.

Table 1. Simulation parameters used in case studies.

Simulation Parameters	
N	200 homes
ϵ_{dsm}	-1
L_{target}	200 kWh

In Figures 6 and 7, we demonstrate a price-load feedback simulation for a period of $T = 48$ h under Goals 1 and 2 with parameters defined in Table 1. Under the Goal 1 scenario when $\kappa_i = 0, \forall_i$, shown in Figure 6a, we see that prices range from \$0.01 to \$0.03 per kWh. As expected with no DSM program to influence demand, the observed load is equal to the base load. The same is observed in Figure 7a when the simulation is run with the Goal 2 RTP model. The only difference under this scenario is the price range which is exceptionally high ranging from \$0.01 to \$0.6 per kWh. This is expected in this scenario since the ISO is attempting to set prices to maximize the effect on DSM customers, of which there are none, but this is unknown to the utility. With no DSM, the mean observed load is 332 kWh.

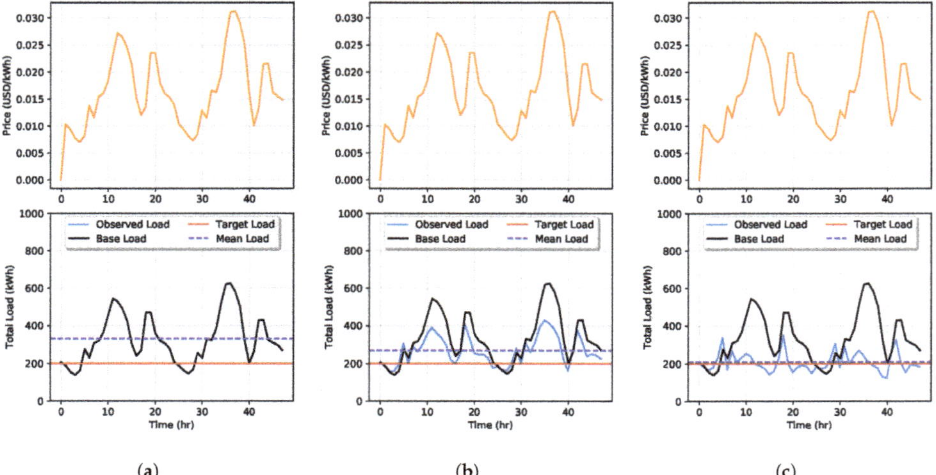

Figure 6. Simulation of price-load interaction with Goal 1 with DSM when $\kappa_i = 0, \forall_i$ (**a**), $\kappa_i = 0.5, \forall_i$ (**b**), and $\kappa_i = 0.99, \forall_i$ in (**c**). Since this simulation is with Goal 1, we see that, while price affects the load demand, load doesn't affect price.

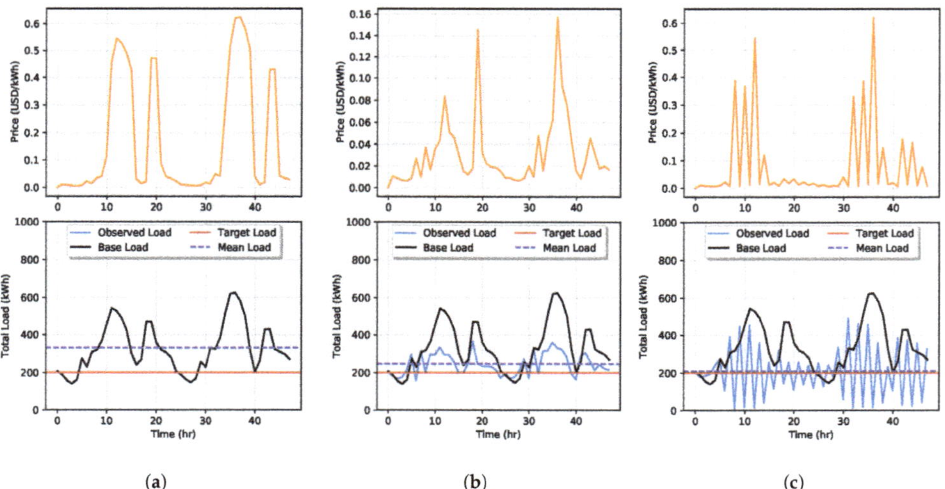

Figure 7. Simulation of price-load interaction with Goal 2 with DSM when $\kappa_i = 0, \forall_i$ (**a**), $\kappa_i = 0.5, \forall_i$ (**b**), and $\kappa_i = 0.99, \forall_i$ in (**c**). In this simulation under with Goal 2, we see that price does affect the load demand and vice versa.

Next, we rerun the simulation setting $\kappa_i = 0.5, \forall_i$. In the Goal 1 scenario in Figure 6b, we see that the base load was reduced with a mean observed load of 269 kWh. In Figure 7b, the same simulation is run with Goal 2. Here, the mean observed load was further reduced to 246 kWh, but large price spikes occur at peak load times. Finally, we run simulations for $\kappa_i = 0.99, \forall_i$ to study the effects of high penetration of DSM. In simulating both goals, shown in Figures 6c and 7c, the mean observed load is reduced to 208 kWh, which is very close to the target load. However, in Goal 2, we see great resonating feedback effects occur when prices spike very high. RTP increases as a response to large values in the observed load. Then, when the load decreases to low levels, a price decrease causes the load to spike more during the next time step, while under Goal 1, this is not observed. The higher prices set in Goal 2 would see a large cost to DSM participating customers.

5. DSM Attack Models

An attacker can exploit the feedback between the customer and utility in determining RTP and load usage by cyber DSM programs by injecting false price or corrupted load data into the feedback loop. The attack exploitations we study here differ from the false data injection attacks studied in other smart grid papers. Most false data injection attack works [29] study the compromise in energy management systems to alter power state estimates by the utility operator. In our case, we study attacks that aim to alter a user's load profile by exploiting cyber DSM vulnerabilities.

For modeling attacks on a cyber DSM managed micro-grid, we assume that the attacker compromises a subset of all the N customers; we denote this subset as $\mathcal{A}$, for an attack period $t \in T_a$. We study two modes of attacks: false pricing data injection attacks in which a compromised user receives manipulated pricing information and a direct load manipulation attack in which the compromised customer's appliances are under the control of the attacker. When communication encryption is broken with an AMI, then a pricing data injection attack can occur. A direct load manipulation can occur by hacking into a cyber DSM load controller, or directly hacking into smart appliances, altering a user's load profile. The two modes of attacks are outlined below.

- **False Pricing Data Injection**: The attacker can manipulate prices P_t received by each compromised costumer $i \in \mathcal{A}$, and the received price P_t^i can be different for various customers in order to achieve the attacker's desired effect:

$$P_{t,i}^a = P_{t,i} + a_{t,i}^P, \forall_i \in \mathcal{A}, t \in T_a$$

This has the effect of compromising the demand response of a customer in the following way:

$$l_{t,i}^{a^P} = (\kappa_i \phi_{t,i})(a_{t,i}^P)^{\epsilon^{dsm}} + (1 - \kappa_i)\phi_{t,i}$$

- **Direct Load Manipulation**: The attacker can manipulate the load of each compromised customer $l_{t,i}, i \in D$ directly:

$$l_{t,i}^{a^L} = l_{t,i} + a_{t,i}^L, \forall_i \in \mathcal{A}, t \in T_a$$

Under both attack modes, we would get a compromised aggregate load which may include one or both attacks occurring simultaneously

$$L_t^a = \sum_{i=1}^{N} l_{t,i}^{a^P} + l_{t,i}^{a^L}.$$

These two modes of attack are equivalent as they both affect a customer's load response as long as part of the load is under cyber DSM control that is sensitive to price changes.

Theorem 1. *Given a set of customers $\mathcal{A}$ compromised by the attacker, there always exists a direct load manipulation attack such that all customers behave the same as a pricing data injection attack and vice versa for* $\kappa_i > 0, \forall_i, t \in T_a$.

Proof. Setting both attacks to have the same load, then the attack load is set as follows:

$$a^L_{t,i} = (\kappa_i \phi_{t,i})(a^P_{t,i})^{\epsilon^{dsm}} + (1-\kappa_i)\phi_{t,i},$$

and the equivalent attacked price is

$$a^P_{t,i} = \left(\frac{a^L_{t,i} - (1-\kappa_i)\phi_{t,i}}{\kappa_i \phi_{t,i}}\right)^{1/\epsilon^{dsm}}$$

If $\kappa_i = 0$, then the two attack modes are not equivalent since a price attack will have no effect on customer load. □

Since false pricing data injection and direct load manipulation attacks are equivalent, we focus only on direct load manipulation attack analysis.

There are different goals an attacker can have to harm the power grid or exploit it. For example, an attacker can cause chaotic metering by messing the metering data transmission, efficiency loss of the energy provided by causing higher load volatility, or the energy system failure by overloading the power lines or devices. The focus of this work is efficiency loss by increasing user loads through direct load manipulation. In this scenario, we introduce three possible types of load attacks: ramp, sudden, and point attacks. These types of attacks are visualized in Figure 8 wherein plot (a) an attacker gradually increases a user's load over time. In plot (b), an attacker suddenly ramps up the power usage to a specified level, and, in plot (c), we demonstrate a point attack where the attacker increases loads only for specific hours.

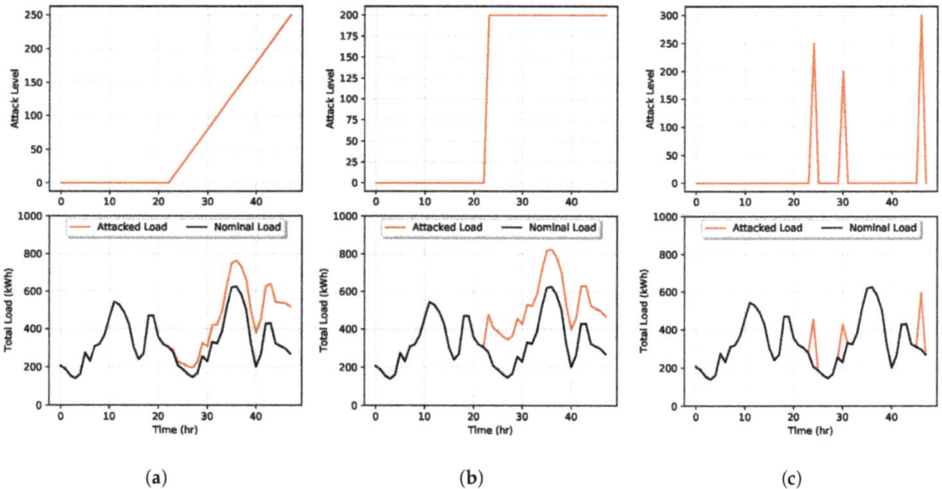

(a) (b) (c)

Figure 8. Examples of different type of DSM attacks: (**a**) ramp attack, (**b**) sudden attack, and (**c**) point attack.

6. Attack Detection

Here, we outline sequential parametric, supervised learning-based, and nonparametric methods for attack detection. The sequential detections methods are the one-sided CUmulative SUM (CUSUM) test [30] and windowed GLRT [31]. Both these methods take as an input a residual time series that is the output of applying a SARIMA filter to load observations. If enough past observations of load data labeled as nominal or under attack are collected, then detection of attacks can be made by training supervised learning classifiers. Supervised learning algorithms have been broadly adopted to the smart grid literature to monitor and detect cyber attacks on power systems [32–34]. Here, we employ several supervised learning methods for detecting attacks on cyber-DSM systems. It is unlikely that an ISO can collect high-quality attack data due to the lack of such attacks occurring. However, such data can be simulated. Using past load data, we simulate direct load manipulation attacks by creating different types of attacks, as shown in Figure 8.

6.1. SARIMA Model

In almost all our detection models, a Seasonal AutoRegressive Integrated Moving Average (SARIMA) forecast of load demand is used. Mathematically, the SARIMA model is denoted as $SARIMA(p,d,q)(P,D,Q)(S)$, where S refers to the number of time units in each season, p and q are the autoregressive and moving average terms for nonseasonal part, d is the degree of differencing (the number of times the data have had past values subtracted), and the uppercase P, D, and Q refer to the autoregressive, differencing, and moving average terms for the seasonal part of the ARIMA model. We can write the model as the following equation:

$$
\begin{aligned}
&(1-\sum_{i=1}^{p}\alpha_i B)(1-\sum_{i=1}^{P}\beta_i B^S)(1-B)^d(1-B^S)^D Y(t+1) \\
&= (1+\sum_{i=1}^{q}u_i B)(1+\sum_{i=1}^{Q}\gamma_i B^S)x(t+1)
\end{aligned}
$$

where B is the back-shift operator of the model, α_i, β_i, u_i, and γ_i are fixed coefficients determined by the model after capturing the behavior of the series. $Y(t+1)$ is the price or load data at time $t+1$ and $x(t+1)$ is the residual of the data. The exponents in the equation represent the times that the back-shift operator is multiplied.

6.2. Sequential Detection Methods

In sequential change-point detection, a series does not have a fixed length. Instead, observations are received and processed sequentially over time. When an observation is received, a decision is made about whether a change has occurred in the state. This is based only on the observations that have been received so far or within a fixed past window size. If no change is detected, then the next observation in the sequence is processed. The sequential formulation allows sequences containing multiple change points to be easily handled. Sequential change-point detection can be applied in the case of attack detection to identify if a load time series has been compromised. If an attack is flagged, an ISO can take appropriate actions to prevent further damage to the grid.

In the parametric detection realm, the Generalized Likelihood Ratio Test (GLRT) is a popular and influential method; it has been widely used in anomaly detection and signal detection. The advantage of the GLRT is that it can provide estimates of the unknown shift in statistical parameters and has consequently been used to distinguish different types of anomalies [35]. However, its applicability relies on knowledge of the statistical parameters when a system is not under attack, and the effectiveness reduces with an increasing number of unknown parameters. In [36], H.B. Mann and D.R. Whitney first presented the theory of creating a new statistic for a sequence of data to detect a probabilistic change without knowing any distribution information of the data. In [37], Ming Yu illustrated several features for nonparametric detection in network anomaly detection: (1) no assumption is made on the distribution of observations, (2) its detection threshold is self-adjusted, and (3) it can react to the end of an anomaly within a required delay time. In [38], Hawkins employed an idea of dividing data into windows to analyze data anomaly in a window with the adjustable window size using a nonparametric method. In this work, we adapted Hawkins's idea and designed a moving window nonparametric detector to detect false data injection attacks in real time.

Under the Sequential Detection paradigm, we collect observations and apply a whitening filter to produce a continuing series with the assumption that it is white Gaussian noise. If an additive attack

$A_t > 0$ is present for observation t, this will cause a definite shift in the mean of the residual series. This detection problem can thus be stated as deciding if a null hypothesis $\mathcal{H}_0$ is true, where the residual series has zero mean and known variance (invariant in time and estimated from a sample population), or if the alternative hypothesis $\mathcal{H}_1$ is correct which states that the examined series has some mean not equal to zero, thus being under attack. This can be modeled as a hypothesis test, and, for the GLRT and CUSUM detectors, this translates to

$$\mathcal{H}_0 : x_t \sim \mathcal{N}(0, \sigma^2), t = 1, ..., N$$
$$\mathcal{H}_1 : x_t \sim \mathcal{N}(A_t, \sigma^2), A_t > 0, t = 1, ..., N$$

For the GLRT detector, to simplify implementation, we model the attack as if it were constant $A > 0$ but unknown. To produce a residual series, a SARIMA multi-step forecast for $t = 1, ..., N$ is made before the detection period. This forecast is conducted using a past window of training data that was not under attack. The forecast is made for time t to $t + k$. These predictions are then subtracted from the incoming observations to produce a residual time series, which is then fed as input into the GLRT and CUSUM detectors.

6.3. Nonparametric Detection Methods

We look at two nonparametric detectors: CUSUM and Mann–Whitney. CUSUM is a classic nonparametric detection method. A CUSUM test is a control chart that is used to monitor the mean of a process based on samples taken from past data at specific time intervals. It is a class of nonlinear stopping rules for structural changes. Given information on current and previous samples, a CUSUM test relies on the specification of a target value h and a known or reliable estimate of the standard deviation σ the process. The CUSUM test typically signals an out of control or anomalous process by an upward or downward drift of the cumulative sum until it crosses the target threshold. If the mean of the load series shifts above the target threshold for attack detection, we assume the grid is under attack.

We define the CUSUM detector as follows. Taking the residual series $x_t = y_t - E_{t-1}[y_t]$, again defined by a SARIMA forecast $E_{t-1}[y_t]$, we define a one-sided CUSUM detector as

$$g_t = \max(0, g_{t-1} + x_t - k)$$

where k is called the reference value (sometimes also called drift) set priori to values such as 0, 0.5, or $A/2$ if the size of A is known in advance. When $g_t = 0$, then we define the change time as $t_c = t$, and, when $g_t > h > 0$, we reset $g_t = 0$ and flag an alarm at time $t_a = T$. The alarm threshold is also set priori to some value based on the sample population standard deviation such as $h = 2\sigma$, where σ is estimated from past training data used to produce the SARIMA forecasts.

Building off the CUSUM, a more sophisticated nonparametric detector is the Mann–Whitney method. Following the approach in [36], we consider a sequence of data samples sorted into descending order. Each of them has k elements from set X and $n - k$ elements from set Y. We label elements from Y as a_1, X as a_0. Let $P_{k,n-k}(U)$ represent the probability of a sequence in which a_1 precedes a_0 U times. We define the following statistic from the data

$$U_{k,n-k} = \sum_{i=k+1}^{n} \sum_{j=1}^{k} \psi(X_i - X_j)$$

where $\psi(t)$ is the unit step function ($\psi(t) = 0, t < 0$ and $\psi(t) = 1, t \geq 0$). Consequently, we can write

$$P_{k,n-k}(U) = \frac{k}{n} P_{k-1,n-k}(U - n + k) + \frac{n-k}{n} P_{k,n-k-1}(U)$$

Following the approach in [36], we get

$$E_{k,n-k}(U) = \frac{k(n-k)}{2}$$
$$Var_{k,n-k}(U) = \frac{k(n-k)(n+1)}{12}$$

Placing the n data points into a window and utilizing k as a moving split point, the pre split sequence is a_0 and the post-split sequence is a_1.

When the data are under attack, by computing $U_{k,n-k}$ for different values of k, the detector can determine a suspect start point of the attack. The algorithm can be illustrated by the following equation:

$$T_{max,k,n-k} = \max_{1 \leq k \leq n-1} \left\{ \frac{|U_{k,n-k} - \frac{k(n-k)}{2}|}{\sqrt{\frac{k(n-k)(n+1)}{12}}} \right\}$$
$$\hat{\mathcal{T}}_T = \arg \max_{1 \leq k \leq n-1} (T_{max,k,n-k})$$

where $T_{k,n-k}$ is the test statistic of the detector, $T_{max,k,n-k}$ is the maximum value of the $T_{k,n-k}$ by traversing k in $1 \leq k \leq n-1$ and $\hat{\mathcal{T}}_T$ is the corresponding k when the $T_{max,k,n-k}$ is located. With all previous discussion, we can introduce a specified control limit h_n and create a hypothesis test:

- Compute $T_{max,k,n-k}$, the maximized split statistic.
- If $T_{max,k,n-k} \leq h_n$, then conclude that the data in this window is in control without attack and continue to move the window forward along the whole sequence of data.
- If $T_{max,k,n-k} > h_n$, then conclude that the data in this window is under attack and stop the process of moving the window for diagnosis. Estimate the epoch of attack start point by the maximizing k, which is $\hat{\mathcal{T}}_T$.

6.4. Supervised Learning Methods

Changepoint detection could alternatively be seen as a supervised learning binary classification problem. Under this scheme, all of the change point sequences, or in our case attacks, represent one class, and all of the nominal sequences represent a second class. Supervised learning methods are machine learning algorithms that learn a mapping from input data to a target class label. Given a set of samples $\mathcal{X} = \{x_i\}_{i=1}^N$ and a set of labels $\mathcal{Y} = \{y_i\}_{i=1}^N$, then the supervised learning detection problem is defined as a hypothesis function that captures the relationship between samples and labels $f : \mathcal{X} \longrightarrow \mathcal{Y}$. A sliding window moves through the data, considering each difference between two data points as a possible change point.

An advantage of treating attack detection as a supervised learning problem is that it has a more straightforward training phase. However, a sufficient amount and diversity of training data need to be provided to represent all of the classes of attacks. To ensure enough training data, and to prevent class label imbalance, we simulate all attack data to train our algorithms. Machine learning methods have successfully been applied several times in data injection attacks in power systems [33,34], so we analyze here their ability to detect data attacks on cyber-DSM systems. The binary classification problem for attack detection can be defined as

$$y_i = \begin{cases} 1, \text{ if } A_i > 0 \\ 0, \text{ if } A_i = 0 \end{cases}$$

where $y_i = 1$ if the i-th observation is under attack, and $y_i = 0$ if there is no attack. A variety of classifiers can be used for this learning problem. For detection of DSM attacks, we examine Logistic Regression (LR), Random Forests (RF), Gaussian Naive Bayes (GNB), Gradient Boosting Classifier (GBC), and an Artificial Neural Network (ANN). We chose these classifiers because they are all compelling and have widespread use in both industry and academia; model descriptions of these methods can be found in [39,40]. All these models were implemented using the Python package scikit-learn [41] with each model using default parameters from the package.

6.5. Performance Analysis

For the security of cyber DSM systems, the major concern is not just the detection of attacks, but also the detection of nominal data with high reliability. We want a detection system that can predict not only with high accuracy but also with high precision and recall to avoid false alarms. Therefore, we measure the True Positives (TP), the True Negatives (TN), the False Positives (FP), and the False Negatives (FN). Definitions of these measures are visually shown in Figure 9. We use these measures to calculate several main performance indicators of accuracy, precision, and recall.

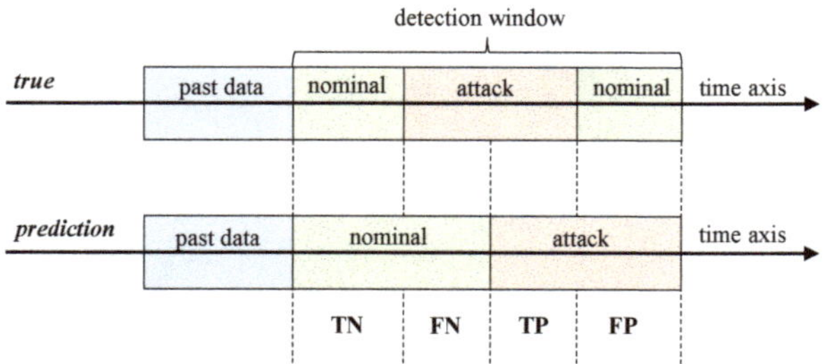

Figure 9. Confusion matrix imposed on a time axis of attack predictions vs. true observations.

We calculate accuracy as the ratio of correctly classified data points to total data points

$$Accuracy = \frac{TP + TN}{TP + TN + FP + FN}$$

This measure provides the total classification success of the models. However, accuracy is not enough to get a full picture of performance alone. Precision is calculated as the ratio of true positive data points (attacks) to total points classified as attacks:

$$Precision = \frac{TP}{TP + FP}$$

On the other hand, recall, also known as the True Positive Rate (TPR), refers to the portion of attacks that were recognized correctly

$$Recall = \frac{TP}{TP + FN}$$

Precision values give information about the prediction performance of the algorithms, whereas recall values measure the degree of attack retrieval. For instance, a recall value equal to 1 signifies that none of the attacked measurements were misclassified as nominal.

We use one more final measure of total performance of our detectors, the Receiver Operating Characteristics (ROC) curve. The ROC curve is an assessment that enables visual analysis of the trade-off between TPR and False Positive Rate (FPR). This can also be seen as the trade-off between the probability of detection and the probability of false alarm. FPR is defined as follows:

$$FPR = \frac{FP}{FP + TN}.$$

The ROC curve is constructed by plotting a two-dimensional graph with false positive rate on the x-axis and true positive rate on the y-axis at various threshold settings. A detection algorithm produces a (TPR, FPR) pair that corresponds to a single point in the ROC space. An ROC curve that's close to the upper axes boundary is reflective of a good detector. The best possible detection method would produce a point in the upper left corner, coordinate (0,1) of the ROC space. A random prediction would give a point along a diagonal line from (0,0) to (1,1). Points above this line are considered to have performance, while points below are considered with performance worse than guessing.

7. Detection Experiments and Results

For attack simulation and detection experiments, we use all the same parameters from Table 1, but we varied the levels of κ and attacks a_t. We simulate each of the three types of attacks, as visualized in Figure 8, each in the form of direct load manipulation attack, under DSM participation levels $\kappa = 0.1$ and 0.9. This creates a total of six experiment scenarios. For each scenario, we simulate 28 days of training data (672 observations) and two proceeding days of test data (48 observations), both at a resolution of 1 hr. All training data are created nominally with no attacks. In the test data sets, the first 24 hrs of the test data are not under attack. In the last 24 h of the test data, we add one of the three types of attacks. For ramp attacks, for each time step, we add an attack $a_t = a_{t-1} + 5$ where the first 24 h of the test set $a_{t=0...23} = 0$. The sudden attacks are similar except the attack is an additive constant $a_t = 150$. The last type of attack, point attacks, are $a_{t=24} = 250, a_{t=29} = 200, a_{t=34} = 300, a_{t=37} = 100, a_{t=46} = 150$, and $a_t = 0$ everywhere else. To showcase a wide variety of detection strategies, we conduct two main studies. In the first, we compare sequential detectors such as GLRT and CUSUM to five machine learning-based classifiers using accuracy, recall, precision, and ROC curves. In the second study, we do a deep analysis comparing GLRT to nonparametric based detection with ROC curves.

7.1. Sequential vs Supervised Learning Detection

For the sequential detectors, we use multi-step SARIMA forecasts to predict the next 48 h and then use those forecasts to filter incoming test data for detection. The training was conducted on the training time series of 28 days of nominal data. SARIMA hyperparameters were chosen by examining lag one differenced AutoCorrelation Function (ACF) and Partial AutoCorrelation Function (PACF) plots of the training data. These forecasts where then subtracted from the incoming test data to obtain a residual series that is input to the sequential detectors. The assumption for both detectors is that the residual series is Gaussian white noise, where the series has zero mean, and each observation is independent identically distributed from a Gaussian distribution.

To ensure the residuals are white noise, we apply the Augmented Dickey–Fuller (ADF) test to check for stationarity, and we examine the ACF/PACF plots to check for independence and run a Jarque–Bera test, and we examine a Q-Q plot to check if the residual series has a Gaussian probability distribution. We conduct these checks on all scenario training datasets, where we first train a SARIMA fit on them and then subtract that fit to produce the residual series. In all the training data sets, the residuals were stationary from the ADF test and independent from ACF/PACF plots. However, the Jarque–Bera test and Q-Q plots showed that the observations did not come from a Gaussian distribution. Example, ACF/PACF, and Q-Q plots for $\kappa = 0.1$ are shown in Figures 10 and 11. Despite the residuals not being Gaussian, we still run the sequential detectors and examine their performance.

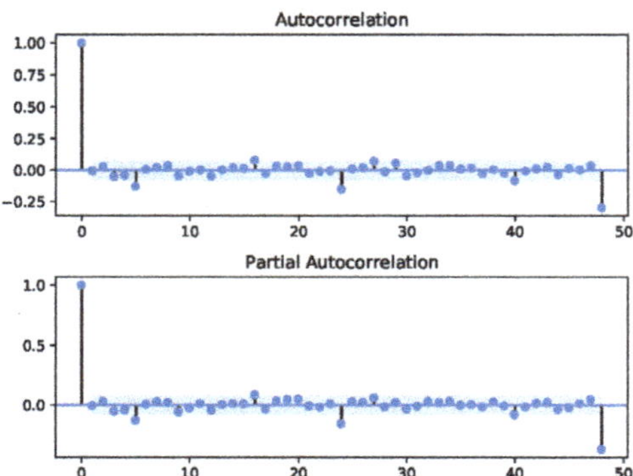

Figure 10. ACF and PACF plots of the residual series between the SARIMA fit tp training data. From the plots, we observe that the residuals are stationary.

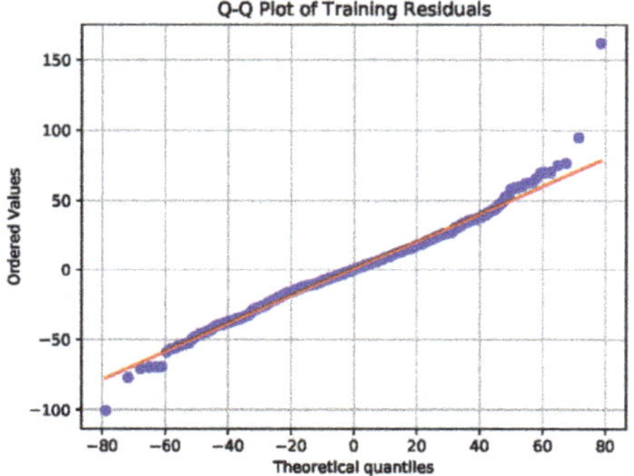

Figure 11. Q-Q plot of the residual series between the SARIMA fit tp training data. From the plot, we observe that, for extreme quantiles, the distribution is not Gaussian.

For training, the supervised learning methods for each of the six scenarios, we simulate more data to ensure proper class learning. We keep the test sets the same, but we extend each training set doubling its size. We keep the original training set, labeling it as nominal, and then make a copy of it. In the copy, we split it into three parts; for each part, we add one of the types of attacks at random levels. We label each observation of this set as under attack. We added the nominal and attacked training sets to form a new training set with a total of 1344 observations. With our training and test time series, we then create a set $\mathcal{X}$ features and $\mathcal{Y}$ labels that could be fed into the supervised learning classifiers. Each training sample $x_i \in \mathcal{X}$ is composed of 24 h of lagged data, each hour is one feature, and each label $y_i \in \mathcal{Y}$ corresponds to a class 0 (not under attack) and class 1 (under attack). Together, we get N =1298 pairs of (x_i, y_i) training samples.

We run our experiments on a computer with an Intel i7 6700 2.6 GHz and 16 GB of RAM. Implementation of the simulations, experiments, and sequential detectors were done in Python 3.6. Implementation for SARIMA forecasts was done using the Python package Statsmodels [42], the supervised learning methods, and confusion evaluation metrics were implemented using the Scikit-Learn Python package [41], and ROC curves were created using the Matplotlib Python package [43]. All classifiers used default hyperparameters from Scikit-Learn. We compiled all code and data used in our work into a Python package titled LehighDSM, which is publicly available on GitHub [44].

Experimental results from the six scenarios are reported in Table 2, in the form of accuracy, recall, and precision metrics, and Figure 12 which showcases six ROC curves, one for each scenario. All performance measures in Table 2 were calculated using the best thresholds found in the ROC curves by searching for the shortest distance from each curve to the corner (0,1). As seen in Figure 12c,f type 3 attacks, under any DSM level, had the lowest detection rates across thresholds with ANN being the worst performer. In Figure 12a,d, LR yielded the best performance with GNB, and again ANN being the worst. Type 2 attacks in Figure 12b,e resulted in the best detection with CUSUM, LR, GNB, and GBC having perfect performance.

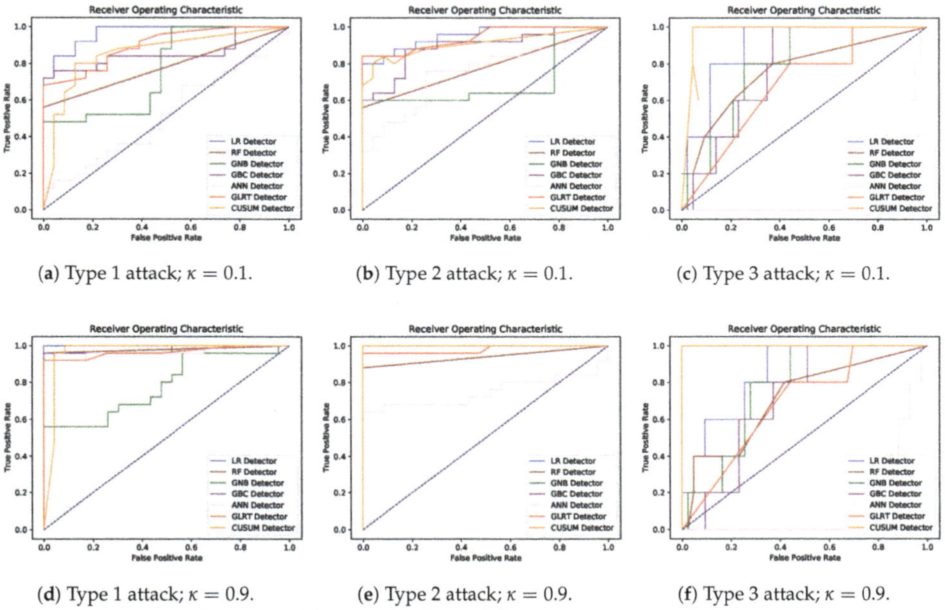

(**a**) Type 1 attack; $\kappa = 0.1$. (**b**) Type 2 attack; $\kappa = 0.1$. (**c**) Type 3 attack; $\kappa = 0.1$.

(**d**) Type 1 attack; $\kappa = 0.9$. (**e**) Type 2 attack; $\kappa = 0.9$. (**f**) Type 3 attack; $\kappa = 0.9$.

Figure 12. ROC curves for the six different experiment scenarios.

With the results from Table 2, we have the following conclusions: demonstrated findings imply that detection of attacks has a higher accuracy with higher levels of DSM participation. This occurs since higher DSM penetration results in more aggregated attacks. However, we note that an attacker, if they have perfect knowledge of the grid, could decrease their intensity and make detection more difficult. Furthermore, the supervised learning classifiers' performance on average was on par or better than sequential detection methods. From a detector comparison perspective, the parametric and nonparametric detectors showcased higher performance on most of the attacks and varying parameters. In the following section, we do a deeper dive on a comparison between the GLRT and the nonparametric detectors to derive further inferences on operations paradigms for the attack detection problem.

Table 2. Evaluation metrics for attack detection. Best results in bold.

	κ	0.1			0.9		
	Attack Type	**1**	**2**	**3**	**1**	**2**	**3**
Accuracy	LR	**89.6**	**100.0**	87.5	87.5	**100.0**	75.0
	RF	77.1	97.9	64.6	77.1	93.8	60.4
	GNB	70.8	77.1	75.0	79.2	**100.0**	72.9
	GBC	85.4	97.9	66.7	85.4	**100.0**	64.6
	ANN	56.3	95.8	20.8	75.0	79.2	14.6
	GLRT	79.2	95.8	58.3	**91.7**	97.9	58.3
	CUSUM	83.3	95.8	**95.8**	87.5	**100.0**	**100.0**
Recall	LR	**92.0**	**100.0**	80.0	**89.0**	**100.0**	80.0
	RF	56.0	96.0	80.0	56.0	88.0	80.0
	GNB	88.0	56.0	80.0	60.0	**100.0**	80.0
	GBC	76.0	96.0	**100.0**	88.0	**100.0**	80.0
	ANN	68.0	92.0	60.0	76.0	68.0	80.0
	GLRT	76.0	92.0	80.0	84.0	96.0	80.0
	CUSUM	80.0	96.0	**100.0**	84.0	**100.0**	**100.0**
Precision	LR	88.5	**100.0**	44.4	88.0	**100.0**	26.7
	RF	**100.0**	**100.0**	20.0	**100.0**	**100.0**	18.2
	GNB	66.7	**100.0**	26.7	**100.0**	**100.0**	25.0
	GBC	95.0	**100.0**	23.8	84.0	**100.0**	20.0
	ANN	56.7	**100.0**	7.7	76.0	89.5	9.1
	GLRT	82.6	**100.0**	17.4	**100.0**	**100.0**	17.4
	CUSUM	87.0	96.0	**71.4**	91.3	**100.0**	**100.0**

7.2. GLRT Detection versus Nonparametric Detection

For the comparison between the nonparametric detector and parametric GLRT detector, we conduct our numerical experiments using the block bootstrap approach on real demand data (unmodulated by DSM) available from the New England ISO, to generate demand for 7 to 200 users. We apply the price-elasticity model with the DSM participating rate of k and pricing information from the New England ISO to generate the DSM modulated price and demand for multiple time periods. We first compare the performance of the GLR and nonparametric detectors on identical data windows in Figures 13 and 14. P denotes the DSM participation rate, k means the percentage of the data in the selected window is under attack. Using twin attacks, we compare the detectors using ROC curves.

There are a few key observations from Figures 13 and 14. First, as expected, the performance of both detectors improves with attack magnitude. Second, the nonparametric detector outperforms the GLRT for small user sizes, whereas the comparison yields the opposite conclusion in the 200 user simulations. The reasoning behind this observation is that: when fewer users' data are analyzed, the residuals based on the SARIMA forecasts are far less likely to behave like white noise, and, consequently, the GLRT optimized for Gaussian residuals underperforms. The nonparametric detector, aside from not requiring the legitimate statistics, purely checks for statistical shifts and performs better in a smaller user pool. In a larger user pool, we enter a law of large numbers regime, where the aggregate residuals increasingly resemble Gaussian variables, and thus the optimized GLRT has a better ROC curve.

Figure 15 plots the performance comparison as a function of attack magnitude and window sizes for a fixed false alarm probability (P_{FA}). As expected, the GLRT outperforms the nonparametric detector for larger window sizes regardless of attack magnitude. With smaller window sizes and, consequently, shorter delay in detection, the nonparametric detector outperforms the GLRT, which illustrates that, when statistical variations are small (due to fewer users, smaller attack magnitude or lower delay), nonparametric detectors are appropriate.

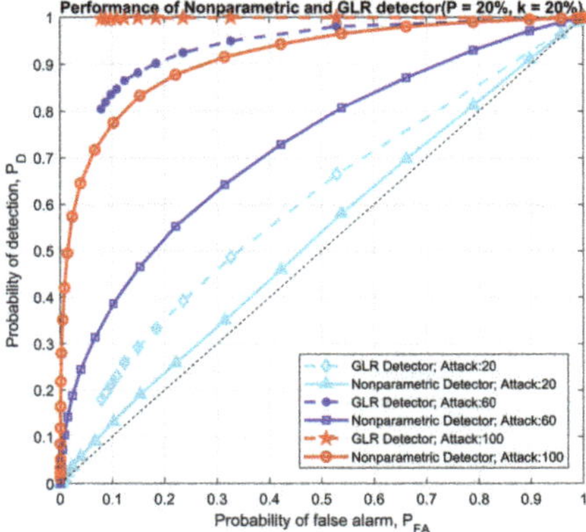

Figure 13. Performance comparison between parametric and nonparametric detector under 200 users.

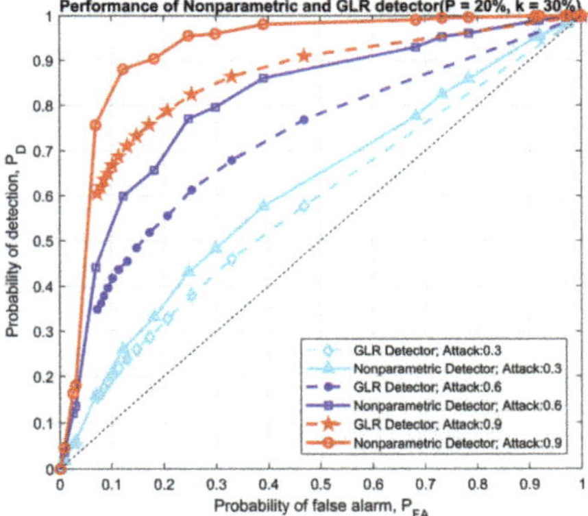

Figure 14. Performance comparison between parametric and nonparametric detectors with two users only.

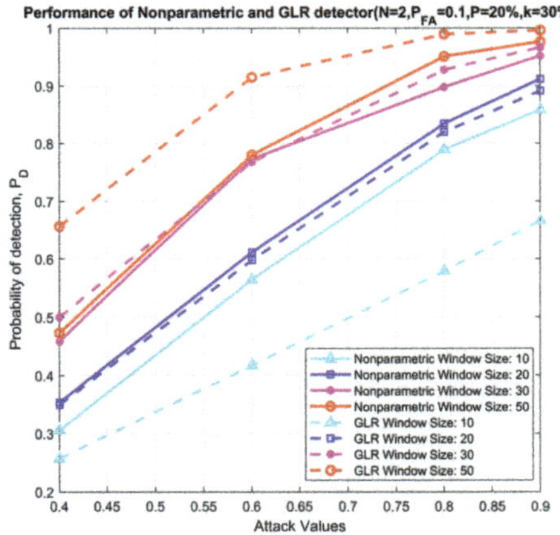

Figure 15. Performance comparison between nonparametric and GLRT detector, N = 2, $P_{FA} = 0.1$, $P = 20\%$, k = 30%.

8. Conclusions

In this work, we study the exploitation of the hypothetical feedback between cyber-enabled DSM programs, on the consumer side, and dynamic RTP on the utility side. An attacker with exploitive economic or nefarious intentions, such as causing efficiency loss of energy provision, can take advantage of the dependency between dynamic pricing and DSM load control. The utility modifies prices in response to forecasted demand in order to push realized load up or down. This is done to achieve some target load level to reduce peak load or achieve some other DSM objective. On the user side, cyber DSM programs then autonomously respond to prices to adjust certain portions of a user's load up or down with some given elasticity.

We propose two modes of attacks, false price data injections, and direct load manipulation. Under a false price data injection, an attacker modifies the RTP that users receive to alter their demand. Through a direct load manipulation attack, attacker hacks and alters a user's load profile directly. In both of these attacks, aggregate load from the grid is modified, which can alter future prices or demand. We showcase how these two modes of attacks are equivalent and introduce three ways an attack can occur. The first type is a ramp attack, the second is a sudden attack, and the third is a point attack. We simulate these types of attacks and review several methods to detect them.

We simulate and examine load-price data under different DSM participation levels with three types of additive random attacks: ramp, sudden, and point attacks. We then conduct two tutorial-style detection studies. First, we compare sequential change point and supervised learning methods. Moreover, in a second study, we compare parametric and nonparametric detectors under different conditions. These two studies are meant to provide a broad review of various possible detection approaches. The proposed approaches are pure data driven approaches which can be implemented without additional hardware assuming that the data are collected at a demand response management system. The computational cost of the proposed detection methods would depend on the specific methodologies adopted. Statistical tests can be implemented without incurring significant computational cost whereas more complex detectors such as ANNs require hardware acceleration for reasonable performance.

Results conclude that higher amounts of DSM participation can exacerbate attacks and lead to better detection of such attacks, point attacks are the hardest to detect, and supervised learning methods produce par or better results than sequential detectors. Amongst the GLRT based parametric and the nonparametric detector, we show that, with growing window size, the performance of both the detectors will increase. The performance of the GLRT detector increases faster when the detection horizon increases due to the MLE mechanism. Thus, if the detection horizon is broad and the user number is fixed, both the detectors become more potent due to the more information and larger size of the attack data.

Due to the speculative nature of these types of attacks, we plan to look into many different issues for future studies. All our detection approaches for attacks depend on predictions. When an accurate detection is made, the ISO would take measures to mitigate that attack by adjusting price levels appropriately. In the event of false negatives or false positives, for part of the future work, we envision to study how an ISO could have a risk cost matrix that would define the amount of price adjustments for which it would feel comfortable to mitigate attacks taking into account probability confidence levels on the accuracy of the attacks. Additional new directions include expanding our model to incorporate day ahead pricing and DSM scheduling, adding models of cost and utility on for a customer and ISO, expanding our model taking into account grid topology, and modeling more advanced attack goals such as system failure by power line overload. Additionally, in this work, we only looked at the DSM goal of strategic load conservation. More complex strategies, such as load shifting, could result in more complex types of cyber attacks; this also makes for compelling future work.

Author Contributions: Conceptualization, methodology, investigation, software K.H.; nonparametric detection investigation, C.Z.; writing—original draft preparation, K.H. and C.Z.; writing—review and editing, P.V., S.K., and R.S.B.; supervision, P.V. and S.K.; funding acquisition, P.V., S.K., R.S.B., and L.S. All authors have read and agreed to the published version of the manuscript.

Funding: This work is supported by the Department of Energy through Award Number DE-OE0000779.

Acknowledgments: The authors thank the anonymous reviewers for their careful reading of our manuscript and their many insightful comments and suggestions.

Conflicts of Interest: The authors declare no conflict of interest.

References

1. Faruqui, A.; Sergici, S. Household Response to Dynamic Pricing of Electricity-a Survey of The Empirical Evidence. 2010. Available online: https://papers.ssrn.com/sol3/papers.cfm?abstract_id=1134132 (accessed on 5 April 2019).
2. Grochocki, D.; Huh, J.H.; Berthier, R.; Bobba, R.; Sanders, W.H.; Cárdenas, A.A.; Jetcheva, J.G. AMI threats, intrusion detection requirements and deployment recommendations. In Proceedings of the 2012 IEEE Third International Conference on Smart Grid Communications (SmartGridComm), Tainan, Taiwan, 5–8 November 2012; pp. 395–400.
3. Khan, A.R.; Mahmood, A.; Safdar, A.; Khan, Z.A.; Khan, N.A. Load forecasting, dynamic pricing and DSM in smart grid: A review. *Renew. Sustain. Energy Rev.* **2016**, *54*, 1311–1322. [CrossRef]
4. Xie, L.; Mo, Y.; Sinopoli, B. Integrity data attacks in power market operations. *IEEE Trans. Smart Grid* **2011**, *2*, 659–666. [CrossRef]
5. Tan, R.; Badrinath Krishna, V.; Yau, D.K.; Kalbarczyk, Z. Impact of integrity attacks on real-time pricing in smart grids. In Proceedings of the 2013 ACM SIGSAC Conference on Computer & Communications Security, Berlin, Germany, 4–8 November 2013; ACM: New York, NY, USA, 2013; pp. 439–450.
6. Palensky, P.; Dietrich, D. Demand side management: Demand response, intelligent energy systems, and smart loads. *IEEE Trans. Ind. Inform.* **2011**, *7*, 381–388. [CrossRef]
7. Logenthiran, T.; Srinivasan, D.; Shun, T.Z. Demand side management in smart grid using heuristic optimization. *IEEE Trans. Smart Grid* **2012**, *3*, 1244–1252. [CrossRef]
8. Mohsenian-Rad, A.H.; Wong, V.W.; Jatskevich, J.; Schober, R.; Leon-Garcia, A. Autonomous demand-side management based on game-theoretic energy consumption scheduling for the future smart grid. *IEEE Trans. Smart Grid* **2010**, *1*, 320–331. [CrossRef]
9. Gelazanskas, L.; Gamage, K.A. Demand side management in smart grid: A review and proposals for future direction. *Sustain. Cities Soc.* **2014**, *11*, 22–30. [CrossRef]
10. Batista, A.C.; Batista, L.S. Demand Side Management using a multi-criteria epsilon-constraint based exact approach. *Expert Syst. Appl.* **2018**, *99*, 180–192. [CrossRef]
11. Dudek, G. Pattern-based local linear regression models for short-term load forecasting. *Electr. Power Syst. Res.* **2016**, *130*, 139–147. [CrossRef]
12. Pappas, S.S.; Ekonomou, L.; Karampelas, P.; Karamousantas, D.; Katsikas, S.; Chatzarakis, G.; Skafidas, P. Electricity demand load forecasting of the Hellenic power system using an ARMA model. *Electr. Power Syst. Res.* **2010**, *80*, 256–264. [CrossRef]
13. Christiaanse, W. Short-term load forecasting using general exponential smoothing. *IEEE Trans. Power Appar. Syst.* **1971**, *PAS-90*, 900–911. [CrossRef]
14. Taylor, J.W. Triple seasonal methods for short-term electricity demand forecasting. *Eur. J. Oper. Res.* **2010**, *204*, 139–152. [CrossRef]
15. Kavousi-Fard, A.; Samet, H.; Marzbani, F. A new hybrid modified firefly algorithm and support vector regression model for accurate short term load forecasting. *Expert Syst. Appl.* **2014**, *41*, 6047–6056. [CrossRef]
16. Baliyan, A.; Gaurav, K.; Mishra, S.K. A review of short term load forecasting using artificial neural network models. *Procedia Comput. Sci.* **2015**, *48*, 121–125. [CrossRef]
17. Kong, W.; Dong, Z.Y.; Jia, Y.; Hill, D.J.; Xu, Y.; Zhang, Y. Short-term residential load forecasting based on LSTM recurrent neural network. *IEEE Trans. Smart Grid* **2017**, *10*, 841–851. [CrossRef]

18. Dudek, G. Short-term load forecasting using random forests. In *Intelligent Systems'2014, Proceedings of the 7th IEEE International Conference Intelligent Systems IS'2014, Warsaw, Poland, 24–26 September 2014*; Springer: Berlin/Heidelberg, Germany, 2015; pp. 821–828.
19. Li, S.; Goel, L.; Wang, P. An ensemble approach for short-term load forecasting by extreme learning machine. *Appl. Energy* **2016**, *170*, 22–29. [CrossRef]
20. Chakraborty, S.; Ito, T.; Senjyu, T. Smart pricing scheme: A multi-layered scoring rule application. *Expert Syst. Appl.* **2014**, *41*, 3726–3735. [CrossRef]
21. Nazar, N.; Abdullah, M.; Hassan, M.; Hussin, F. Time-based electricity pricing for Demand Response implementation in monopolized electricity market. In Proceedings of the 2012 IEEE Student Conference on Research and Development (SCOReD), Pulau Pinang, Malaysia, 5–6 December 2012; IEEE: Piscataway, NJ, USA, 2012; pp. 178–181.
22. Zimmerman, R.D.; Murillo-Sánchez, C.E.; Gan, D. *MATPOWER: A MATLAB Power System Simulation Package*; Manual, Power Systems Engineering Research Center: Ithaca, NY, USA, 1997; Volume 1.
23. Chassin, D.P.; Schneider, K.; Gerkensmeyer, C. GridLAB-D: An open-source power systems modeling and simulation environment. In Proceedings of the 2008 IEEE/PES Transmission and Distribution Conference and Exposition, Chicago, IL, USA, 21–24 April 2008; IEEE: Piscataway, NJ, USA, 2008; pp. 1–5.
24. Barker, S.; Mishra, A.; Irwin, D.; Cecchet, E.; Shenoy, P.; Albrecht, J. Smart*: An open data set and tools for enabling research in sustainable homes. *Sustkdd August* **2012**, *111*, 108.
25. Brockwell, P.J.; Davis, R.A.; Calder, M.V. *Introduction to Time Series and Forecasting*; Springer: Berlin/Heidelberg, Germany, 2002; Volume 2.
26. Gamerman, D.; Lopes, H.F. *Markov Chain Monte Carlo: Stochastic Simulation for Bayesian Inference*; Chapman and Hall/CRC: Boca Raton, FL, USA, 2006.
27. Politis, D.N.; White, H. Automatic block-length selection for the dependent bootstrap. *Econom. Rev.* **2004**, *23*, 53–70. [CrossRef]
28. Thimmapuram, P.R.; Kim, J.; Botterud, A.; Nam, Y. Modeling and simulation of price elasticity of demand using an agent-based model. In Proceedings of the 2010 Innovative Smart Grid Technologies (ISGT), Gothenburg, Sweden, 19–21 January 2010; IEEE: Piscataway, NJ, USA, 2010; pp. 1–8.
29. Liang, G.; Zhao, J.; Luo, F.; Weller, S.R.; Dong, Z.Y. A review of false data injection attacks against modern power systems. *IEEE Trans. Smart Grid* **2017**, *8*, 1630–1638. [CrossRef]
30. Tartakovsky, A.; Nikiforov, I.; Basseville, M. *Sequential Analysis: Hypothesis Testing and Changepoint Detection*; Chapman and Hall/CRC: Boca Raton, FL, USA, 2014.
31. Key, S. *Fundamentals of Statistical Signal Processing*; Volume ii: Detection Theory; Prentice Hall PTR: Upper Saddle River, NJ, USA, 1993.
32. Hink, R.C.B.; Beaver, J.M.; Buckner, M.A.; Morris, T.; Adhikari, U.; Pan, S. Machine learning for power system disturbance and cyber-attack discrimination. In Proceedings of the 2014 7th International Symposium on Resilient Control Systems (ISRCS), Denver, CO, USA, 19–21 August 2014; pp. 1–8.
33. Ozay, M.; Esnaola, I.; Vural, F.T.Y.; Kulkarni, S.R.; Poor, H.V. Machine learning methods for attack detection in the smart grid. *IEEE Trans. Neural Netw. Learn. Syst.* **2016**, *27*, 1773–1786. [CrossRef]
34. Esmalifalak, M.; Liu, L.; Nguyen, N.; Zheng, R.; Han, Z. Detecting stealthy false data injection using machine learning in smart grid. *IEEE Syst. J.* **2017**, *11*, 1644–1652. [CrossRef]
35. Apley, D.W.; Shi, J. The GLRT for statistical process control of autocorrelated processes. *IIE Trans.* **1999**, *31*, 1123–1134. [CrossRef]
36. Mann, H.B.; Whitney, D.R. On a test of whether one of two random variables is stochastically larger than the other. *Ann. Math. Stat.* **1947**, 18, 50–60. [CrossRef]
37. Yu, M. A nonparametric adaptive CUSUM method and its application in network anomaly detection. *Int. J. Adv. Comput. Technol.* **2012**, *4*, 280–288.
38. Hawkins, D.M.; Deng, Q. A nonparametric change-point control chart. *J. Qual. Technol.* **2010**, *42*, 165–173. [CrossRef]
39. Bishop, C. Pattern Recognition and Machine Learning. In *Pattern Recognition and Machine Learning*; Springer: Berlin/Heidelberg, Germany, 2006.
40. James, G.; Witten, D.; Hastie, T.; Tibshirani, R. *An Introduction to Statistical Learning*; Springer: Berlin/Heidelberg, Germany, 2013; Volume 112.

41. Pedregosa, F.; Varoquaux, G.; Gramfort, A.; Michel, V.; Thirion, B.; Grisel, O.; Blondel, M.; Prettenhofer, P.; Weiss, R.; Dubourg, V.; et al. Scikit-learn: Machine learning in Python. *J. Mach. Learn. Res.* **2011**, *12*, 2825–2830.
42. Seabold, S.; Perktold, J. Statsmodels: Econometric and statistical modeling with python. In Proceedings of the 9th Python in Science Conference, SciPy Society Austin, Austin, TX, USA, 28 June–3 July 2010; Volume 57, p. 61.
43. Hunter, J.D. Matplotlib: A 2D graphics environment. *Comput. Sci. Eng.* **2007**, *9*, 90–95. [CrossRef]
44. Hatalis, K. LehighDSM Python Package. 2019. Available online: https://github.com/hatalis/DSM (accessed on 1 August 2020).

Article

Simulation of Incidental Distributed Generation Curtailment to Maximize the Integration of Renewable Energy Generation in Power Systems

Ingo Liere-Netheler *, Frank Schuldt, Karsten von Maydell and Carsten Agert

DLR Institute of Networked Energy Systems, Carl-von-Ossietzky-Str. 15, 26129 Oldenburg, Germany; Frank.Schuldt@dlr.de (F.S.); Karsten.Maydell@dlr.de (K.v.M.); Carsten.Agert@dlr.de (C.A.)

* Correspondence: Ingo.Liere-Netheler@dlr.de; Tel.: +49-441-99906-457

Received: 20 July 2020; Accepted: 11 August 2020; Published: 12 August 2020

Abstract: Power system security is increasingly endangered due to novel power flow situations caused by the growing integration of distributed generation. Consequently, grid operators are forced to request the curtailment of distributed generators to ensure the compliance with operational limits more often. This research proposes a framework to simulate the incidental amount of renewable energy curtailment based on load flow analysis of the network. Real data from a 110 kV distribution network located in Germany are used to validate the proposed framework by implementing best practice curtailment approaches. Furthermore, novel operational concepts are investigated to improve the practical implementation of distributed generation curtailment. Specifically, smaller curtailment level increments, coordinated selection methods, and an extension of the n-1 security criterion are analyzed. Moreover, combinations of these concepts are considered to depict interdependencies between several operational aspects. The results quantify the potential of the proposed concepts to improve established grid operation practices by minimizing distributed generation curtailment and, thus, maximizing power system integration of renewable energies. In particular, the extension of the n-1 criterion offers significant potential to reduce curtailment by up to 94.8% through a more efficient utilization of grid capacities.

Keywords: power system operation; power system security; renewable energy integration; load flow analysis; congestion management; distributed generation curtailment

1. Introduction

The integration of distributed generators (DGs), such as wind turbines or photovoltaics, causes formerly unknown load flow situations that can endanger the operational security of power systems. In Germany, particularly in rural areas, the increasing installation of DG systems coincides with low hosting capacities due to the historical dimensioning of the network [1]. Hence, violations of operational security limits, such as overvoltage problems or thermal overloadings of grid components, can occur more frequently [2]. To prevent this, grid operators are required to enhance existing congestion management regimes. In the long term, grid expansion and reinforcement measures are required to ensure larger DG penetrations without jeopardizing the operational security of power systems. However, several studies have analyzed the grid expansion demand in Germany and agreed that the provision of a system without any bottlenecks solely through grid expansion or reinforcement is neither the most efficient solution at the distribution level (≤110 kV) [3] nor at the transmission level (≥220 kV) [4]. Consequently, an efficient utilization of existing grid capacities through operational congestion management approaches will become more relevant in the future in times of high renewable feed-in [5].

In the literature, a wide range of congestion management regimes is investigated. Market design approaches, such as nodal pricing or market splitting [6], consider network restrictions in the trading stage and therefore prevent congestions from the outset. Technical solutions, such as control of generation [1], consumption [7], or power flow controlling devices [8], are used to solve congestions in short-term network security management. These can further be subdivided into market-oriented and direct congestion management measures [9]. Market-oriented approaches require contractual agreements between grid operator and flexible consumers or generators, respectively. Local flexibility markets are a possible organizational form of these agreements. However, these markets face challenges regarding illiquidity, lack of customer participation, and regulatory problems [10]. Direct congestion management measures are executable by the grid operator without the necessity of additional agreements. These can either be the application of controllable grid components, such as switching actions, or the control of generation or consumption in emergency situations [9].

In Germany, both transmission system operators (TSOs) and distribution system operators (DSOs) are allowed to temporarily reduce the power injections of DGs to maintain grid security in critical situations. DG operators are then legally required to reduce their power generation output. Hence, this measure constitutes a direct emergency action and is called feed-in management (FIM). Although DG curtailment can only be exploited as a last resort in Germany [1], a significant historical increase of these measures can be observed. In 2017, FIM caused annual costs of approximately 610 M€ because each curtailed DG was financially compensated [11]. Consequently, reducing DG curtailment offers the potential to both significantly reduce the costs for congestion management and increase the hosting capacity of power systems for renewable energy generation. The remainder of the paper will thus focus on optimizing the practical implementation of DG curtailment as a congestion management measure.

Therefore, this work presents a simulation framework for DG curtailment based on load flow analysis. Different operational aspects to enhance grid capacity utilization compared to the recent best practice are investigated. Consequently, this work demonstrates possibilities to improve the practical implementation of DG curtailment and, thus, maximize the integration of renewable energy generation into power systems.

Several studies have examined selected aspects of DG curtailment as an operational congestion management measure. The effect of increasing the controllability of DGs on the resulting curtailment demand is analyzed in [1]. The authors show that the amount of curtailed energy can be reduced by controlling small scale DGs individually and thus increasing the granularity of curtailment measures. In [12] a linear state estimation algorithm and a remedial action scheme are developed and used to determine the optimal amount of DG curtailment based on the DC power flow model. The algorithms enable the substitution of preventive DG curtailment by offering fast automated control actions. The authors of [13] show that a curtailment strategy based on power flow sensitivities can reduce the amount of power adjustments compared to established principles of access in the United Kingdom. The substitution of DG curtailment by fast-reacting curative congestion management measures is investigated in [5]. The authors show that capacities of the transmission network can be utilized more efficiently, if fast-reacting measures, such as phase shifting transformers, are implemented. Reference [14] proposes a congestion management regime including multiple voltage levels simultaneously. The results indicate optimization potentials regarding the coordination between grid operators of different voltage levels. In [15], potentials for reducing the demand for redispatch and DG curtailment in the transmission network through dynamic line rating are depicted.

The above articles analyze selected aspects and their influence on DG curtailment. However, to the best of the authors' knowledge, neither a comparative study taking into account multiple operational aspects simultaneously nor the validation of curtailment simulations based on real grid data have been carried out yet. One of the main contributions of this paper is thus the development of a curtailment simulation framework and its application to the model of a real 110 kV distribution network from a region located in Germany. To validate the simulated amount of DG curtailment, recent best practice approaches are first implemented into the framework and the results derived from load flow simulations are compared to curtailment measures from real grid operation. Another main contribution of this work is the proposal of three novel operational concepts to improve grid capacity utilization in critical operating states. These concepts include increasing the grid operator's ability to control DG units by implementing smaller curtailment level increments, improving DG selection methods with a special focus on increasing the coordination between different grid operators and adapting the deterministic n-1 security criterion to maximize DG integration. Furthermore, combinations of the proposed concepts are investigated to demonstrate possible interdependencies. The concepts are implemented into the simulation framework and their potential to reduce the incidental amount of renewable energy curtailment compared to the best practice is quantified. Hence, by combining the proposal of different operational concepts and their implementation into the validated model of a real distribution network, this work extends the existing research in the field by quantifying the influence of several operational aspects on DG curtailment and by depicting potentials to maximize DG integration into power systems by enhancing the utilization of existing transmission capacities.

The remainder of the paper is structured as follows. Section 2 introduces the simulation framework and discusses different operational aspects influencing DG curtailment. The current best practice regarding these aspects in grid operation and concepts to improve the practical implementation are discussed. Section 3 includes a description of the 110 kV distribution grid model and the validation based on the simulation of best practice curtailment as a reference scenario. Furthermore, scenarios for the implementation of the proposed concepts are defined and simulation results for each of the scenarios are presented, summarized, and discussed. Section 4 concludes the most relevant findings of the paper.

2. DG Curtailment Simulation Framework

The process of DG curtailment is influenced by different operational aspects. This section describes the consideration of different aspects in the developed simulation framework. Because the model of a real distribution network of a region located in Germany is investigated as a case study in this paper, recent best practice approaches applied there are described as an exemplary specification of the considered aspects. Furthermore, inefficiencies of these approaches are discussed and used to derive concepts to improve the practical implementation of DG curtailment.

2.1. DG Controllability

Grid operators can request the reduction of power generation of DG units but are not allowed to directly control these. Instead, generation units receive curtailment requests and are required to realize the requested power output reduction. The maximum permitted power injection $P_{inj,max}$ of a DG unit requested by a grid operator can be formulated as:

$$P_{inj,max} = l_{curt} \cdot P_{nom} \text{ with } l_{curt} \in \Omega_{curt} \quad (1)$$

where P_{nom} describes the nominal capacity of the DG unit and l_{curt} the requested percentage level included in the feasible set of all possible curtailment levels Ω_{curt}. Recently, the established granularity of curtailment levels is four discrete output levels [1]:

$$\Omega_{curt,established} = [1,\ 0.6,\ 0.3,\ 0]$$

Here, the curtailment level of 1 is used to cancel previously requested commands and allow the maximum possible feed-in again. The ability of grid operators to request the most efficient amount of DG curtailment is thus limited due to their limited control opportunities. As a consequence, the practically curtailed power feed-in can exceed the theoretically necessary amount of power reductions to solve congestions in certain situations, which can result in an inefficient utilization of transmission capacities [1]. Therefore, implementing an extended feasible set of curtailment levels with smaller increments can potentially enable the request of more efficient measures and reduce the amount of incidental renewable energy curtailment compared to the best practice.

2.2. DG Selection Procedure

In addition to requesting the most suitable curtailment level of a DG unit, usually multiple generation units in different locations can potentially terminate grid congestions. Thus, grid operators have to select which DG units to curtail. This selection process has to satisfy all technical constraints, reach the practical minimum of curtailment, and avoid discrimination of specific generation units [1]. However, several studies have shown that minimizing curtailment and sharing the amount of power adjustments between multiple DGs equally are contrary objectives and a trade-off between both aspects has to be found [16,17].

2.2.1. DG Selection Methods

The recent best practice in Germany is to curtail DGs iteratively based on their sensitivity to congested grid components [18]. Following this curtailment strategy, the most sensitive unit is curtailed first. If congestions remain after completely reducing the power output of the first unit, the second most sensitive unit is selected, etc.

The sensitivities can either be derived from simplified DC load flow models or from nonlinear AC models. The advantage of DC models is the independence of the recent grid state. Hence, these sensitivities can be calculated once and used permanently. In contrast, the AC formulation has a higher accuracy but requires separate calculation of sensitivity values for each grid state [19] In this work, the sensitivities are derived from the AC load flow model and calculated for each grid state separately. Sensitivities enable the approximation of the relation between changes in real and reactive branch power flows and changes in nodal real and reactive power injections as follows [20]:

$$\begin{bmatrix} \Delta \mathbf{P}_{\text{branch}} \\ \Delta \mathbf{Q}_{\text{branch}} \end{bmatrix} = \mathbf{M} \cdot \begin{bmatrix} \Delta \mathbf{P}_{\text{inj}} \\ \Delta \mathbf{Q}_{\text{inj}} \end{bmatrix} \tag{2}$$

where $\Delta \mathbf{P}_{\text{branch}}$ and $\Delta \mathbf{Q}_{\text{branch}}$ are the vectors of real and reactive power flow changes over all considered grid components, such as lines or transformers, $\Delta \mathbf{P}_{\text{inj}}$ and $\Delta \mathbf{Q}_{\text{inj}}$ are the vectors of changes in nodal real and reactive power injections, and $\mathbf{M}$ is the sensitivity matrix of the system. Each value of $\mathbf{M}$ approximates the effect of a change in power injection in a certain node in the network on a certain grid component for one specific power flow solution. Accordingly, power flow sensitivities can be used to determine the most sensitive DG unit to certain congested grid components.

However, the iterative selection scheme in combination with discrete curtailment level increments minimizes the number of necessary curtailment requests but does not necessarily lead to the minimum possible amount of curtailed power. This is due to the fact that the most sensitive unit can, e.g., be a large wind farm and only be curtailed in relatively large discrete steps, whereas smaller units with smaller possible curtailment levels with an equally large sensitivity are not considered.

To address this issue and enable the request of more efficient curtailment measures, the problem of selecting DG units can also be implemented in the simulation framework as an optimization problem aiming at minimizing the total amount of power reductions [20]:

$$\min_{\mathrm{x}} \mathrm{f(x)} = \sum_{\mathrm{i=1}}^{\mathrm{n_u}} \Delta \mathrm{P_i} \tag{3}$$

where $\mathrm{n_u}$ describes the total number of available DG units and $\Delta \mathrm{P_i}$ the curtailed power of unit *i*, which amounts to:

$$\Delta \mathrm{P_i} = \mathrm{P_{i,init}} - \mathrm{P_{inj,max}} \tag{4}$$

where $P_{i,init}$ describes the initial power infeed and $P_{inj,max}$ the requested maximum power injection from Equation (1). Accordingly, the decision variables are discrete due to the discrete curtailment level increments which results in a mixed integer programming problem. Real power flow restrictions are displayed by the addition of constraints for each grid component:

$$P_{comp,init} + \Delta P_{comp} \leq P_{comp,max} \tag{5}$$

where $P_{comp,init}$ describes the real power flow over the considered grid component in the initial grid state, ΔP_{comp} the change in the real power flow due to DG curtailment, and $P_{comp,max}$ the maximum permitted power flow over the grid component. The power flow change can be written in terms of the power flow sensitivities of each DG unit on the respective grid component:

$$\Delta P_{comp} = \sum_{i=1}^{n_u} \frac{\partial P_{comp}}{\partial P_i} \cdot \Delta P_i \tag{6}$$

This optimization approach solely minimizes the amount of curtailed power, whereas the number of curtailment requests is not directly considered in the objective function. On the contrary, the best practice iterative approach curtails the least number of DGs and neglects the amount of reduced power output. However, grid operators might practically want to reach a certain objective, such as minimizing curtailment, with a limited number of control actions [21]. Therefore, a hybrid approach is also proposed as a candidate selection method in this paper, which aims at reducing the resulting renewable energy curtailment while limiting the complexity of the practical implementation by limiting the number of curtailment requests. The three considered selection methods are compared schematically regarding the amount of curtailment and the number of requests in Figure 1.

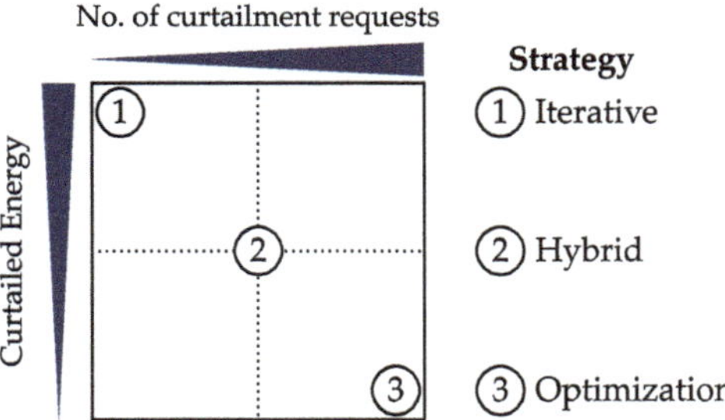

Figure 1. Schematic comparison of distributed generator (DG) selection methods regarding the resulting amount of curtailed energy and number of curtailment requests.

The hybrid selection method takes the number of curtailment requests into account by implementing a penalty term for each curtailment level leading to adjustments compared to the initial grid state ($\Delta P_i > 0$) in the optimization objective (Equation (3)). This can be interpreted as minimizing the total costs of curtailment consisting of a variable part depending on the amount of curtailment and a fixed part for each measure. The variable costs describe the financial reimbursement of curtailed DGs whereas the fixed proportion represents organizational expenses of grid operators (e.g., costs of billing curtailment). The optimization objective is thus adapted as follows:

$$\begin{gathered} \min_x f(x) = \sum_{i=1}^{n_u} \Delta P_i \cdot c_{var} \cdot \Delta t + p_c(\Delta P_i) \\ \text{with } p_c = \begin{cases} 0 \text{ if } \Delta P_i = 0 \\ c_{fix} \text{ else} \end{cases} \end{gathered} \tag{7}$$

The variable costs c_{var} are estimated based on the total amount of curtailed energy in Germany in 2017 and the respective amount of financial reimbursements, which amount to 610 M€ for 5518 GWh of curtailed energy [11], resulting in variable costs of 110 €/MWh. The penalty term p_c is set to 0 if no curtailment is requested and to a fixed cost parameter for curtailment requests. This fixed cost parameter c_{fix} directly influences the solution of the optimization problem. The higher this parameter, the higher the solution would prioritize the number of curtailment requests towards the amount of curtailed energy and vice versa. Due to the available data set in this work, the time interval Δt is set to 15 min or 0.25 h, respectively.

2.2.2. Influence of Grid Operator Coordination

Another issue reducing the efficiency of requested curtailment measures is the separate assessment of different power system levels. DG units can affect load flow situations both on the local distribution grid level and the superordinate transmission level. Consequently, DG curtailment is requested both by DSOs for preventing congestions at the distribution grid level and by TSOs to support grid operation at the superordinate transmission level in the course of an operative cascade [18]. TSO and DSO are only responsible for their own grid and request curtailment measures independently. Practically, measures requested by the TSO can also resolve congestions at the distribution level and vice versa. However, the separate security assessment and derivation of suitable curtailment measures carried out by each grid operator independently can cause redundant requests. Hence, this work proposes a coordinated curtailment procedure, which aims at finding the global optimal amount of DG curtailment. Therefore, grid restrictions at multiple system levels are considered simultaneously. Figure 2 visualizes the comparison of the established operative cascade and the proposed curtailment framework in this work.

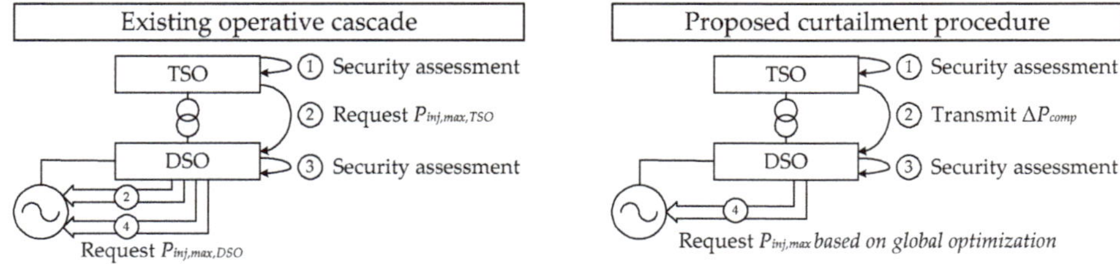

Figure 2. Comparison of established operative cascade and proposed curtailment procedure.

Following the existing operative cascade (left side of Figure 2), in the first step, the TSO assesses security criteria in its responsibility area (Step 1). If congestions, e.g., overloadings of transformers linking the transmission and distribution system, are identified, the TSO is allowed to send curtailment requests to the respective DSO. The DSO is then responsible for forwarding these signals to the respective DG units (Step 2). Additionally, the DSO determines the security state in its own responsibility area (Step 3), which can lead to additional curtailment requests (Step 4).

To avoid redundant or unnecessary curtailment requests, the proposed procedure aims at finding the global optimal amount of DG curtailment (right side of Figure 2). Therefore, grid restrictions of multiple system levels are considered simultaneously. Consequently, the TSO does not request DG curtailment directly but rather transmits the necessary power adjustments on all congested grid components ΔP_{comp} to the DSO (Step 2). These are added to the optimization problem as constraints and complemented by additional security constraints due to distribution network restrictions (Step 3). As a consequence, the optimization problem can be solved once considering all necessary security constraints simultaneously (Step 4). The operative cascade is also extensible for multiple grid operators on the distribution grid level by adding additional hierarchy levels. Because this work focuses on the 110 kV distribution level, subordinate grid levels are not represented in detail and the operational cascade only consists of two system levels.

2.3. Operational Security Assessment

The first step of the curtailment process is the security assessment carried out by the respective grid operator (Figure 2). Accordingly, the necessity and the extent of practical DG curtailment measures are also strongly influenced by the way power system security is assessed. This is done based on load flow calculation of the undisturbed base case and several credible contingency conditions [22]. Contingencies include the outage of a grid component or generation unit. The European Network of Transmission System Operators for Electricity (ENTSO-E) defines five different steady-state security states, namely normal, alert, emergency, blackout, and restoration [23]. The monitored security parameters contain the thermal loading of grid components, such as transmission lines and transformers, and the nodal voltages in the network.

The normal state is also referred to as n-1-secure and is reached if all parameters are within allowed operational limits both in the base case and in any contingency state. In this case, no correcting congestion management measures are necessary. The alert state is reached if at least one of the monitored parameters exceeds security limits in a certain contingency event. This state is also referred to as n-secure, since no limit violations are present in the undisturbed base case [24]. In the alert state, preventive congestion management measures are activated with the aim to return the system to a n-1 secure grid operation. This guarantees an a priori security margin for contingency events, which, however, leads to an inefficient utilization of transmission capacities in the undisturbed grid state since the considered contingencies are possible but not certain events. The emergency state is reached if operational limit violations occur in the undisturbed base case. Then, curative congestion management measures have to be realized. These include fast-reacting measures, which terminate congestions as quickly as possible. On the other hand, these enhance the efficiency of grid capacity utilization in the base case since they decrease preventive security margins [25].

Recently, best practice is to apply DG curtailment as a preventive congestion management measure [5]. Accordingly, it is activated in the alert state to maintain a security margin for possible fault situations. Hence, the utilization of grid components in the base case can be increased by increasing the loadability of grid components in contingency events and, thus, the amount of preventive renewable energy curtailment can be reduced [25].

The implementation of this aspect into the developed curtailment simulation framework can be realized by adapting the network security constraints in Equation (5). A distinction can be made between the maximum permissible active power transfer over grid components in the undisturbed grid state (n-0) and in possible contingency states (n-1):

$$P_{comp,init,n-0} + \Delta P_{comp,n-0} \leq P_{comp,max,n-0} \tag{8}$$

$$P_{comp,init,n-1} + \Delta P_{comp,n-1} \leq P_{comp,max,n-1} \tag{9}$$

Due to the uncertain occurrence of contingency events, the increase of the maximum permissible real power flow in n-1 events $P_{comp,max,n\text{-}1}$ above the physical limit of the respective grid component is not necessarily accompanied by actual overloadings.

2.4. Dynamic Line Rating

The developed simulation framework furthermore considers changes in the power transfer capability of overhead transmission lines dependent on environmental factors. The factor limiting the transmission capacity of overhead lines is usually the sag of the line, which results from its temperature [26]. However, this is not only dependent on the transmitted current but rather on several environmental factors such as the ambient temperature, wind speed, and solar radiation [27]. The transmission capacity of overhead lines is thus dynamically changing with variations in these factors. Generally, the rated current of overhead lines is determined by assuming normative values with an ambient temperature of 35 °C, a wind speed of 0.6 m/s, and solar radiation of 900 W/m^2 [28].

Particularly in times of high wind occurrence accompanied by a high amount of DG feed-in, the demand for congestion management measures can be reduced due to higher admissible currents on transmission lines resulting from additional cooling effects [27].

The application of dynamic line rating is already well established in practice. Therefore, it is not considered as a novel operational concept in this work but rather integrated into the simulation environment to increase the accuracy of simulation results in comparison to real data from grid operation practice because dynamic line rating has a significant impact on the amount of incidental DG curtailment. Based on a meteorological dataset, the maximum transferable current of a transmission line is calculated dynamically using the approach proposed by the Institute of Electrical and Electronics Engineers (IEEE) [29]. Thus, the maximum real power transfer capabilities in Equations (5), (8), and (9) change dynamically.

3. Implementation and Simulation Results

The developed simulation framework is applied to the model of a real 110 kV distribution network of a region located in Germany, which is characterized by a high share of installed DG capacities. Firstly, the model and data sources are described. Then, best practice curtailment approaches are implemented as a reference simulation scenario to validate the model regarding its capability to reproduce historical curtailment measures based on load flow simulations. Finally, different scenarios including the proposed novel concepts are defined and implemented, and the resulting reduction potential compared to the reference scenario is quantified and discussed.

3.1. Grid Simulation Model and Data Sources

The model of a real 110 kV high voltage (HV) distribution system with 100 nodes and a ring topology located in Germany is used as a case study in this paper. Grid topology data of the 110 kV level are provided by the respective grid operator and further extended by a simplified representation of the superordinate transmission system. Accordingly, the power flows at the interconnection points between transmission and distribution system are also modelled to simulate possible congestions of transformers operated by the TSO. The investigated 110 kV system is supplied by three interconnection points and operated by the DSO.

The amount of curtailment due to congestions on the subordinate medium voltage (MV) and low voltage (LV) levels is negligibly small, which can be derived from a dataset including historical curtailment requests in the region. Thus, grid restrictions on the MV and LV level are not considered. Instead, DG units and loads on the MV and LV level are modelled in aggregate at the interconnection point to the HV network. Historical measurements of active and reactive power flows of all HV/MV transformers with a resolution of 15 min are available, which enable time series load flow simulations. Since these measurements are aggregated and the wind feed-in needs to be known separately to determine curtailment potentials, the method presented in [30] is used to disaggregate the measured time series into aggregated power injections of wind energy generators and the remaining part. Consequently, the simulation of curtailment is limited to wind turbines whereas other DG technologies are neglected. This assumption is expected as reasonable since FIM is mainly carried out using wind turbines [11]. Additionally, the time series are distorted by historical curtailment measures since the feed-in of DGs is reduced in these situations. The disaggregation enables the simulation of times with historical curtailment measures since the potential DG feed-in can be estimated based on meteorological data. A clear sky model provides the solar radiation [31] whereas the wind speed and ambient temperature are provided by the COSMO-DE analysis of the DWD (German Meteorological Service) [32]. These are geographically interpolated to the coordinates of all MV grid districts. Measured time series of wind farms directly connected to the 110 kV level are not available. Therefore, these are created synthetically by using the meteorological data set and the performance curve of a typical wind turbine [33]. The meteorological data set is also used to implement dynamic line rating for the respective 110 kV transmission lines. For this, each transmission line is assigned to the substation it has the shortest geographical distance to.

The balance of the total real and reactive power within power flow simulations in the region is ensured by the reference node on the 220 kV level. The parameters of transmission lines on the 110 kV level are provided by the respective grid operator (GO). Furthermore, the line lengths of the 220 kV lines are derived from a publicly available GIS database and the electrical parameters are extracted from a typical AL/ST 265/35 220 kV overhead line model [34]. Capacities of 220 kV/110 kV transformers are also given through grid operator information, whereas the short-circuit voltage for power flow calculations is specified to 13% based on [35]. A schematic graphical representation of the model is given by Table 1 including an overview of all numbered model objects and the respective data sources.

Table 1. Schematic visualization of grid simulation model including the data sources of different model parameters.

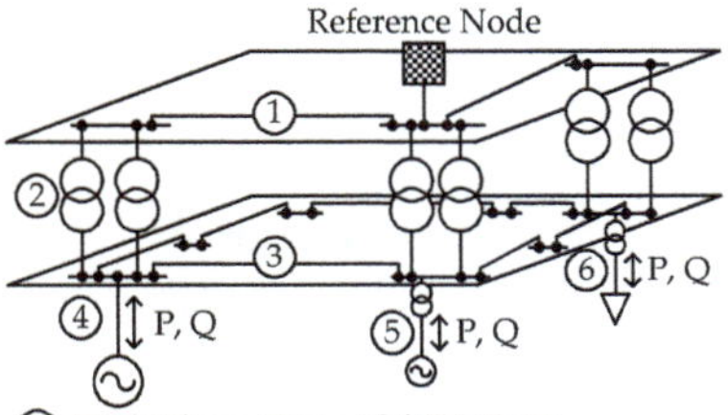

Reference Node

① 220 kV lines (Type AL/ST 265-35)
② 220 kV/110 kV Transformers
③ 110 kV lines
④ DG units on 110kV level
⑤ Aggregated DG units on MV/LV level
⑥ Aggregated Loads on MV/LV level

Objects	Parameter	Source
①	Line length Electrical parameters	GIS [34]
②	Capacity Short-circuit voltage	GO [35]
③	Line length Electrical parameters	GO
④	Installed capacity Wind speed Performance curve	[36] [32] [33]
⑤ ⑥	Active and reactive power time series	GO

3.2. Validation of Simulated Curtailment Demand

The aim of the model validation is to assess the extent to which the developed simulation framework is capable of displaying realistic curtailment measures. Therefore, the recent best practice approaches for DG curtailment are implemented into the framework and the simulated curtailment measures are compared to historical measures, which are derived from published data by the respective grid operators in the model region.

Firstly, the best practice scenario setup is implemented into the simulation framework. The feasible set of curtailment levels is considered as:

$$\Omega_{\text{curt}} = [1,\ 0.6,\ 0.3,\ 0]$$

Furthermore, the iterative curtailment method is implemented and the curtailment requests by TSO and DSO are derived independently. In addition, the maximum loading of grid components in contingency situations is set to 100%. In the following, this scenario configuration is also considered as the reference scenario to evaluate the effectiveness of the proposed concepts.

The published data set of historical curtailment measures includes starting time, level (1, 0.6, 0.3, 0), ending time, and the receiving substation of each request. Due to the frequent occurrence of curtailment requests in the historical data set and the availability of comprehensive time series data, February 2016 was selected as the investigated time period in this work.

Hence, the historical curtailment of DG units is determined using the respective time series and the historical data set with a resolution of 15 min. For the same time horizon, the incidental curtailment using the best practice approaches is simulated. Therefore, in each time step, a contingency analysis is executed to assess the security state of the network. In the alert state, DG curtailment is requested as a preventive congestion management measure. In case of multiple critical contingencies, grid operators usually initiate countermeasures ordered by the severity of operational limit violations [22]. Subsequently, the amount of curtailment is determined for the most severe contingency at first. This is iteratively repeated until all congestions in each possible n-1 event are terminated.

Firstly, according to the cascading grid operation approach (Figure 2), congestions on the TSO level are identified by performing a contingency analysis of all 220/110 kV transformers in the grid area. For all insecure grid states, the resulting amount of DG curtailment is derived by the simulation framework. Since most curtailment measures contained in the historical data set result from congestions of transformers connecting the transmission and distribution system it is assumed that the TSO requests FIM measures only to prevent overloadings of transformers. Thus, congestions of 220 kV transmission lines are not considered in this work. After deriving curtailment requests by the TSO, another contingency analysis is executed to determine the security state of the 110 kV system and derive curtailment requests by the DSO if necessary. Although the curtailment demand is determined for each installed DG unit individually, simulation results are validated for the total amount of curtailment in the investigated region to assess the overall amount of incidental renewable energy curtailment.

3.2.1. Validation Metrics

In the first step of the validation process, the accordance of simulated curtailment with the historical measures in each of the 15-min intervals is analyzed. Therefore, each simulation step can be assigned to one of the four binary states visualized in Table 2. In each time step, it is examined if the historical data set contains any curtailment requests and if curtailment requests are simulated based on the best practice approach. Consequently, four different combinations (a, b, c, d) of these aspects are possible [37].

Table 2. Depiction of different binary states to assess time steps resulting from combinations of the presence or absence of simulated and historical curtailment measures.

	Simulated Curtailment	
Historical Curtailment	**True**	**False**
True	a	b
False	c	d

To increase the comparability of the simulation results achieved in this work, these are put in relation to the study of [2] as a reference. There, the authors developed a congestion forecast model based on autonomous neural networks for a German transmission network and validated it based on historical data about congestion management measures. Although neither these measures are quantified nor multiple system levels are considered in [2], the results can be compared to the simulations carried out in this work regarding the ability to represent historical congestion situations. The precision (p), recall (r), and f1-score (f) are used as similarity indices between simulation and historical values in [2], which are defined as follows:

$$p = \frac{a}{a + c} \tag{10}$$

$$r = \frac{a}{a + b} \tag{11}$$

$$f = 2 \cdot \frac{p \cdot r}{p + r} \tag{12}$$

For each of those indices, a value of 1 indicates total similarity. Additionally, in this work the mean absolute error (MAE) and the root mean square error (RMSE) are used as error metrics to assess the accuracy of curtailment simulations, which are defined as follows [38]:

$$\mathrm{MAE} = \frac{1}{n} \cdot \sum_{k=1}^{n} |e_i| = \frac{1}{n} \cdot \sum_{k=1}^{n} \left| P_{curt,historic} - P_{curt,simulated} \right| \tag{13}$$

$$\mathrm{RMSE} = \sqrt{\frac{1}{n} \cdot \sum_{k=1}^{n} |e_i|^2} = \sqrt{\frac{1}{n} \cdot \sum_{k=1}^{n} \left| P_{curt,historic} - P_{curt,simulated} \right|^2} \tag{14}$$

where n describes the number of simulated time steps, $P_{curt,historic}$ the amount of power reduction in the historical data set, and $P_{curt,simulated}$ the simulated power curtailment per time step.

Due to the independent simulation of curtailment requests by the TSO and the DSO, the similarity indices and error metrics are calculated separately for both grid operators. In addition, the aggregated results, defined as the sum of TSO and DSO curtailment requests, are determined. Table 3 depicts the resulting similarity indices from reference [2] compared to those achieved in this paper. Furthermore, the resulting MAE and RMSE from this work are depicted. In order to visualize the significance of the absolute values for the error metrics, the maximum amount of simulated power curtailment is also depicted.

Table 3. Resulting similarity and error metrics between historical and simulated DG curtailment in comparison to results published in [2].

	[2]	This Work		
		DSO	**TSO**	**Aggregated**
Precision p	0.699	0.462	0.812	0.930
Recall r	0.636	0.263	0.748	0.760
f1-score f	0.648	0.335	0.779	0.837
Maximum curtailment [MW]	/	162.844	389.393	389.393
MAE [MW]	/	7.892	19.985	22.484
RMSE [MW]	/	26.353	51.396	51.948

According to Table 3, regarding all similarity values, the simulation of TSO curtailment is more accurate compared to [2], whereas DSO measures are depicted less accurately. The results also show that the aggregated consideration has the highest values in all similarity indices. Both error metrics have the lowest values for the DSO curtailment and the highest values for the aggregated consideration.

Consequently, the similarity metrics indicate a lack of accuracy regarding the simulation of curtailment requests by the DSO. This is due to several reasons. In particular, the recall r of only 26.3% indicates a poor model performance. Since this index is dependent on the inaccurate state b, it can be concluded that a large number of historical DSO requests do not appear in the simulation results, because the corresponding congestions are already resolved by the prior determined TSO curtailment demand. In 86% of the times assigned to state b, TSO requests are simulated. Hence, congestions in the distribution network are often already resolved and no additional DSO curtailment is necessary. On the contrary, 58% of times assigned to state c are accompanied by an underestimation of TSO curtailment requests. Consequently, the precision p is also influenced by this dependency but to a smaller extent. Accordingly, this index indicates a larger similarity of 46.2%. The MAE values are relatively small compared to the maximum occurring power reductions in all three cases. The RMSE values are considerably larger than the MAE values. Hence, it can be concluded that the curtailment simulation values occasionally include larger deviations from the historical values, since the RMSE is more sensitive to large errors than the MAE [38]. However, both error metrics are relatively small compared to the maximum occurring values. The differences between simulation results and historical data are further visualized by the time series comparison of requests by the TSO, DSO and the aggregated amount of DG curtailment in Figure 3. The time series confirm the high degree of temporal coincidence between simulation and historical values. Occasionally, larger deviations are apparent, which contribute to the larger RMSE values.

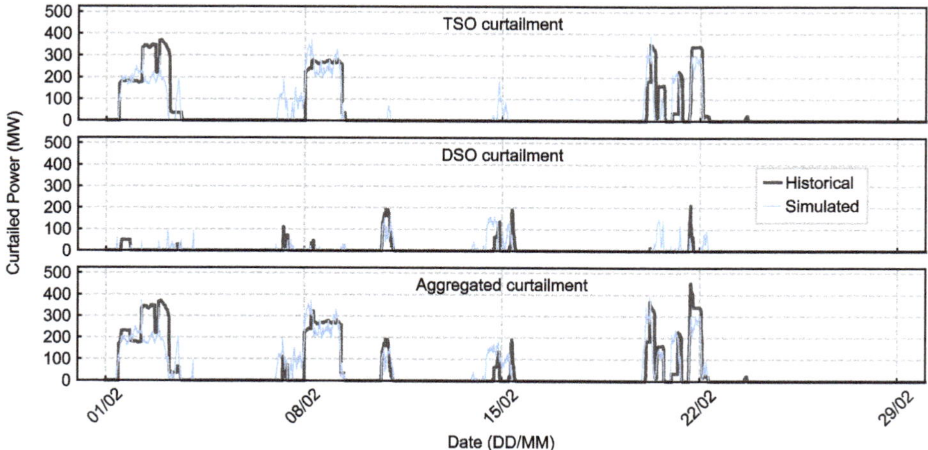

Figure 3. Time series comparison of simulated and historical curtailment measures requested by the transmission system operators (TSO) (**top**), distribution system operators (DSO) (**middle**) and aggregated (**bottom**).

Overall, the simulation of curtailment measures requested by the DSO is highly influenced by the simulation of the TSO demand beforehand, which also contains errors, as can be seen in Table 3. Nevertheless, the aggregated results gain the highest similarity values and only slightly higher values in the error metrics. This confirms that some measures historically requested by the DSO are mistakenly simulated as measures by the TSO and vice versa. The relatively low values of MAE and RMSE and the graphical time series comparison in Figure 3 confirm the accuracy of simulated DG curtailment, particularly considering the aggregated results.

Possible sources of errors decreasing the accuracy of curtailment simulations are the meteorological data which affect both the wind power time series and the dynamic line rating approach. As stated in [39], the quality of the dynamic line rating approaches could be improved by further increasing the maximum geographical resolution of 2.8 km of established meteorological models, which are also used in this work. This is especially relevant for the 110 kV lines and, thus, for the curtailment requests by the DSO. Additionally, due to the available data basis, the simulation was carried out for 15-min time intervals. However, grid security is typically assessed more frequently (e.g., in intervals of 2 [40] or 5 [41] min) by state estimation algorithms. Therefore, the practical security assessment fluctuates more than in the simulations, which can also affect the derivation of curtailment measures.

3.2.2. Comparison of Historical and Simulated Renewable Energy Curtailment

In addition, the total amount of renewable energy curtailment in the considered time horizon is derived from the data of simulated and historical curtailed power per time step. The resulting amounts of energy curtailment requested by TSO, DSO, and aggregated are depicted in Figure 4.

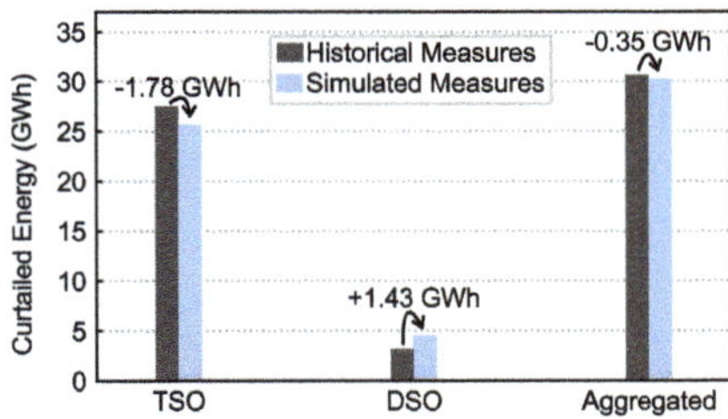

Figure 4. Comparison of renewable energy curtailment requested by the TSO, DSO, and aggregated.

The curtailed energy due to requests by the TSO is slightly underestimated by 1.78 GWh (6.5%), whereas the demand from the DSO is overestimated by 1.43 GWh (45.6%). Consequently, the combined energy amount is underestimated by 0.35 GWh or 1.1%. Accordingly, the model is able to project the historical curtailment measures regarding the total amount of curtailed energy very well. The share of TSO requests in the aggregated amount is comparably large in the simulation (84.89%) and the historical (89.74%) dataset. Consequently, most curtailment requests are due to congestions of transformers linking the distribution and transmission system.

Overall, the similarity measures, time series analysis, and evaluation of the curtailed energy indicate the capability of the simulation framework to represent the demand and number of curtailment measures in the investigated region to a reasonable extent. Despite inaccuracies regarding the allocation of curtailment measures to the requesting grid operator, the aggregated simulation results are particularly promising. Therefore, the developed simulation framework is assumed to achieve reasonable results regarding the simulation of DG curtailment. The resulting amount of renewable energy curtailment of 30.27 GWh in the reference scenario is thus further used to assess the extent to which the proposed operational concepts can increase grid capacity utilization and, thereby, decrease the incidental amount of DG curtailment.

3.3. Scenarios for the Implementation of Novel Operational Concepts

Section 2 discusses relevant aspects influencing the demand for DG curtailment including the description of recent best practice approaches and possible concepts to improve the practical implementation on a theoretical basis. Based on different scenarios, these concepts are implemented into the simulation framework to quantify their benefit in terms of maximizing the integration of renewable energy generation into the investigated 110 kV distribution network. The concepts include increasing the DG controllability by extending the feasible set of curtailment levels. In addition, the implementation of the proposed novel DG selection procedures is investigated. These include the coordination of curtailment requests between TSO and DSO by implementing both the optimization (Equation (3)) and the hybrid selection method (Equation (7)) into the simulation framework. Finally, the concept of improving the utilization of transmission capacities by increasing the maximum permissible loading of grid components in possible contingency events is considered. However, the implementation of one of these concepts does not exclude the coincident implementation of the others. Instead, combinations can be also realized to improve the DG curtailment process regarding different aspects simultaneously. Table 4 gives an overview of the investigated scenarios regarding the considered adaptations compared to the best practice reference scenario.

Table 4. Overview of investigated scenarios and adaptations regarding the specification of curtailment level increments, DG selection method, and maximum permissible loading in contingency states compared to best practice approaches.

Scenario Description	Specification of Aspect		
	Feasible Set of Curtailment Levels Ω_{curt}	DG Selection Method and Grid Operator Coordination	Maximum Permissible Loading in n-1 Events
Best practice reference scenario	[1,0.6,0.3,0]	Iterative selection method, independent curtailment requests by TSO and DSO	100%
	Adaptations compared to best practice approaches		
Scenario 1: Extended feasible set	X		
Scenario 2: Coordinated DG selection methods	X	X	
Scenario 3: Increased loadability in contingency events		X	X

Firstly, the reference scenario includes four discrete curtailment levels, the iterative selection scheme accompanied by independent curtailment request of TSO and DSO, and a maximum loadability of grid components of 100% in contingency events. The first scenario quantifies the benefit of decreasing curtailment level increments and thereby enabling the grid operator to control DG units to a larger extent. The DG selection procedure and maximum permissible loading in contingency events remain unmodified in comparison to the reference scenario. The second scenario supplements the larger control opportunities by also implementing the optimization and hybrid approaches as candidate DG selection methods and thereby improving the coordination between TSO and DSO. Finally, the last scenario quantifies reduction potentials of DG curtailment dependent on increasing the maximum loadability of grid components in n-1 events. This scenario also considers the different DG selection methods. In the following, simulation results for each of the described scenarios are presented.

3.4. Curtailment Reduction Potentials of Proposed Concepts

3.4.1. Scenario 1: Extended Feasible Set of Curtailment Levels

Smaller curtailment level increments enable more efficient requests due to the enhanced DG control capability of grid operators. Consequently, feasible sets of curtailment levels with smaller increments compared to the best practice approach are implemented into the simulation framework to quantify the benefit of this concept. The incremental sizes investigated in this scenario are 10% and 1% related to the nominal capacity of DG units. Table 5 depicts the resulting amount of DG energy curtailment within the considered time horizon of one month requested by DSO, TSO, and the aggregated sum of both in comparison to the reference scenario with only four discrete levels.

Table 5. Resulting renewable energy curtailment for different incremental sizes of curtailment levels.

Incremental Size of Curtailment Levels	Energy Curtailment [GWh]		
	DSO	TSO	Aggregated
Reference scenario ([1,0.6,0.3,0])	4.57	25.7	30.27
10%	4.16	25.19	29.35
1%	3.99	24.98	28.97

Curtailment requested by the DSO can be reduced by 0.41 GWh (9%) for 10% and 0.58 GWh (12.7%) for 1% increments. Moreover, TSO curtailment can be reduced by 0.51 GWh (2%) and 0.72 GWh (2.8%). Accordingly, the aggregated savings amount to 0.92 GWh (3%) and 1.3 GWh (4.3%) for 10%

and 1% increments, respectively. The results show that the reduction of the total amount of curtailment is comparable between DSO and TSO. Consequently, the relative savings achievable by the DSO are distinctly higher. Due to the iterative selection scheme applied in this scenario, the benefit of smaller curtailment level increments is limited to the least sensitive DG unit affected by the respective request, because each of the other affected units is fully curtailed. Hence, the relative reduction of curtailed energy is reduced significantly only in the case of less severe congestions that induce a small amount of total power curtailment.

This effect is illustrated in Figure 5, which shows the relative reduction of the aggregated amount of DG curtailment requested by TSO and DSO gained by 10% and 1% increments per time step depending on the respective amount of total power reduction. The relative savings are related to the best practice increments. It becomes clear that the highest relative savings are achieved at the less severe congestions. Smaller increments tend to only have minor advantages in situations with large amounts of power reductions. As depicted by Table 5, the relative reduction potential is thus considerably lower for TSO requests, since these exhibit larger amounts of power adjustments compared to DSO curtailment.

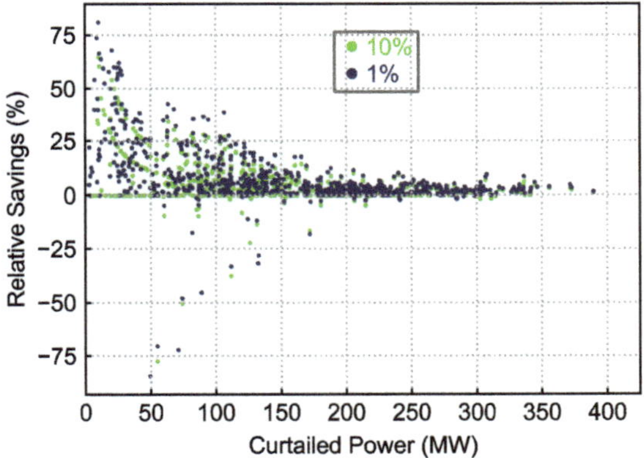

Figure 5. Relative reduction of curtailed DG power feed-in achieved by 10% and 1% increments in comparison to established curtailment levels.

Furthermore, a few measures show negative savings for 10% and 1% increments in Figure 5. This is due to the fact that TSO curtailment also reduces congestions at the distribution level. Therefore, the avoidance of TSO requests can inevitably lead to a higher amount of curtailment demand at the distribution grid level. Consequently, the aggregated amount resulting from smaller increments can even exceed the demand induced by larger increments in certain situations.

In total, the advantage of decreasing the curtailment level increments is its easy implementation into existing grid control strategies, since modern communication standards and technologies already enable a finer granularity in DG controllability compared to FIM [20]. However, the simulation results in this scenario imply that the potential to reduce the curtailed energy is limited to a few percent due to the frequent occurrence of severe congestions in real grid operation practice, especially in combination with the iterative DG selection approach. Therefore, the benefit of this concept is also assessed in combination with implementing the novel DG selection methods in the following scenario.

3.4.2. Scenario 2: Coordinated DG Selection Methods

In this scenario, the optimization approach (Equation (3)) and the hybrid approach (Equation (7)) are implemented as candidate DG selection methods to increase the coordination between TSO and DSO. The benefit of this concept is assessed based on both the potential to reduce the resulting amount of renewable energy curtailment and the complexity of its practical implementation compared to the best practice iterative approach. Therefore, in addition to the resulting amount of DG curtailment, the number of necessary control signals is also considered to evaluate the communication demand associated with the implementation of this concept.

To implement the hybrid approach (Equation (7)), firstly a suitable choice for the fixed cost parameter c_{fix} in the objective function has to be determined. Therefore, this parameter is varied iteratively between 100 € and 500 € in steps of 100 € and analyzed regarding its influence on the curtailed energy and the number of requested measures. Both variables are depicted in Table 6 dependent on the respective fixed cost parameter. In addition, the results obtained from the reference scenario are shown. As expected, the lowest fixed cost parameter (100 €) causes the lowest amount of curtailed energy but the largest number of measures. In contrast, the highest fixed cost parameter of 500 € prioritizes the number of measures over the amount of curtailed energy. Compared to the reference case, an increase in the number of curtailment measures for all parameters is present, whereas the amount of curtailed energy is lower. For the further analyses in this work, the parameter is set to a value of 300 €, since this causes a considerable reduction of curtailed energy (2.79 GWh) while increasing the number of curtailment measures acceptably (49.43%) compared to the iterative approach.

Table 6. Renewable energy curtailment and number of curtailment requests for different fix cost parameter specifications.

		Curtailed Energy [GWh]	Curtailment Requests [#]
Reference Scenario		**30.27**	**437**
c_{fix} [€]	100	27.46	989
	200	27.48	738
	300	27.49	653
	400	27.77	542
	500	27.87	497

The three different DG selection methods are compared regarding the incidental amount of renewable energy curtailment and the number of resulting curtailment requests. In addition, the approaches are combined with the concept of smaller curtailment level increments to show possible interdependencies between both concepts. Therefore, nine different scenario specifications are investigated. The resulting energy curtailment and the number of measures are visualized in Figure 6. The different increments are represented by varying colors, and the number and the amount of curtailment are reflected by different markers ('o','x'), respectively.

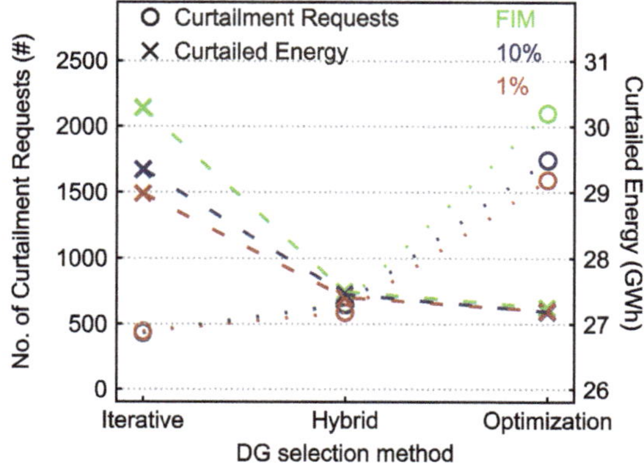

Figure 6. Resulting renewable energy curtailment and number of curtailment requests for different DG selection methods and incremental sizes of curtailment levels.

The iterative approach yields the least number of measures and the highest amount of curtailed energy. Decreasing the curtailment increments mainly reduces the curtailed energy (Section 3.4.1.), whereas the number of measures nearly remains constant. In contrast, the implementation of the optimization approach reduces the amount of curtailed energy by up to 3 GWh (10%) at the expense of 1660 additional curtailment requests. In contrast to the iterative approach, the energy amount remains constant with decreasing increments whereas the number of measures can be reduced considerably. This is due to the fact that smaller increments result in a higher flexibility regarding the controllability of DGs. Therefore, it is less probable that curtailment measures are firstly cancelled and, subsequently, reactivated if congestions return. Smaller increments allow the grid operator to carefully return the DGs feed-in to the initial value while monitoring the operational security limits of the network. Finally, the hybrid approach nearly obtains the same amount of curtailed energy compared to the optimization approach (dashed trend line) and a similar number of curtailment measures compared to the iterative approach (dotted trend line). Moreover, the dependence of this approach on the incremental size is relatively small.

Overall, the hybrid approach yields the best trade-off between the amount of curtailed energy and the complexity of its practical implementation. Combining this strategy with smaller increments only has a minor effect on both of these aspects. However, compared to the iterative approach, the curtailed energy can be reduced by 2.78 GWh at the expanse of 216 additional curtailment measures. Considering the average financial reimbursement of DG operators, reducing curtailment by 2.78 GWh could save 305,800 € of compensation payments within the investigated time period of one month. Concurrently, 1444 measures can be avoided compared to the optimization approach, which reduces the communication demand and the complexity of the implementation into grid operation significantly. This is realized at the expense of increasing the amount of curtailed energy by only 0.24 GWh.

3.4.3. Scenario 3: Increased Loadability in Contingency Events

Increasing the loadability of grid components in contingency events requires the availability of fast-reacting curative congestion management measures, such as fast power flow controlling devices, to terminate the occurring overloading in case of n-1 events as soon as possible. In this work, these measures are not explicitly investigated. Rather, the focus is on analyzing the resulting demand for preventive curtailment measures in the alert security state, if grid components can be loaded more heavily in case of contingency events.

This concept is implemented into the simulation framework by increasing the maximum permissible real power flow $P_{comp,max,n-1}$ in the security constraints (Equation (9)). Therefore, the maximum thermal loading of grid components is enhanced iteratively in steps of 5% until a maximum value of 130% related to the physical limit of grid components in this scenario. Hence, a value of 130% assumes that the loading of grid components in any contingency situation is preventively reduced to a maximum value of 130%. If this contingency actually occurred during grid operation, curative measures would have to ensure the return of the thermal loading to a value below 100% as quickly as possible. Concurrently, a maximum thermal loading of grid components of 100% in the base case is assured and, thus, the system remains in the alert state and does not reach the emergency state. Furthermore, the concept is applied for the iterative, optimization, and hybrid DG selection methods. The resulting amount of renewable energy curtailment for different maximum loading values and the different selection methods is depicted in Figure 7.

The results clearly illustrate the potential to reduce the incidental DG curtailment significantly by increasing the thermal loadability of grid components in n-1 situations. By enhancing the thermal loadability to 130%, the curtailed energy amounts to only 1.56 GWh, which is 5.2% of the initial value. According to Figure 7, the differences between different DG selection methods decrease as the loadability is increased. Hence, the potential to reduce DG curtailment by implementing novel selection methods is lowered due to the massive reductions obtained by increasing the loadability of grid components in contingency events.

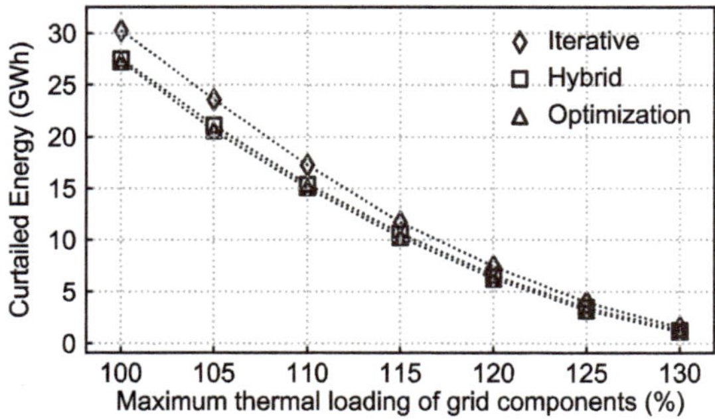

Figure 7. Resulting renewable energy curtailment for different maximum permissible grid component loadings in contingency events and different DG selection methods.

In addition to curative congestion management measures, another possibility to increase the loadability of grid components in contingency events is to consider their thermal inertia, as presented in [15]. Using more detailed thermal models of components, the temperature rise in contingency situations can be simulated dynamically and, thus, a critical time can be derived until the maximum temperature of transmission lines or transformers is reached to derive timely congestion management measures.

3.4.4. Summary and Discussion

To summarize the presented simulation results from the investigated scenarios, Table 7 presents the curtailment reduction potentials of the proposed concepts and evaluates the complexity of implementing these concepts into grid operation practice. The potential to reduce the incidental amount of DG curtailment by implementing smaller curtailment level increments of 10% and 1% was proven to be small due to the frequent occurrence of severe congestions. However, the implementation of this concept can be realized without large expanses since modern communication standards and technologies already enable a finer granularity in DG controllability compared to best practice [20]. Furthermore, simulation results show that the amount of DG curtailment can be reduced by implementing novel DG selection methods. The hybrid approach was proven to be the most suitable approach for the practical implementation since it nearly reaches the same reduction of renewable energy curtailment (9.48%) compared to the optimization approach (10.14%), while reducing the complexity of operational implementation by reducing the number of curtailment requests significantly. The coordinated curtailment procedure requires an extended data exchange between different grid operators regarding the demand for power flow adjustments and power flow sensitivities considering multiple system levels. Due to the low complexity of their practical implementation, both of the aforementioned concepts could be realized in the near term to maximize the integration of DG units into power systems. However, in the long-term perspective the concept of utilizing existing grid capacities by allowing heavier loadings of grid components in uncertain contingency events should be intensified since it clearly shows the highest potential to reduce incidental renewable energy curtailment, by up to 94.8%. On the contrary, this concept requires the implementation of curative congestion management measures or dynamic thermal models of components into grid operation. Therefore, future research should address the question of whether investments in novel curative technologies can be justified by their potential to reduce the amount of renewable energy curtailment quantified in this work.

Table 7. Overview of resulting renewable energy curtailment of proposed concepts and challenges regarding the practical implementation.

Proposed Concept	Minimum Resulting Renewable Energy Curtailment [GWh]	Challenges Regarding Practical Implementation
Best practice	30.27	/
Increased DG controllability	28.97	Adjustment of applied ICT standards
Improved DG selection scheme	27.20 (Optimization approach) 27.40 (Hybrid approach)	Extended data exchange between grid operators
Increased loadability in contingency events	1.56	Implementation of curative congestion management measuresImplementation of dynamic thermal grid component models

4. Conclusions

The on-going integration of DGs into the energy system causes an increase in curtailment measures executed by grid operators due to limited transmission capacities. To address this problem, this paper presents a simulation framework for optimizing the practical implementation of DG curtailment based on load flow analysis. The framework proved to be sufficient for simulating the curtailment demand in the investigated region in Germany, although the allocation of curtailment measures between DSO and TSO shows some inaccuracies. The framework is further used to analyze the ability of three operational concepts and combinations of these to reduce renewable energy curtailment. Firstly, an enhancement of DG controllability is investigated. Furthermore, novel DG selection methods are proposed that enable a simultaneous consideration of multiple voltage levels. Additionally, an extension of the deterministic n-1 criterion is considered by enhancing the maximum thermal loading of grid components in contingency events.

The simulation results obtained indicate that the amount of renewable energy curtailment can be reduced both by implementing smaller curtailment level increments by up to 4.3% and by establishing coordinated novel DG selection methods by up to 10%. To limit the complexity of the practical implementation, the hybrid approach proves to be superior to solely minimizing the amount of power adjustments. Both concepts could be a temporary solution to optimize the curtailment process since the effort for their realization is rather small. However, the concept of increasing the utilization of transmission capacities by enhancing operational limits in contingency situations has clearly proven to be the most efficient concept to reduce preventive DG curtailment. Enhancing the maximum permissible loadings of grid components to 130% reduces the amount of renewable energy curtailment by up to 94.8% and, thus, nearly leads to a complete prevention of these congestion management measures. Concurrently, this is also the most challenging concept regarding its implementation. Therefore, future research should address possibilities to extend operational limits, such as considering the thermal inertia of grid components or establishing fast-reacting, curative congestion management measures, and compare the additional costs of these measures with their benefit regarding the reduction of renewable energy curtailment.

Author Contributions: Conceptualization I.L.-N.; Writing-original draft preparation I.L.-N.; Writing-review and editing I.L.-N., F.S. and K.v.M.; and Supervision C.A. All authors have read and agreed to the published version of the manuscript.

Funding: This research received no external funding.

Acknowledgments: The authors would like to thank all project partners of the research project "enera". In particular, the authors thank Avacon Netz GmbH for providing their grid model. Furthermore, the authors thank EWE NETZ GmbH for the provision of the aggregated load time series and the German Meteorological Service (DWD) for providing meteorological data through the COSMO-DE model. The project enera is funded by the Federal Ministry for Economic Affairs and Energy (BMWi, grant no. 03SIN317).

Conflicts of Interest: The authors declare no conflict of interest.

References

1. Gross, T.; Reese, S.; Petters, B.; Cupelli, M.; Mildt, D.; Monti, A.A. Novel Approach to DG Curtailment in Rural Distribution Networks—A Case Study of the Avacon Grid as Part of the InterFlex Field Trial. In Proceedings of the 2018 IEEE 16th International Conference on Industrial Informatics (INDIN), Porto, Portugal, 18–20 July 2018; pp. 667–672.
2. Staudt, P.; Rausch, B.; Gärttner, J.; Weinhardt, C. Predicting Transmission Line Congestion in Energy Systems with a High Share of Renewables. In Proceedings of the 2019 IEEE Milan PowerTech, Milan, Italy, 23–27 June 2019; pp. 1–6.
3. Büchner, J.; Katzfey, J.; Flörcken, O.; Moser, A.; Schuster, H.; Dierkes, S.; Leeuwen, T.V.; Verheggen, L.; Amelsvoort, M.V.; Uslar, M. Smart Grids in Germany: How much costs do distribution grids cause at planning time? In Proceedings of the 2015 International Symposium on Smart Electric Distribution Systems and Technologies (EDST), Vienna, Austria, 8–11 September 2015; pp. 224–229.
4. Gunkel, D.; Möst, D. The German transmission grid expansion in long-term perspective—What is the impact of renewable integration? In Proceedings of the 11th International Conference on the European Energy Market (EEM14), Kracow, Poland, 28–30 May 2014; pp. 1–6.
5. Hoffrichter, A.; Kollenda, K.; Schneider, M.; Puffer, R. Simulation of Curative Congestion Management in Large-Scale Transmission Grids. In Proceedings of the 2019 54th International Universities Power Engineering Conference (UPEC), Bucharest, Romania, 3–6 September 2019; pp. 1–6.
6. Yusoff, N.I.; Zin, A.A.M.; Khairuddin, A.B. Congestion management in power system: A review. In Proceedings of the 2017 3rd International Conference on Power Generation Systems and Renewable Energy Technologies (PGSRET), Johor Bahru, Malaysia, 4–6 April 2017; pp. 22–27.
7. Huang, S.; Wu, Q. Real-Time Congestion Management in Distribution Networks by Flexible Demand Swap. *IEEE Trans. Smart Grid* **2017**, *9*, 4346–4355. [CrossRef]
8. Berizzi, A.; Delfanti, M.; Marannino, P.; Pasquadibisceglie, M.; Silvestri, A. Enhanced Security-Constrained OPF With FACTS Devices. *IEEE Trans. Power Syst.* **2005**, *20*, 1597–1605. [CrossRef]
9. Huang, S.; Wu, Q.; Liu, Z.; Nielsen, A.H. Review of congestion management methods for distribution networks with high penetration of distributed energy resources. In Proceedings of the IEEE PES Innovative Smart Grid Technologies, Europe, Istanbul, Turkey, 12–15 October 2014; pp. 1–6.
10. Bouloumpasis, I.; Steen, D.; Tuan, L.A. Congestion Management using Local Flexibility Markets: Recent Development and Challenges. In Proceedings of the 2019 IEEE PES Innovative Smart Grid Technologies Europe (ISGT-Europe), Bucharest, Romania, 29 September–2 October 2019; pp. 1–5.
11. Bundesnetzagentur für Elektrizität, Gas, Telekommunikation, Post und Eisenbahnen Monitoring Report 2018. Berlin, Germany, 29 May 2019. Available online: https://bundesnetzagentur.de/SharedDocs/Downloads/EN/Areas/ElectricityGas/CollectionCompanySpecificData/Monitoring/MonitoringReport2018.pdf?__blob=publicationFile&v=3/ (accessed on 20 July 2020).
12. Liu, R.; Srivastava, A.K.; Bakken, D.E.; Askerman, A.; Panciatici, P. Decentralized State Estimation and Remedial Control Action for Minimum Wind Curtailment Using Distributed Computing Platform. *IEEE Trans. Ind. Appl.* **2017**, *53*, 5915–5926. [CrossRef]
13. Jupe, S.; Taylor, P.; Michiorri, A. Coordinated output control of multiple distributed generation schemes. *IET Renew. Power Gener.* **2010**, *4*, 283–297. [CrossRef]
14. Müller, B.; Salzinger, M.; Lens, H.; Enzenhöfer, R.; Gutekunst, F. Coordinated Congestion Management Across Voltage Levels Using Load Flow Sensitivities. In Proceedings of the 2018 15th International Conference on the European Energy Market (EEM), Lodz, Poland, 27–29 June 2018; pp. 1–6.
15. Tschampion, M.; Bucher, M.A.; Ulbig, A.; Andersson, G. N−1 security assessment incorporating the flexibility offered by dynamic line rating. In Proceedings of the 2016 Power Systems Computation Conference (PSCC), Genoa, Italy, 20–24 June 2016; pp. 1–7.
16. Meier, F.; Többermann, C.; Braun, M. Retrospective Optimal Power Flow for Low Discriminating Active Power Curtailment. In Proceedings of the 2019 IEEE Milan PowerTech, Milan, Italy, 23–27 June 2019; pp. 1–6.
17. Tonkoski, R.; Lopes, L.A.C.; El-Fouly, T.H.M. Coordinated Active Power Curtailment of Grid Connected PV Inverters for Overvoltage Prevention. *IEEE Trans. Sustain. Energy* **2010**, *2*, 139–147. [CrossRef]

18. *VDE-AR-N 4140 Kaskadierung von Maßnahmen für Die Sicherheit von Elektrischen Energieversorgungsnetzen*; VDE Verlag GmbH: Berlin, Germany, February 2017.
19. Stott, B.; Jardim, J.; Alsac, O. DC Power Flow Revisited. *IEEE Trans. Power Syst.* **2009**, *24*, 1290–1300. [CrossRef]
20. Liere-Netheler, I.; Schuldt, F.; Von Maydell, K.; Agert, C. Optimised curtailment of distributed generators for the provision of congestion management services considering discrete controllability. *IET Gener. Transm. Distrib.* **2020**, *14*, 735–744. [CrossRef]
21. Capitanescu, F.; Wehenkel, L. Redispatching Active and Reactive Powers Using a Limited Number of Control Actions. *IEEE Trans. Power Syst.* **2011**, *26*, 1221–1230. [CrossRef]
22. Stott, B.; Alsac, O.; Monticelli, A. Security analysis and optimization. *Proc. IEEE* **1987**, *75*, 1623–1644. [CrossRef]
23. *ENTSO-E Network Code on Operational Security*; European Network of Transmission System Operators for Electricity (ENTSO-E): Brussels, Belgium, February 2013.
24. Schlegel, S.; Westermann, D. Determination of remedial actions taking into account various operational rules. In Proceedings of the 2017 IEEE PES Innovative Smart Grid Technologies Conference Europe (ISGT-Europe), Esslingen, Germany, 26–29 September 2017; pp. 1–6.
25. Westermann, D.; Schlegel, S.; Sass, F.; Schwerdfeger, R.; Wasserrab, A.; Haeger, U.; Dalhues, S.; Biele, C.; Kubis, A.; Hachenberger, J. Curative actions in the power system operation to 2030. In Proceedings of the International ETG-Congress 2019 ETG Symposium, Esslingen, Germany, 8–9 May 2019; pp. 1–6.
26. Kundur, P. AC Transmission. In *Power System Stability and Control*, 1st ed.; McGraw-Hill: New York, NY, USA, 1994; p. 226.
27. Alvarez, D.L.; Rosero, J.A.; Silva, F.F.; Bak, C.L.; Mombello, E.E. Dynamic line rating—Technologies and challenges of PMU on overhead lines: A survey. In Proceedings of the 2016 51st International Universities Power Engineering Conference (UPEC), Coimbra, Portugal, 6–9 September 2016; pp. 1–6.
28. *DIN EN 50341-2-4 Freileitungen über AC 1kV*; VDE Verlag GmbH: Berlin, Germany, September 2019.
29. *738-2012-IEEE Standard for Calculating the Current-Temperature Relationship of Bare Overhead Conductors*; IEEE SA: New York, NY, USA, 2013.
30. Memmel, E.; Peters, D.; Völker, R.; Schuldt, F.; Von Maydell, K.; Agert, C.; Voelker, R. Simulation of vertical power flow at MV/HV transformers for quantification of curtailed renewable power. *IET Renew. Power Gener.* **2019**, *13*, 3071–3079. [CrossRef]
31. Hammer, A. Anwendungsspezifische Strahlungsinformationen aus Meteosat-Daten. In *Dissertation, Fachbereich Physik*; Carl-von-Ossietzky-Universität: Oldenburg, Germany, 2000.
32. Baldauf, M.; Förstner, J.; Klink, S.; Reinhardt, T.; Schraff, C.; Seifert, A.; Stephan, K. Kurze Beschreibung des Lokal-Modells Kürzestfrist COSMO-DE (LMK) und seiner Datenbanken auf dem Datenserver des DWD. Offenbach, Germany, November 2016. Available online: https://dwd.de/SharedDocs/downloads/DE/modelldokumentationen/nwv/cosmo_de/cosmo_de_dbbeschr_version_2_4_161124.pdf?__blob=publicationFile&v=4 (accessed on 20 July 2020).
33. ENERCON. *GmbH ENERCON Produktübersicht*; ENERCON: Aurich, Germany, 2015.
34. Brakelmann, H. Netzverstärkungs-Trassen zur Übertragung von Windenergie: Freileitung oder Kabel? Rheinberg, Germany, October 2004. Available online: https://www.wind-energie.de/fileadmin/redaktion/dokumente/pressemitteilungen/2018/20180103-brak-studie-2004.pdf (accessed on 11 August 2020).
35. Heuck, K.; Dettmann, K.-D.; Schulz, D. Aufbau und Ersatzschaltbilder der Netzelemente. In *Elektrische Energieversorgung—Erzeugung, Übertragung und Verteilung für Studium und Praxis*, 9th ed.; Springer Vieweg: Wiesbaden, Germany, 2013; p. 156.
36. Neon Neue Energieökonomik, Technische Universität Berlin, ETH Zürich, DIW Berlin Open Power System Data. March 2018. Available online: https://open-power-system-data.org/ (accessed on 20 July 2020).
37. Choi, S.-S.; Cha, S.-H.; Tappert, C.C. A Survey of Binary Similarity and Distance Measures. *J. Syst. Cybern. Inform.* **2009**, *8*, 43–48.
38. Cort, J.W.; Kenji, M. Advantages of the mean absolute error (MAE) over the root mean square error (RMSE) in assessing average model performance. *Clim. Res.* **2005**, *30*, 79–82.
39. Molinar, G.; Fan, L.T.; Stork, W. Ampacity forecasting: An approach using Quantile Regression Forests. In Proceedings of the 2019 IEEE Power & Energy Society Innovative Smart Grid Technologies Conference (ISGT), Washington, DC, USA, 18–21 February 2019; pp. 1–5.

40. Primadianto, A.; Lu, C.-N. A Review on Distribution System State Estimation. *IEEE Trans. Power Syst.* **2016**, *32*, 1. [CrossRef]
41. Grigsby, L.L. State Estimation. In *Power System Stability and Control*, 3rd ed.; CRC Press: Boca Raton, FL, USA, 2012.

Article

Recloser-Based Decentralized Control of the Grid with Distributed Generation in the Lahsh District of the Rasht Grid in Tajikistan, Central Asia

Anvari Ghulomzoda [1], Aminjon Gulakhmadov [2,3,4,*], Alexander Fishov [1], Murodbek Safaraliev [5], Xi Chen [2,3], Khusrav Rasulzoda [6], Kamol Gulyamov [7] and Javod Ahyoev [8]

1. Department of Automated Electric Power Systems, Novosibirsk State Technical University, Novosibirsk 630073, Russia; anvar_4301@mail.ru (A.G.); fishov9@gmail.com (A.F.)
2. Research Center for Ecology and Environment of Central Asia, Xinjiang Institute of Ecology and Geography, Chinese Academy of Sciences, Urumqi 830011, China; chenxi@ms.xjb.ac.cn
3. State Key Laboratory of Desert and Oasis Ecology, Xinjiang Institute of Ecology and Geography, Chinese Academy of Sciences, Urumqi 830011, China
4. Ministry of Energy and Water Resources of the Republic of Tajikistan, Dushanbe 734064, Tajikistan
5. Department of Automated Electrical Systems, Ural Federal University, Ekaterinburg 620002, Russia; murodbek_03@mail.ru
6. Andritz Hydro GmbH, Vienna 1120, Austria; khusrav.rasulzoda@andritz.com
7. Department of Electric Drive and Electric Machines, Tajik Technical University named after Academic M. S. Osimi, Dushanbe 734042, Tajikistan; kamol-1804@mail.ru
8. Department of Electric Power Station, Tajik Technical University named after Academic M. S. Osimi, Dushanbe 734042, Tajikistan; javod_66@mail.ru

* Correspondence: aminjon@ms.xjb.ac.cn; Tel.: +86-173-9777-5315

Received: 27 May 2020; Accepted: 15 July 2020; Published: 16 July 2020

Abstract: Small-scale power generation based on renewable energy sources is gaining popularity in distribution grids, creating new challenges for power system control. At the same time, remote consumers with their own small-scale generation still have low reliability of power supply and poor power quality, due to the lack of proper technology for grid control when the main power supply is lost. Today, there is a global trend in the transition from a power supply with centralized control to a decentralized one, which has led to the Microgrid concept. A microgrid is an intelligent automated system that can reconfigure by itself, maintain the power balance, and distribute power flows. The main purpose of this paper is to study the method of control using reclosers in the Lahsh district of the Rasht grid in Tajikistan with distributed small generation. Based on modified reclosers, a method of decentralized synchronization and restoration of the grid normal operation after the loss of the main power source was proposed. In order to assess the stable operation of small hydropower plants under disturbances, the transients caused by proactive automatic islanding (PAI) and restoration of the interconnection between the microgrid and the main grid are shown. Rustab software, as one of the multifunctional software applications in the field of power systems transients study, was used for simulation purposes. Based on the simulation results, it can be concluded that under disturbances, the proposed method had a positive effect on the stability of small hydropower plants, which are owned and dispatched by the Rasht grid. Moreover, the proposed method sufficiently ensures the quality of the supplied power and improves the reliability of power supply in the Lahsh district of Tajikistan.

Keywords: decentralized control; small hydropower plants; microgrid; emergency control; recloser; synchronous coupler; power systems stability

1. Introduction

At present, many technologies provide high-quality and reliable power supply to consumers locally, having the opportunity in emergency conditions to get power from backup sources. That is why centralized electrification of power supply to remote rural areas is a hugely complicated problem to be accomplished by grid companies.

The Tajikistan power system faces these difficulties for several reasons being (1) power supply is centralized, and (2) there is power shortage during the autumn-winter period.

These two factors force the regulation of power generation, which is strongly reflected in the rural areas making up about 70% of Tajikistan's population [1].

Tajikistan is a country where that has a huge stock of hydro resources. Consequently, Tajikistan ranks first among Central Asian countries, and eighth in the world, as noted by Zhan-Kristof et al. [2], for hydropower research. The country possesses 4% of all world hydro-energy potential [1]. Fishov et al. [3] analyzed the condition and progress of Tajikistan's hydro-energy potential, which exceeds 527 billion kWh per year, and the country exploits only 6.6% of it [4]. Potential reserves of small hydropower are 184.46 billion kWh per year [1].

In the field of small hydropower, the CASA-1000 (Central Asia South Asia) [5] project is of paramount importance to Tajikistan. The role envisioned for the project is to limit the export of excess power to neighboring countries during the summer. It is also important to recognize the effects not only of large-scale hydropower (LHPP) plants but also small-scale hydropower plants (SHPPs) on power generation. Implementation of the project will allow the efficient use of SHPPs and the development of this energy sector in Tajikistan.

The 0.4–10 kV rural distribution grids are characterized by longline length and complex structure. The key disadvantages of such grids are low reliability, inadmissible values of power quality indices, and significant power losses. Cherkasova [6] has explored energy efficiency issues, including the reliability and safety of rural electrical grids. The transmission outages can make up 40–90% of the total number of outages. Moreover, due to the significant degradation of rural distribution grids, the reliability of power supply is also deteriorating. For example, in the 6–20 kV grids, 30 blackouts/year × 100 km of line length occurs on average, and in the 0.4 kV grids, 100 blackouts/year × 100 km [7]. It is possible to reduce the number of blackouts using auto-reclosing in transmission lines. As Cherkasova [8] shows in her research on the monitoring and analysis of failures, about 60% of power line damage is unsustainable and eliminated by auto-reclosing. However, since rural distribution grids were built in the 1960s and 1970s in Tajikistan, it is impossible to use auto-reclosing based on old oil circuit breakers at the distribution center. A significant part of outages and interruptions can be reduced, and as a consequence, the reliability of power supply will increase. The use of sectionalizing points when the feeder is divided by switching devices is known to be an effective means of reliability increase. With the development of engineering and technologies based on automatic sectionalizing points (ASP), it is possible to significantly increase the reliability of power supply, especially in rural areas. The use of reclosers can serve as a good opportunity for that purpose.

With the progress of power electronics and control theory, the Microgrid concept emerged, which reflected in the research studies conducted in the USA, Europe, and Asia. Zhou et al. [9] have considered key problems and prospects of microgrid progress. Barnes et al. [10] have considered a principle of work versions of microgrid management. Zaidi et al. [11] discussed a new approach to modeling intellectual microgrids with automatic load control consumers.

The microgrid is a small and independent system that combines small-scale generation (SSG), consumers, energy storage systems, as well as control devices, forming an integrated controlled power supply system. The microgrid is an intelligent automated system that can reconfigure itself, manage power generation and load (maintain power balance), and distribute power flows.

In the present work, the problem of connecting the microgrid based on small generation to an external power grid is solved using its direct connection to the grid. We applied the methods in Zhdanov [12], namely the fixed-point iteration method, equal area criterion, and the method of superposition for determining the stability of power plants operating in parallel with the grid. To simulate the method used in the present article, the Rustab software based on RastrWin3 (2.3, JSC "Scientific and Technical Center of Unified Power System", Yekaterinburg, Russia) was used. It is a multifunctional software devoted to the study of electromagnetic and electromechanical transients during different load flow calculations [13]. The RastrWin3 software includes Eurostag and Mustang libraries, which allow us to calculate, analyze, and optimize the load flow states; to monitor the power generation/consumption; to calculate short circuit currents and to carry out asymmetrical load flow.

Currently, the transition from a centralized to a decentralized power supply system based on the microgrid is relevant. It allows us to efficiently use to distribute power sources over the grid, including renewable energy sources (RES). Therefore, decentralized synchronization and control of the normal operation restoration during the parallel operation of the separated grid and generators is considered in the present work. The control method using modified reclosers with distributed small generation in the Lahsh district of the Rasht grid of Tajikistan is studied. Based on the methodology above, the stability of all available SHPPs under disturbances in the Rasht grid was estimated.

Frequent emergency disturbances in the external network will lead to emergency shutdowns of the SHPPs power units, which feed consumers located in the local network of the Lahsh district. Consequently, consumers of this local network remain without power, and the available capacity of SHPPs is underutilized. Therefore, the goal of this work is to study ways of decentralized management using reclosers of a local network with SHPPs in the rural Lahsh district of Rasht electric networks of Tajikistan.

The paper is structured as follows: Section 2 provides an overview of the literature; Section 3 describes information about the object of study and a problem description; Section 4 discusses methods and tools for solving the problems under consideration; Section 5 presents the results obtained, as well as plans for further research; Section 6 gives the conclusions.

2. Literature Review

A review of research works devoted to the decentralization of grids control is presented in several publications. In Wang et al. [14] study, a decentralized power control system consisting of three integrated microgrids and working both independently and in parallel with the main power grid is considered. The microgrid may contain such power sources as a microturbine unit (under distribution grid operator control), wind turbines, solar panels operating stochastically. When the microgrid is in interconnected mode, the distribution grid operator and each microgrid connected to the grid are considered as separate objects with individual goals, which are minimization of operating costs. In an islanded mode, each microgrid's goal is to provide a reliable power source for its consumers. Tani et al. [15,16] study power control in decentralized power systems using wind turbines with photovoltaic panels and energy storage systems. An energy storage system is needed for a quick response to correct the imbalance between generation and load caused by sudden fluctuations in wind, sun, or consumers. In this work, the energy consumption control method based on a frequency approach using a fluctuating distribution of wind power/load is shown. The main advantage of the control method in Tani et al. [16] is that the used frequency approach takes the dynamic characteristics of the power sources into account and allows them to increase the operating time of the main power sources and reduce the size of the energy storage system. Kargarian et al. [17] considered the decentralized algorithms to determine the optimum power flow in electric power systems. The algorithms offered in their work can be used for the accelerated process of deciding large-scale problems of optimization that are difficult to solve centrally, as well as to control systems with several independent operating objects. Dou et al. [18] proposed double-tier strategies for decentralized online optimization management for an isolated microgrid without communication network channels to ensure frequency security and minimize operating costs. Qi et al. [19] considered a decentralized approach to optimal performance, based on analytical cascading of goals, to optimize the performance of AC/DC hybrid distribution networks. Yin et al. [20] have offered real-time decentralized energy management for multi-source hybrid energy systems that adapts to sudden changes in system configuration. Koukoula et al. [21] offer a new fully decentralized method of controlling congestion in radial distribution networks. The authors propose an algorithm for detecting and eliminating local restrictions on power flows without the need for centralized control. Shi et al. [22] developed a distributed energy management strategy for optimal microgrid performance, considering the distribution network and its associated limitations.

Reclosers are widely used in various countries around the world. Examples can be drawn from several studies. Sazykin et al. [23] considered the decentralized control of 6–10 kV distribution grids and planned to achieve this by installing reclosers that would act like sectionalizers. The placement of reclosers for different versions of the grid design was determined analytically. Reliability indices of the grid under consideration were calculated. Andrianova et al. [24] analyzed the use of reclosers in 6–35 kV rural distribution grids for decentralized control. Reliability indices such as SAIFI (System Average Interruption Frequency Index) and SAIDI (System Average Interruption Duration Index), as well as power supply quality indices, were calculated before and after installation of reclosers. It was noted that the number of load shedding events reduced by 15 times, and the duration of load loss was reduced by 26 times when reclosers were installed in distribution grids. Popovic et al. [25] presented the methodology for optimizing and coordinating placement of reclosers and generators in distribution grids. It was noted that optimal placement contributes to the improvement of system reliability and voltage profile and the reduction of power losses. The locations of reclosers and generators were determined using a genetic algorithm, which helped to improve the overall reliability of the grid. The study [26] was devoted to finding the optimal placement of reclosers in the grid. The reliability index SAIFI was calculated for various grid designs using the developed method. De Bruyn et al. [27] studied the behavior of recloser in the case of distributed wind energy presence in the grid. Misoperations and dead zones of relay protection were found and appropriately addressed using the developed improved version of the method. Sazykin et al. [28] examined the criteria of optimal recloser placement in 6–10 kV distribution grids. The results indicated a reduction in the number and duration of load loss events, with the total energy not served value being reduced from 69.7% to 83.5%.

A review of research on divisive protection and emergency automation can be found in some papers. Onisova [29] has reviewed the basic functions and principles of divisive protections in electric power systems, formulated new requirements for divisive protections, and has proposed improvements in devisive voltage automation. Fishov et al. [30] considered the problem of integrating small generation into electric networks. They offer new technical solutions aimed at ensuring the reliability and cost-effectiveness of systems with distributed small generation through highly automated emergency control in line with the modern SMART GRID concept. Ilyushin et al. [31] presented the features of separation of power plants from the power system with their own needs and load at a voltage of 6–20 kV. Formulated requirements for the speed of unloading during the forced separation of the power plant from the power system. Ilyushin [32] formulated possible limitations in the application of multi-parameter dividing automation related to the peculiarities of active power regulation and the settings of relay protection devices of generating installations of distributed generation facilities. Shabad [33] considered schemes and calculations of the settings of divisive protection, designed to prevent various emergencies in distribution electric networks and power plants. Kalentionok et al. [34] examined the problem of creating divisive automation in industrial enterprises that have their own generating units. The authors' proposed algorithms for the operation of dividing automation, which ensure the minimum possible power imbalances in the allocation of generating units for stand-alone operation, as well as possible parameters for its operation.

3. Data and Materials

Small generation in Tajikistan includes SSG plants with a capacity of 10–10,000 kW. From 1990–2013, many SHPPs of various capacities were built and commissioned in Tajikistan. Gulomzoda et al. [35]

reflect this in their analysis of the development of small hydropower in Tajikistan. The construction dynamics of these SHPPs is shown in Figure 1.

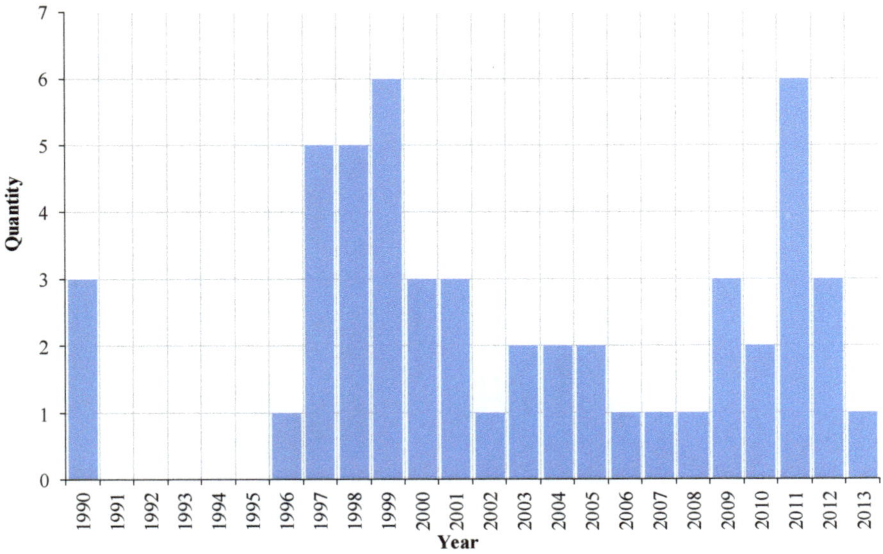

Figure 1. Construction dynamics of the small-scale hydropower plants (SHPPs) from 1990–2013.

The country continues to build SHPPs under the "Long-term program for the construction of SHPPs during 2009–2020." Construction of 189 SHPPs is part of the program, with a total installed capacity of 103 MW (Table 1).

Table 1. The prospect of SHPP building according to the program.

Stages/Capacity	Short-Term Stage (2009–2011), Units	Medium-Term Stage (2012–2015), Units	Long-Term Stage (2016–2020), Units
Up to 100 kW	20	21	21
100 to 1000 kW	34	37	25
Over 1000 kW	12	12	7
Total	66	70	53
Total power, kW	43,350	32,850	26,801

The "Barqi Tojik" Open Joint-Stock Holding Company (OJSHC) energy company is responsible for power supply in the majority of the country's regions, including the Lahsh district of the Rasht grids. The Rasht region has 6 SHPPs [3], and the generation retrospective is shown in Figure 2.

As can be seen from the figures above, the plant utilization factor (which means the ratio of the total actual energy produced or supplied by SHPPs in the Rasht grid over a definite period, to the energy that would have been produced if the plant had operated continuously at the maximum rating) differs from the assigned one, since the capacity of SHPPs is not used correctly. The reasons for the decrease in power generation are mainly frequent disturbances in the grid leading to the shutdown of power units, the operation of the generation units only in stand-alone mode, the impact of external factors, and others.

Frequent disturbances in the Lahsh grid strongly affect consumers, since they lose power both from the main grid and from the local SHPPs. Consequently, it is crucial to solve the problem of customer outage, reducing energy not served, ensuring the quality of supplied power, and improving the reliability of the power system in the Lahsh district of the Rasht grid in Tajikistan. Figure 3 shows a geographical map of the grid under study in the Lahsh district.

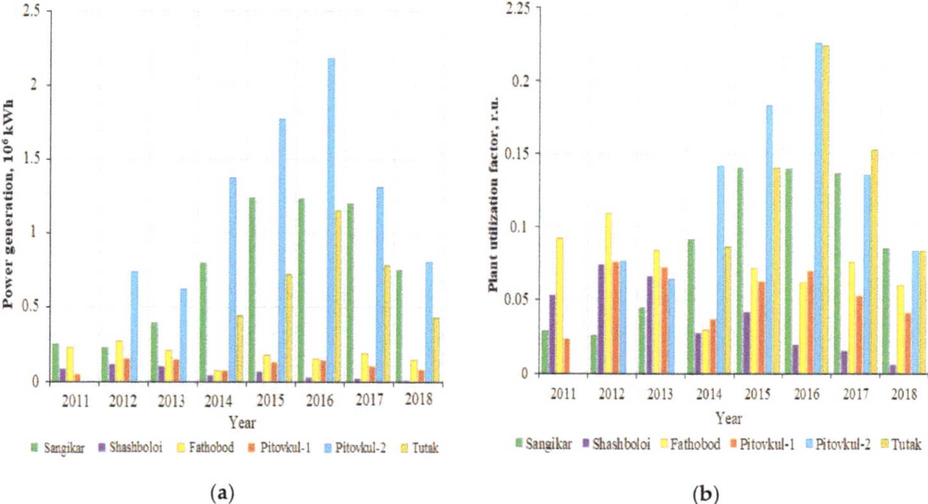

Figure 2. (**a**) The power generation and (**b**) the plant utilization factor of small-scale hydropower plants (SHPPs) in the Rasht grids.

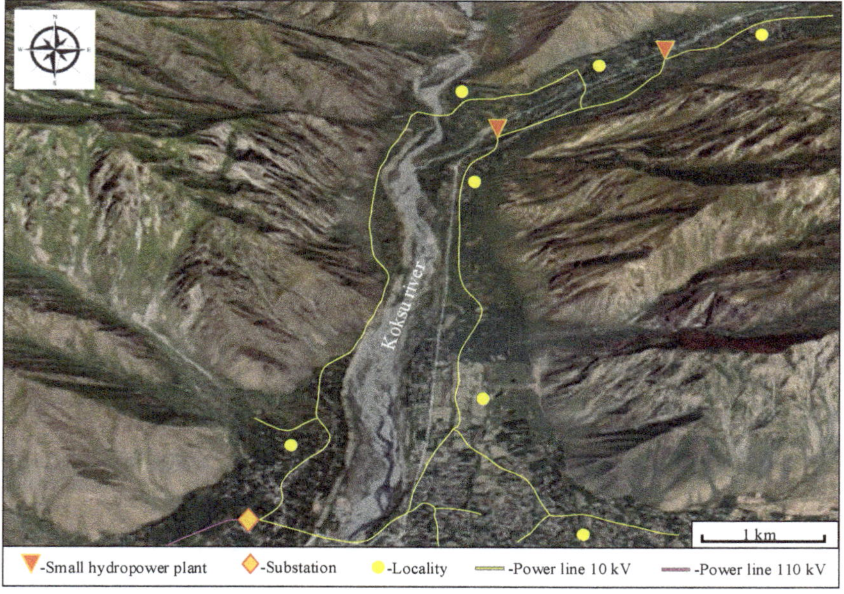

Figure 3. The geographical map of the grid under study of the Lahsh district in Tajikistan.

4. Methodology

4.1. Microgrids Control and Operation

At present, the transition from a centralized control of power supply to a decentralized one, especially for remote consumers, is highly relevant. In this case, it allows the use of electric power sources distributed over the grid, including RES, called distributed generation (DG). Gulomzoda et al. [36] and Ismoilov et al. [37] considered technologies for controlling the modes of local power supply systems with a low generation, as well as their system effects. Along with the development of power electronics and control theory, the concept of the Microgrid [9–11] arose. The microgrid is a small and independent system that combines small-scale generation (SSG), consumers, energy storage systems, as well as control devices, forming an integrated controlled power supply system. Figure 4 shows the basic structure of a hybrid microgrid.

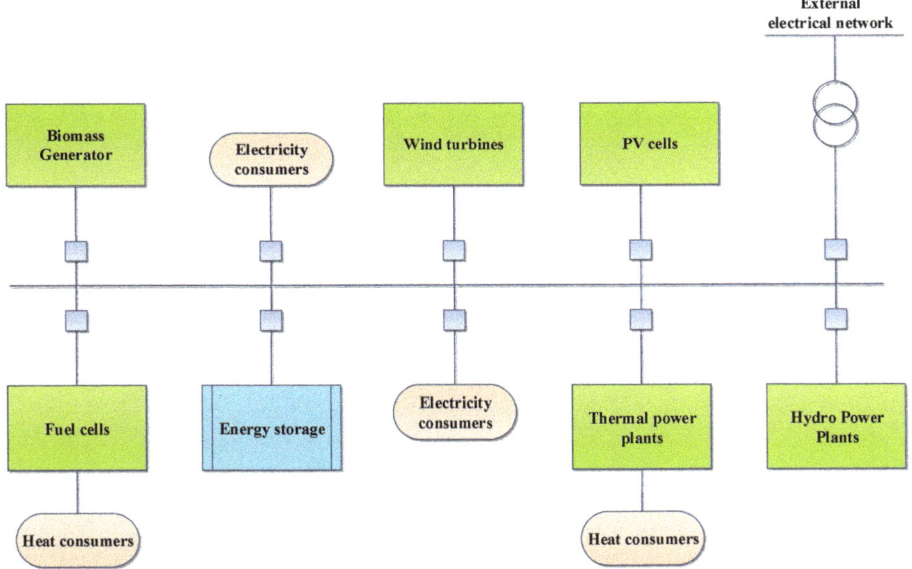

Figure 4. The basic structure of a hybrid microgrid.

The microgrid can be either in interconnected mode, where the microgrid exchanges power with the main grid through the distribution substation transformer, or in the islanded mode, where the microgrid is disconnected from the main grid, and operates autonomously, serving its own local demand using DG and other distributed energy resources. During contingency Microgrid can be independently disconnected from the main grid and operate autonomously. After eliminating the contingency, microgrid can again reconnect to the main grid. Thus, microgrid is an intelligent automated system that is able to reconfigure by itself, maintain power balance and distribute power flows.

This study presents the solution of microgrid (on the base of small generation) connection problem by means of synchronous coupling or the use of microgrid in the grid without centralized control. A prerequisite is the integration of SHPPs into the grid, the integration of several local power supply systems (on the base of SHPPs) into islanded or interconnected power systems.

4.2. Combined Use of Emergency Control and Reclosers

Proactive automatic islanding (PAI) is one of the most important elements for solving the problems under consideration. Gezha et al. [38] considered automation to solve the problem of safe direct connection to the main network of a local power supply system based on small generation. This type of emergency control has proven its efficiency, and it can be applied both to future SSG projects and to existing facilities, which were not initially considered for use in that way. In order to ensure reliable parallel operation of the microgrid in terms of emergency control, a unique method of emergency control was suggested and implemented. Fishov et al. [39] developed a method for emergency control of the mode of parallel operation of synchronous generators in electric networks. The concept behind it is the proactive islanding of the microgrid at selected points when a contingency occurs. After that, the transition to the island mode is made with the synchronization and transition to normal operating conditions with a proper equipment loading schedule. This type of emergency control has already proved its efficiency in test conditions. It was tested in the Berezovoe district, Pervomayskiy area of Novosibirsk, in the Russian Federation.

Another key element of decentralized operation and control of Microgrids is an automatic recloser. Fishov et al. [40] considered a decentralized reconfiguration of the electrical network with a microgrid using advanced reclosers. Reclosers perform the functions of isolation and redundancy in 6–10 kV distribution grids of radial and network types with long transmission lines. The recloser includes a vacuum circuit-breaker, power conversion systems, an auxiliary DC control power system, a microprocessor-based protection and control system, and SCADA communications ports and software. Previously Buzin et al. [41] investigated modern relay protection and automation of 6–10 kV overhead power lines.

The reclosing is that the lines connecting the microgrid to the power system are restricted for grids with generation facilities (SSG in this case). The synchronization conditions should be met in order for these lines to be switched on. If not, out-of-step switching may result in significant dynamic torque, which represents a possible threat to generators and their auxiliaries. Hachaturov [42] presented the principles of synchronization and the consequences of the non-synchronous inclusion of power plant generators in the network. It should be considered when reclosers are to be installed in grids with generation facilities. This work examines whether it is possible to use reclosers for the recovery of power generators parallel operation.

Modern reclosers do not possess synchronization properties needed in grids with SSG penetration. Since reclosers are located away from generators, the synchronization is not carried out using the generator circuit breakers but the reclosers. The following modification was made: the synchronization package was added to the control cabinet of a vacuum circuit-breaker. The functional control diagram is shown in Figure 5.

In Figure 5, the high-voltage equipment includes the Q—vacuum circuit-breaker; ST1, ST2—auxiliary transformers; TV1, TV2—voltage transformers, and a TA—current transformer. The control cabinet is the power supply unit; the control system unit; the relay protection and the control unit (RP and A); the measurement unit; the battery; the GSM modem (in case if SCADA connection is not provided); the monitor and the synchronization package.

The recloser operation modification allows extending the range of control capabilities by adding the synchronization package to the cabinet. The synchronization package receives such measurement signals as voltage module and phase measured by voltage transformers (TV1, TV2) and frequency on both sides of a vacuum circuit-breaker.

The following synchronization conditions are to be met:

$$\left.\begin{array}{l} |U_i - U_j| < dU_{ph.} \\ |\delta_i - \delta_j| < \delta_{perm.} \\ |f_i - f_j| < s_{perm.} \end{array}\right\} \quad (1)$$

where U_i—root-mean-square voltage value in the "from" bus of the transmission line, kV; U_j—root-mean-square voltage value in the "to" bus of the transmission line, kV; $dU_{ph.}$—permissible voltage drop of the transmission line, kV; δ_i—voltage phase angle in the "from" bus of the transmission line, el.; δ_j—voltage phase angle in the "from" bus of the transmission line, el.; $\delta_{perm.}$—permissible phase angle difference of the transmission line, el.; f_i—frequency in the "from" bus of the transmission line, Hz; f_j—frequency in the "to" bus of the transmission line, Hz; $s_{perm.}$—permissible slip of the transmission line, Hz.

If based on values of the mentioned parameters, the synchronization conditions are met, then the synchronizer transmits a signal to the control unit passing through the relay protection and control unit in order to switch on the vacuum circuit-breaker. The process of synchronization can be seen on the monitor.

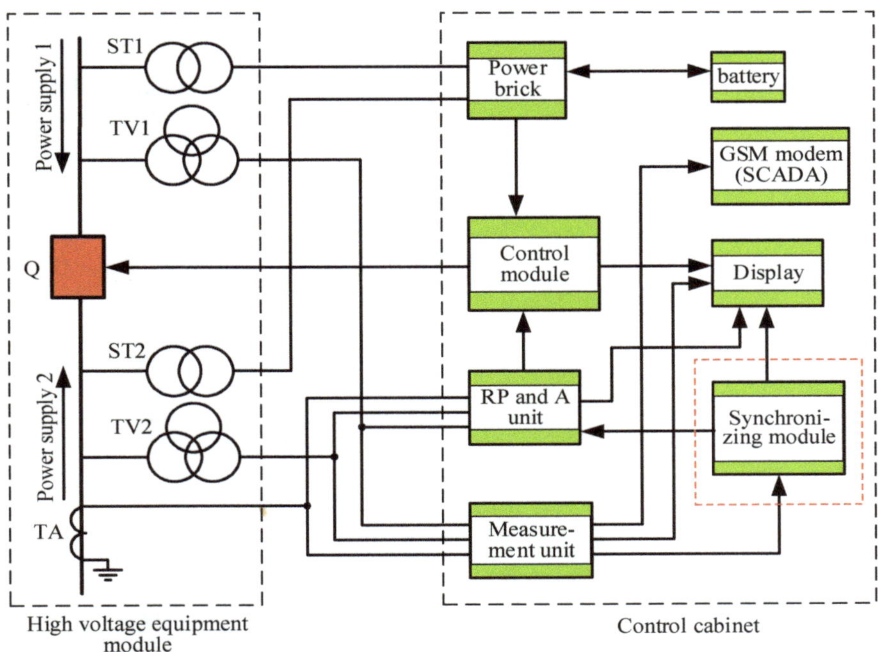

Figure 5. The functional control diagram of the modified recloser.

Such microprocessor-based devices as «AC-M» [43], «Sprint-M» [44] and «SYNCHROTACT» [45] are used as synchronizers. These devices provide compliance with all synchronization conditions. They carry out the functions of measurement, analysis, and, if necessary, synchronization parameters adjustment. When all conditions are met, the synchronizer sends a signal to switch on the recloser. Additionally, the synchronizers are capable of self-testing and displaying state data, which can be considered as an advantage.

4.3. Methods of Ensuring Generators Stability

Parallel operation of power plants in a local grid or connection with a bulk power system exerts a considerable impact on such things as improvement of adequacy and operational reliability of power supply, power losses reduction, optimization of generation reserves to increase export interconnection power flows, among others.

However, any unforeseen disturbance of the power system state (for example, any kind of fault) causes swinging of synchronous machines, and if strong enough, machines or entire power plants may fall out of synchronism, threatening the whole power system stability. Therefore, stability studies of power plants are still of high relevance and importance, regardless of the way a power system is controlled.

Swinging of synchronous generators might be caused by sudden sharp load drop or rise, outage of a transmission line, a transformer, or a generator, with a fault being the most dangerous one. Generally, transmission outages are followed by sudden changes in the output of power plants. Rotors of generators and turbines respond to disturbances with some time delay due to significant inertia. As a result, the power balance between generator and turbine shifts, resulting in excess torque of the generator, in turn causing rotor speed increase.

According to the fixed-point iteration method [12], the rotor acceleration is:

$$\alpha = \frac{\Delta M}{T_j} = \frac{\Delta P}{T_j} \tag{2}$$

where T_j—inertia constant, kg·m^2/s^2; $\Delta M = M_{mech} - M_{el}$—excess torque of turbine, determined by the difference between mechanical torque and electrical torque, Nm; ΔP—excess power of the generator, MW.

Taking into account that angular acceleration of the rotor is the second derivative of rotor angle with respect to time $\alpha = d^2\delta/dt^2$; and the difference between turbine power and generator power output $\Delta P = P_0 - P_m \cdot \sin\delta$, the swing equation of rotor is:

$$T_j \frac{d^2\delta}{dt^2} = P_0 - P_m \sin\delta \tag{3}$$

Solving the Equation (3) shows how to load angle changes over time and allows the estimation of generator stability.

The stable parallel operation of two power plants can be estimated using the fixed-point iteration method and equal area criterion. The power output of each power plant is:

$$\left.\begin{aligned} P_1 &= E_1^2 y_{11} \sin\alpha_{11} + E_1 E_2 y_{12} \sin(\delta_{12} - \alpha_{12}) \\ P_2 &= E_2^2 y_{22} \sin\alpha_{22} - E_1 E_2 y_{12} \sin(\delta_{12} + \alpha_{12}) \end{aligned}\right\} \tag{4}$$

where P_1, P_2—the power output of the first and the second power plant, respectively; E_1, E_2—EMF of the first and the second power plant, respectively; y_{11}, y_{22}, y_{12}—self-admittances of both power plants and mutual admittance, respectively; δ_{12}—angle difference between power plants; α_{11}, α_{22}, α_{12}—absolute acceleration of the first power plant generator rotor, the second power plant generator rotor, and their relative acceleration, respectively.

Absolute rotor acceleration of both power plants, and also their relative acceleration is:

$$\left.\begin{aligned} \alpha_1 &= \tfrac{360f}{T_{j1}} \Delta P_1 \\ \alpha_2 &= \tfrac{360f}{T_{j2}} \Delta P_2 \end{aligned}\right\} \tag{5}$$

$$\alpha_{12} = \alpha_1 - \alpha_2 = 360f\left(\frac{\Delta P_1}{T_{j1}} - \frac{\Delta P_2}{T_{j2}}\right) \tag{6}$$

where f—utility frequency;

ΔP_1, ΔP_1—excess power of generators:

$$\left.\begin{aligned} \Delta P_1 &= P_0 - P_1 \cdot \sin\delta \\ \Delta P_2 &= P_0 - P_2 \cdot \sin\delta \end{aligned}\right\} \tag{7}$$

Hence, the stability of power plants working in parallel can be estimated based on the equal area criterion (equality of deceleration and acceleration areas).

The stability of systems with complex structures can be analyzed using the superposition technique. The power equation for a system with n—generators is:

$$\left.\begin{array}{l} P_1 = E_1^2 y_{11} \sin\alpha_{11} + E_1E_2y_{12}\sin(\delta_{12}-\alpha_{12}) + E_1E_3y_{13}\sin(\delta_{13}-\alpha_{13}) + \ldots + E_1E_ny_{1n}\sin(\delta_{1n}-\alpha_{1n}); \\ P_2 = E_2E_1y_{21}\sin(\delta_{21}-\alpha_{21}) + E_2^2y_{22}\sin\alpha_{22} + E_2E_3y_{23}\sin(\delta_{23}-\alpha_{23}) + \ldots + E_2E_ny_{2n}\sin(\delta_{2n}-\alpha_{2n}); \\ \ldots \\ P_{(n-1)} = E_{(n-1)}E_1y_{(n-1)1}\sin(\delta_{(n-1)1}-\alpha_{(n-1)1}) + E_{(n-1)}E_2y_{(n-1)2}\sin(\delta_{(n-1)2}-\alpha_{(n-1)2}) + \ldots \\ \ldots + E_{(n-1)}^2y_{(n-1)(n-1)}\sin\alpha_{(n-1)(n-1)} + E_{(n-1)}E_ny_{(n-1)n}\sin(\delta_{(n-1)n}-\alpha_{(n-1)n}); \\ P_n = E_nE_1y_{n1}\sin(\delta_{n1}-\alpha_{n1}) + E_nE_2y_{n2}\sin(\delta_{n2}-\alpha_{n2}) + \ldots \\ \ldots + E_nE_{(n-1)}y_{n(n-1)}\sin(\delta_{n(n-1)}-\alpha_{n(n-1)}) + E_n^2y_{nn}\sin\alpha_{nn} \end{array}\right\} \quad (8)$$

Consequently, the excess power of each generator can be found using the methods mentioned above, despite the complexity of the system structure. Expressions (5)–(7), and angle-time curves form the basis of generators stability analysis.

4.4. Power Flow Studies Software

The Rustab software (package of RastrWin3) was used in this study. This software is widely used in Russia and CIS countries for transient studies in power systems [13]. RastrWin3 includes Eurostag and Mustang libraries, which makes power flow and optimization studies possible, as well as adequacy estimation and faults analysis (including asymmetrical load flow).

The huge range of Rustab capabilities was used in the research in order to create a model of the small hydro of the Lahsh Microgrid and to study its transient stability and power flow states. The examples of models considered were a fault, substation relay protection, recloser-based PAI, auto-reclosing with synchronism check, and others. In addition, seasonal variations of load were simulated as well.

5. Results and Discussions

Simulations were carried out for a grid of rural areas in the Lahsh district in Tajikistan, Central Asia. The two small hydro plants are located in the area. Power is supplied in either island mode or using the external 10 kV grid interconnection. This is due to frequent disturbances in the external grid, which may lead to generator outages in case of parallel operation and, therefore, the whole feeder interruption. Reclosers installation in different parts of the feeder is suggested (Figure 6), considering the high values of outage rates, loss of load, and energy not supplied.

Four reclosers were suggested to be placed, two of which (R1, R2) separated the microgrid from the external one (highlighted part of the grid in Figure 6) and reconnected them if synchronization conditions were met. Recloser R3 is used to connect (switched-on state) two parts of the microgrid in the island mode and to avoid shunt connection (switched-off state). Recloser R4 was used for under-frequency load shedding (UFLS).

Without those reclosers, generators of small hydropower plants would fall out of synchronism each time a disturbance occurs. An example of such an event is shown in Figure 7.

Figure 7 shows that after a disturbance, the generator falls out of synchronism since its rotor angle constantly increases. In this case, the power plant has lost its stability.

As suggested, reclosers are placed according to the power balance of power generators capacity and customer load (Figure 6). Such placement allows automatic rapid separation of the microgrid from the external grid during disturbances. The PAI was used to perform this function [20,21]. The improved recloser was used for disturbance recovery. The combined use of all mentioned devices and methods has prevented generators from falling out of synchronism and has maintained stable operation in the island mode. In addition, the smooth transition from the latter to parallel operation with the external grid was achieved (Figures 8–11).

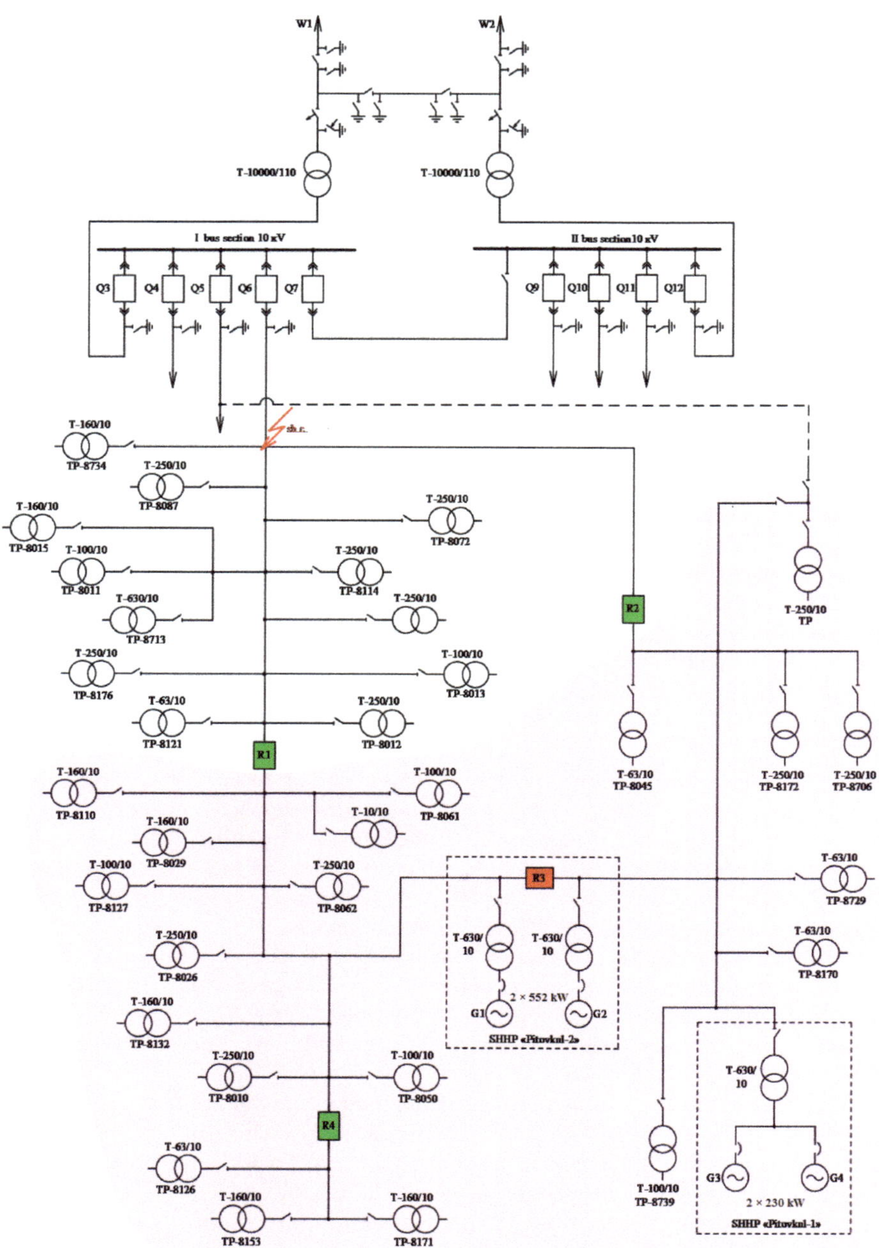

Figure 6. Model of the rural area grid of the Lahsh region of Tajikistan, Central Asia.

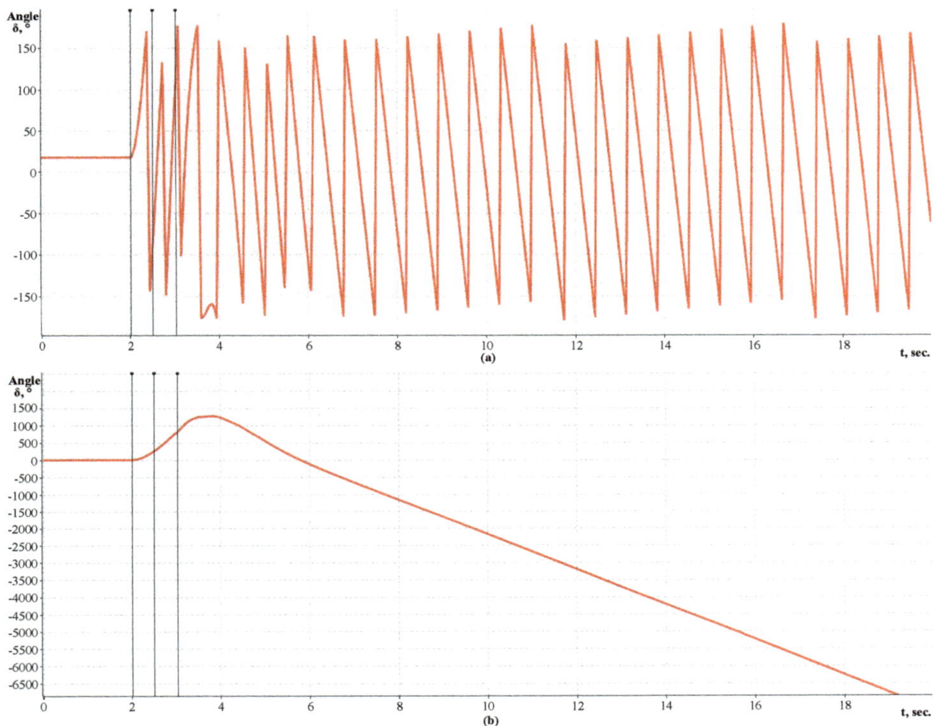

Figure 7. The angle-time curve in case of disturbance: (**a**) ±180° curve; (**b**) incremental angle curve.

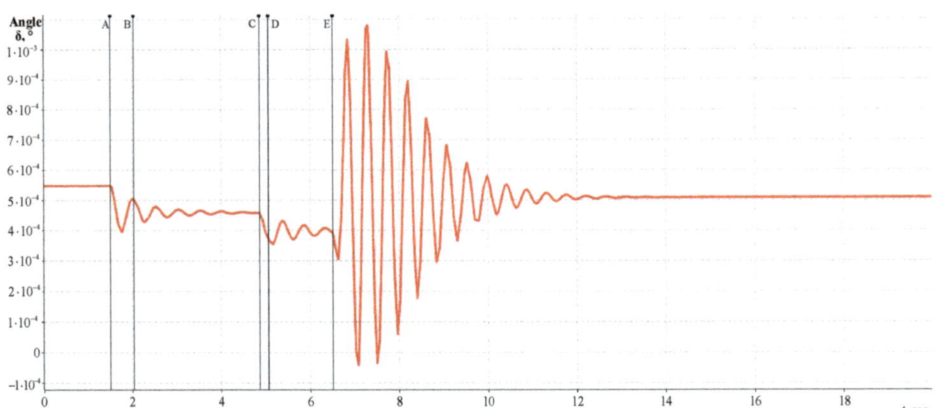

Figure 8. The angle-time curve in disturbance conditions while using the suggested method.

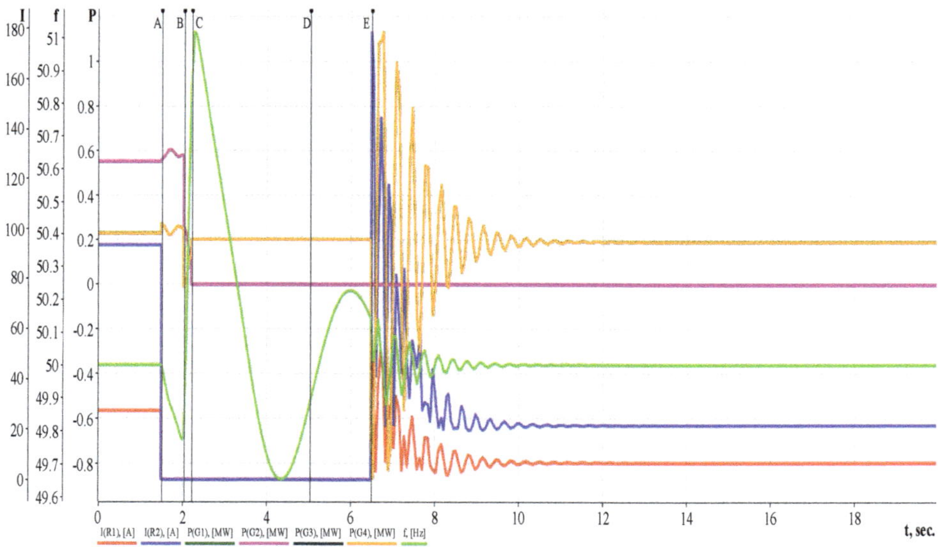

Figure 9. The microgrid transient response with 20% of the maximum load.

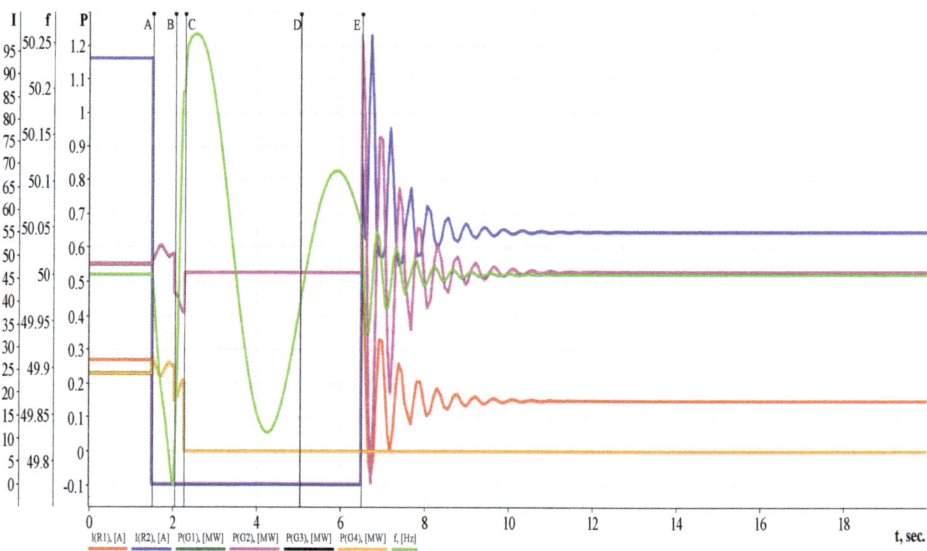

Figure 10. The microgrid transient response with 60% of the maximum load.

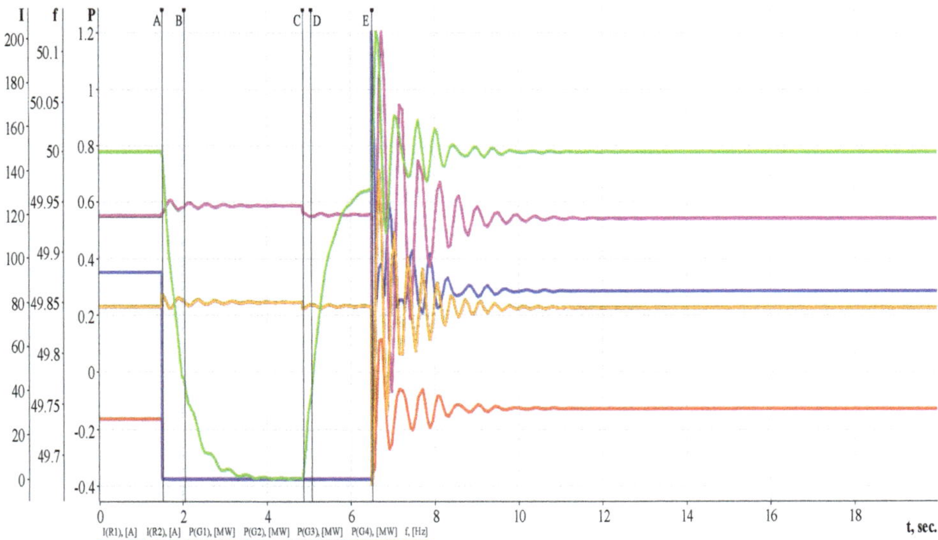

Figure 11. The microgrid transient response with 100% of the maximum load.

Figure 8 shows a curve of the change in the angle of the generator's rotor, and Figures 9–11 show the curves of the transition process on the microgrid network when an emergency disturbance occurs. On the listed curves, the vertical bars indicated by the letters A–E are action indicators. Curves 9–11 show the following dependencies: changes in currents through reclosers R1, R2—I (R1), I (R2), A; power of generators G1–G4—P (G1), P (G2), P (G3), P (G4), MW; Microgrid frequency—f, Hz. The PAI acted according to the algorithms aimed at forcing generator outages depending on the value of the microgrid load. The simulation scenario was as follows: when a fault occurs (Figure 6), PAI acts ahead of current relay protection and sends signals to reclosers R1, R2 to switch off and to recloser R3 to switch on (point A, at t = 1.5 s, Figures 9–11) and generators to turn off (point C, at t = 2.27 s, G1, G2 on Figure 9 and G3, G4 in Figure 10). It eliminates excess power in the microgrid being separated. In case of power shortage in the Microgrid UFLS acts and sends a signal to recloser R4 to switch off (point C, at t = 4.85 s, Figure 11); then current relay protection acts (point B at t = 2 s, Figures 9–11) and sends signals to circuit-breaker Q6 to switch off, causing a fault to self-clear and circuit-breaker Q6 to reconnect automatically (point D, at t = 5 s, Figures 9–11). When voltage appears on both sides of reclosers R1, R2, the mechanism of auto-reclosing with a synchronism check is triggered. If synchronization conditions are met, then reclosers R1, R2 switch on, and recloser R3 switches off (point E, at t = 6, 5 s, Figure 9, Figure 10, and Figure 11). The power supply of customers is restored. No more load losses occur due to the available generation in the grid.

The simulation results prove the efficiency of the suggested method, which leads to the improvement of the small hydro plant stability. This, in its turn, provides the required power quality and power supply reliability for customers in the Lahsh district. The suggested method of power system control is planned to be applied in other regions of Tajikistan in Central Asia with distributed generation. In addition, at the next stage of the study, this method will be applied for studying voltage control capabilities, reliability analysis, and the influence of reclosers placement on these parameters.

6. Conclusions

The suggested method of distributed synchronization of the microgrid connection to the external grid was studied. The obtained data of the object under study was analyzed and used in the simulation.

The wide range of Rustab package functions was used to simulate disturbances and study transient stability of the small hydro plants in the Lahsh Microgrid in Tajikistan.

The suggested modified recloser was used in the study. Using such reclosers makes it possible to decentralize emergency control, provide independent operation of the sectionalizing point, minimize human error, ensure fast recovery of supply in areas untouched by disturbance, and, as a consequence, keep the values of energy not served to a minimum. Automation and control in overhead power lines based on reclosers were considered by Vorotnickij et al. [46,47].

According to the results of the simulation, the suggested method has a positive impact on the ability of small hydropower plants in the Rasht grid to withstand disturbances. It provides the required power quality and improves power supply reliability in remote areas of Tajikistan's Lahsh district.

Author Contributions: All authors have made valuable contributions to this work. A.G. (Anvari Ghulomzoda) and A.F. engaged in the development, research of the main part of the work, and wrote the manuscript; M.S. assisted in modeling and software; A.G. (Aminjon Gulakhmadov) and X.C. helped revise the manuscript structure and funding; K.R., K.G., and J.A. helped in the calculation and organization of data. All authors have read and agreed to the published version of the manuscript.

Funding: This research was funded by the Strategic Priority Research Program of the Chinese Academy of Sciences, Pan-Third Pole Environment Study for a Green Silk Road (Grant No. XDA20060303), the International Cooperation Project of the National Natural Science Foundation of China (Grant No. 41761144079), the Xinjiang Tianchi Hundred Talents Program (Grant No. Y848041), and the project of the Research Center of Ecology and Environment in Central Asia (Grant No. Y934031).

Acknowledgments: The authors are thankful to the Ministry of Energy and Water Resources of the Republic of Tajikistan and the Rasht electric networks OJSHC "Barqi Tojik" for providing the data for this research work.

Conflicts of Interest: The authors declare no conflicts of interest.

References

1. Tajikistan. Rapid Assessment and Gap Analysis. Available online: https://www.undp.org/content/dam/tajikistan/docs/library/UNDP_TJK_SE4ALL_Rapid_Assessment_and_gap_analysis_Rus.pdf (accessed on 10 April 2020).
2. Zhan-Kristof, F.; Varasteh, E.; Jako, R.; Chobanova, B.; Bilolov, F. *Tajikistan: In-Depth Energy Efficiency Review*; Spotinov Print LLC: Brussels, Belgium, 2013; pp. 76–79. ISBN 978-905948-138-1.
3. Fishov, A.G.; Gulomzoda, A.H.; Kasobov, L.S. Analysis of the state and direction of development of small hydropower of Tajikistan. *Polytech. Bull. Ser. Eng. Stud.* **2019**, *1*, 13–20.
4. Barki Tojik OJSHC. Available online: http://www.barqitojik.tj (accessed on 15 April 2020).
5. CASA-1000. Available online: http://www.casa-1000.org/indexr.php (accessed on 17 April 2020).
6. Cherkasova, N.I. Fundamentals of Managing Technological Risks and the Efficient Functioning of Power Supply Systems for Agricultural Consumers. Ph.D. Thesis, Altai State Technical University, Barnaul, Russia, 15 December 2017.
7. *Standing Order of PJSC "ROSSETI" on a Unified Technical Policy in the Electric Grid Complex*; Protocol No.138.; PJSC Rosseti: Moscow, Russia, 2013.
8. Cherkasova, N.I. Analysis of the state of rural electric networks of 10 kV in the light of failure monitoring. *Polzunovsky Bull.* **2012**, *4*, 49–54.
9. Zhou, X.; Guo, T.; Ma, Y. An Overview on Microgrid Technology. In Proceedings of the 2015 IEEE International Conference on Mechatronics and Automation (ICMA), Beijing, China, 2–5 August 2015; pp. 76–81.
10. Barnes, M.; Kondoh, J.; Asano, H.; Oyarzabal, J.; Ventakaramanan, G.; Lasseter, R.; Hatziargyriou, N.; Green, T. Real-World MicroGrids-An Overview. In Proceedings of the 2007 IEEE International Conference on System of Systems Engineering, San Antonio, TX, USA, 16–18 April 2007; pp. 1–8.
11. Zaidi, A.A.; Kupzog, F. Microgrid Automation—A Self-Configuring Approach. In Proceedings of the 2008 IEEE International Multitopic Conference, Karachi, Pakistan, 23–24 December 2008; pp. 565–570.
12. Zhdanov, P.S. *Stability Issues of Electrical Systems*; Energy: Moscow, Russia, 1979; pp. 77–105. ISBN 31.27-01-6112.11.
13. RastrWin. Available online: http://www.rastrwin.ru (accessed on 7 May 2020).

14. Wang, Z.; Chen, B.; Wang, J.; Kim, J. Decentralized Energy Management System for Networked Microgrids in Grid-Connected and Islanded Modes. *IEEE Trans. Smart Grid* **2016**, *7*, 1097–1105. [CrossRef]
15. Tani, A.; Camara, M.B.; Dakyo, B. Energy Management in the Decentralized Generation Systems Based on Renewable Energy—Ultracapacitors and Battery to Compensate the Wind/Load Power Fluctuations. *IEEE Trans. Ind. Appl.* **2015**, *51*, 1817–1827. [CrossRef]
16. Tani, A.; Camara, M.B.; Dakyo, B. Energy Management in the Decentralized Generation Systems Based on Renewable Energy Sources. In Proceedings of the 2012 International Conference on Renewable Energy Research and Applications (ICRERA), Nagasaki, Japan, 11–14 November 2012; pp. 1–6.
17. Kargarian, A.; Mohammadi, J.; Guo, J.; Chakrabarti, S.; Barati, M.; Hug, G.; Kar, S.; Baldick, R. Toward Distributed/Decentralized DC Optimal Power Flow Implementation in Future Electric Power Systems. *IEEE Trans. Smart Grid* **2018**, *9*, 2574–2594. [CrossRef]
18. Dou, C.; An, X.; Yue, D.; Li, F. Two-Level Decentralized Optimization Power Dispatch Control Strategies for an Islanded Microgrid without Communication Network. *Int. Trans. Electr. Energy Syst.* **2017**, *27*, e2244. [CrossRef]
19. Qi, C.; Wang, K.; Fu, Y.; Li, G.; Han, B.; Huang, R.; Pu, T. A Decentralized Optimal Operation of AC/DC Hybrid Distribution Grids. *IEEE Trans. Smart Grid* **2018**, *9*, 6095–6105. [CrossRef]
20. Yin, H.; Zhao, C.; Ma, C. Decentralized Real-Time Energy Management for a Reconfigurable Multiple-Source Energy System. *IEEE Trans. Ind. Inform.* **2018**, *14*, 4128–4137. [CrossRef]
21. Koukoula, D.I.; Hatziargyriou, N.D. Gossip Algorithms for Decentralized Congestion Management of Distribution Grids. *IEEE Trans. Sustain. Energy* **2016**, *7*, 1071–1080. [CrossRef]
22. Shi, W.; Xie, X.; Chu, C.-C.; Gadh, R. Distributed Optimal Energy Management in Microgrids. *IEEE Trans. Smart Grid* **2015**, *6*, 1137–1146. [CrossRef]
23. Sazykin, V.G.; Kudrjakov, A.G. *Actual Issues of Technical Sciences: Theoretical and Practical Aspects. Collective Monograph*; Aeterna: Ufa, Russia, 2017; pp. 64–87.
24. Andrianova, L.P.; Kabashov, V.J.; Hajrislamov, D.S. Reliable indicators of SAIFI and SAIDI rural electric networks with intelligent sectional reclosers. *Bull. Bashkir State Agrar. Univ.* **2019**, *3*, 83–92. [CrossRef]
25. Popović, D.H.; Greatbanks, J.A.; Begović, M.; Pregelj, A. Placement of Distributed Generators and Reclosers for Distribution Network Security and Reliability. *Int. J. Electr. Power Energy Syst.* **2005**, *27*, 398–408. [CrossRef]
26. Karpov, A.I.; Akimov, D.A. Integral Indicators Improvement (SAIFI) of Power Supply Reliability in Electric Distribution Systems Based on Reclosers Placement Optimization. In Proceedings of the 2018 IEEE Conference of Russian Young Researchers in Electrical and Electronic Engineering (EIConRus), Moscow, Russia, 29 January–1 February 2018; pp. 663–666.
27. De Bruyn, S.; Fadiran, J.; Chowdhury, S.; Chowdhury, S.P.; Kolhe, P. The Impact of Wind Power Penetration on Recloser Operation in Distribution Networks. In Proceedings of the 2012 47th International Universities Power Engineering Conference (UPEC), London, UK, 4–7 September 2012; pp. 1–6.
28. Sazykin, V.; Kudryakov, A.; Bagmetov, A. Criteria for Optimizing the Place of Reklouser Installation in the 6-10 KV Distribution Network. *Electrotech. Syst. Complexes* **2018**, *1*, 33–39. [CrossRef]
29. Onisova, O.A. Subdivision Protection in Electric Power Systems Containing Low-Power Electric Power Plants. *Power Technol. Eng.* **2014**, *48*, 322–330. [CrossRef]
30. Fishov, A.G.; Landman, A.K.; Serdyukov, O.V. SMART Technologies for Connecting to Electric Networks and Control Small Generation Modes. In Proceedings of the VIII International Scientific and Technical Conference "Electricity through the Eyes of Youth—2017", Samara, Russia, 2–6 October 2017; pp. 27–34.
31. Ilyushin, P.V.; Pazderin, A.V. Requirements for dividing automation of distributed generation facilities, taking into account the influence of adjacent network parameters and load. *Electr. Transm. Distrib.* **2018**, *4*, 42–47.
32. Ilyushin, P.V. Features of the implementation of multi-parameter dividing automation in energy areas with distributed generation facilities. *Relay Prot. Autom.* **2018**, *2*, 12–24.
33. Shabad, M.A. *Fission Protections—Automation of Division During Failures*, 7th ed.; Dyakov, A.F., Ed.; Energoprogress: Moscow, Russia, 2006; pp. 5–37.
34. Kalentionok, E.V.; Filipchik, Y.D. Methodological approaches to the creation of dividing automation in industrial enterprises with generating power plants. *Energy Proc. Higher Educ. Establ. Energy Assoc. CIS* **2010**, *6*, 14–19.

35. Gulomzoda, A.; Fishov, A.G.; Nikroshkina, S.V. Development of Small-Scale Hydropower Generation in Tajikistan. In Proceedings of the 2019 8th International Academic and Research Conference of Graduate and Postgraduate Students, Novosibirsk, Russia, 28 March 2019; pp. 123–126.
36. Gulomzoda, A.; Fishov, A.G.; Nikroshkina, S.V. Technology of Managing the Modes of Local Energy Supply Systems. In Proceedings of the 2018 2nd All Russia Academic and Research Conference of Graduate and Postgraduate Students, Novosibirsk, Russia, 20 December 2018; pp. 70–72.
37. Ismoilov, S.T.; Gulomzoda, A.H.; Rahimov, F.M. Putting distributed generation into the network to provide system services. In Proceedings of the 11th International Conference on Mining, Construction and Energy, Tula, Russia, 5–6 November 2015; pp. 362–367.
38. Gezha, E.N.; Glazyrin, V.E.; Glazyrin, G.V.; Ivkin, E.S.; Marchenko, A.I.; Semendjaev, R.J.; Serdjukov, O.V.; Fishov, A.G. System automation for the integration of local power supply systems with synchronous low generation into electrical networks. *Prot. Eng. 75 Years Dep. RP APS NRU MPEI* **2018**, *2*, 24–31.
39. Fishov, A.G.; Mukatov, B.B.; Marchenko, A.I. Method for Emergency Control of the Mode of Parallel Operation of Synchronous Generators in Electrical Networks. RU Patent 2662728, 30 July 2018.
40. Fishov, A.G.; Ghulomzoda, A.H.; Kasobov, L.S. Decentralised reconfiguration of a Microgrid electrical network using reclosers. *Proc. Irkutsk State Tech. Univ.* **2020**, *24*, 382–395. [CrossRef]
41. Buzin, S.; Vorotnickij, V. *Modern Relay Protection and Automation for the Automation of 6–10 Kv Overhead Electrical Networks*; Technical Report; RK Tavrida Elektrik Ltd.: Krasnodar, Russia, 2010; pp. 1–4.
42. Hachaturov, A.A. *Non-Synchronous Switching and Resynchronization in Power Systems*, 2nd ed.; Energy: Moscow, Russia, 1977; pp. 11–25. ISBN 621-311-018-5.
43. Beljaev, N.A. Synthesis of Adaptive Synchronization Systems for Generators with an Electric Network Based on Automatic Control Methods with a Reference Model. Master's Thesis, Tomsk Polytechnic University, Tomsk, Russia, 7 October 2015. (In Russian).
44. Sprint-M Accurate Automatic Synchronization Device. Available online: https://www.rza.ru/catalog/zahita-i-avtomatika-stancionnogo-oborudovaniya/sprint-m.php?sphrase_id=29938&click=130 (accessed on 7 April 2020).
45. Synchrotact. Sync Devices and Systems. Available online: https://new.abb.com/power-electronics (accessed on 25 March 2020).
46. Reliability of Distribution Electric Networks 6 (10) Kv Automation Using Reclosers. Available online: http://www.news.elteh.ru/arh/2002/17/08.php (accessed on 23 April 2020).
47. Vorotnickij, V.; Buzin, S. The Recloser is a New Level of Automation and Control of Overhead Lines of 6 (10) kV. Available online: http://www.news.elteh.ru/arh/2005/33/11.php (accessed on 27 April 2020).

Article

Location and Sizing of Micro-Grids to Improve Continuity of Supply in Radial Distribution Networks

Fernando Postigo Marcos *, Carlos Mateo Domingo, Tomás Gómez San Román and Rafael Cossent Arín

Institute for Research in Technology (IIT), Comillas Pontifical University, 28015 Madrid, Spain; carlos.mateo@comillas.edu or carlos.mateo@iit.comillas.edu (C.M.D.); tomas.gomez@comillas.edu or tomas.gomez@iit.comillas.edu (T.G.S.R.); rafael.cossent@comillas.edu or rafael.cossent@iit.comillas.edu (R.C.A.)

* Correspondence: fpostigo@comillas.edu or fernando.postigo@comillas.edu; Tel.: +34-91-542-2800 (ext. 2729)

Received: 19 May 2020; Accepted: 1 July 2020; Published: 6 July 2020

Abstract: The steady decline in the prices of distributed energy resources (DERs), such as distributed renewable generation and storage systems, together with more sophisticated monitoring and control strategies allow power distribution companies to enhance the performance of the distribution network, for instance improving voltage control, congestion management, or reliability. The latter will be the subject of this paper. This paper addresses the improvement of continuity of supply in radial distribution grids in rural areas, where traditional reinforcements cannot be carried out because they are located in secluded areas or in naturally protected zones, where the permits to build new lines are difficult to obtain. When a contingency occurs in such a feeder, protection systems isolate it, and all downstream users suffer an interruption until the service is restored. This paper proposes a novel methodology to determine the optimal location and size of micro-grid systems (MGs) used to reduce non-served energy, considering reliability and investment costs. The proposed model additionally determines the most suitable combination of DER technologies. The resulting set of MGs would be used to supply consumers located in the isolated area while the upstream fault is being repaired. The proposed methodology is validated through its application to a case study of an actual rural feeder which suffers from reliability issues due to the difficulties in obtaining the necessary permissions to undertake conventional grid reinforcements.

Keywords: Micro-grids; continuity of supply; power distribution; power system planning

1. Introduction

The digitalization of the energy industry represents a turning point in the sector development [1]. In the case of electricity network planning [2], until the end of the 20th century, networks were mainly composed of a set of passive assets governed by electromechanical protection elements. With the sector digitalization, a wide range of possibilities enabling an increase in global social welfare has been opened. Regarding system planning and operation, these possibilities are translated into new management strategies that allow optimizing the performance of the system.

Nowadays, the levels of monitoring and control of transmission grids are fairly high; however, transmission only accounts for 3% of the total length of the European electricity network [3]. The remaining 97% corresponds to the distribution system [4], whose monitoring levels are significantly lower, especially in the case of rural networks [5]. These grids are mostly radial. Consequently, a failure in any of the network elements usually implies the loss of supply for all customers located downstream of the nearest circuit-breaking device [6]. Moreover, in many cases, orography hampers fast fault detection and repair, leading to long restoration times and deterioration in quality of service.

It is worth stressing that in more than half of the European countries [4], Distribution System Operators (DSOs) are exposed to incentive-penalty schemes associated with the power quality and continuity of supply indexes in their networks [7]. Increasing grid reliability has conventionally been achieved through traditional network reinforcements, such as the installation of switching and control equipment throughout feeders, the construction of new parallel power feeders, or meshing existing feeders to allow network reconfiguration [8]. However, these solutions are not always possible, since rural networks are sometimes placed in secluded and/or protected areas where permits to build this type of infrastructures are very hard to obtain. To overcome these difficulties, DSOs may consider alternative solutions based on micro-grids (MGs) in order to improve quality of supply without causing a major environmental impact.

MGs can be defined as network subsets with a high degree of automation and distributed energy resources (DERs) that can be operated in an islanded manner [9]. As several international experiences show [10], MGs are not a futuristic solution, but current solutions for which new policies and regulations are already being developed [11]. This is mainly due to the sharp reduction in the costs of DERs in recent years [12].

According to the existing literature, MGs can be classified according to different criteria. In [13], two types of MGs are defined from a regulatory perspective: customer microgrids (or true microgrids) which are located downstream of an end-user meter, and utility microgrids (or milligrids) that involve a segment of the common distribution network. While milligrids are not technologically different from customer microgrids, they are radically different from a regulatory viewpoint, largely because milligrids integrate the conventional utility network as in this case. However, other categorizations can be made as indicated in [14], where a taxonomy is presented based on their use, with the main categories being: remote, utility distribution, community, institutional, military, and commercial. Islanded operation of MGs also imposes new technological requirements on the operation and control of DER. If only inverter-connected DER is available, at least one of the DER units must operate in grid-following or voltage-source-based grid-supporting mode [15].

Regarding distribution networks, MGs are able to increase the number of control variables (e.g., islanded operation), and enable new management strategies [16] that can be applied in order to achieve diverse objectives depending on the network needs [17]. Previous research has already addressed several of these applications, including: decreasing technical losses [18], helping voltage control [19], increasing PV hosting capacity [20], rural electrification [21], increasing global efficiency [22], or market services [23]. Nevertheless, one of the most relevant advantages of these solutions comes from the opportunity of improving continuity of supply in the case of network failures. To this end, some publications have developed methodologies for locating generators [24,25] and reclosers [26] in distribution networks.

Nonetheless, as far as we know, to date, no reference in the literature has addressed the same problem tackled in this paper. This paper proposes a novel methodology that allows obtaining optimal MG designs that enable supplying all those consumers that, after a contingency, could not be supplied from the original source (normally the upstream substation). In this paper, the location, the type (storage, PV systems, and diesel groups) and the size of the DER technologies in charge of feeding the MG are chosen according to a multicriteria optimization, in which both the investment in these new technologies and the increase of reliability obtained with them are evaluated.

The remainder of this paper is structured as follows. Section 2 defines the problem addressed and poses the type of solutions that can be obtained. Section 3 presents the nomenclature used in this paper. Section 4 digs into the methodology applied to optimize the location and size of MGs. Section 5 applies the previously developed methodology to a synthetic realistic rural network. Finally, Section 6 outlines the main conclusions.

2. Problem Statement

As presented in the introduction, this paper addresses the problem of improving the reliability of radial distribution networks in rural and secluded areas. Currently, when a contingency occurs in such a feeder, protection systems isolate it, and all downstream users suffer an interruption until the fault is repaired and service is restored. This paper proposes a methodology for determining the optimal size and location of a set of DERs (a combination of generators and storage), considering both reliability improvements and investment minimization. In the event of a contingency, and once the fault is isolated, the installed DERs allow to keep supplying the affected consumers while the fault is being repaired, thus minimizing the duration of the interruption times of those customers. This process is shown in Figure 1. The proposed reliability improvement application can be combined with other potential services that DERs can provide, such as voltage control or congestion management, since the islanded operation will only be used sporadically when network outages occur.

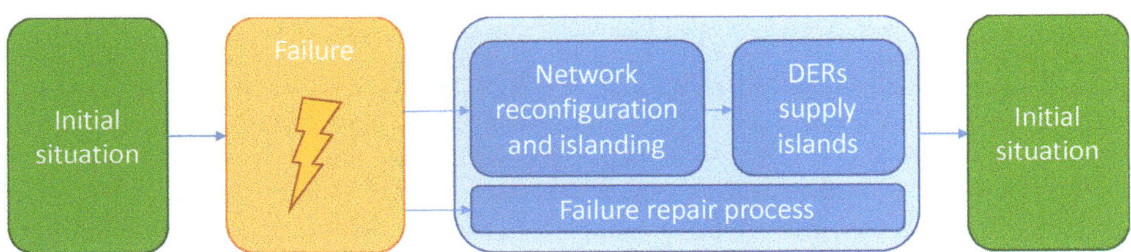

Figure 1. Action plan when a contingency occurs.

For illustrative purposes, this section provides an overview the kind of MG-based solutions that the proposed model would yield, and the main concepts used throughout the paper. However, these concepts are further detailed in Section 3. In this line, Figure 2 shows an example of a network divided into four zones (z_1, z_2, z_3, z_4). These zones are delimited by three smart switches (ss_1, ss_2, ss_3) that enable the reconfiguration of the network when an outage occurs. This figure shows a possible solution to improve the reliability of this network. Two sets of DERs (der_{d1} and der_{d2}) are installed in different zones of the system. The colors used denote that $derd_1$ is designed to supply three zones (z_1, z_2, and z_3), while $derd_2$ is dimensioned to supply just z_4.

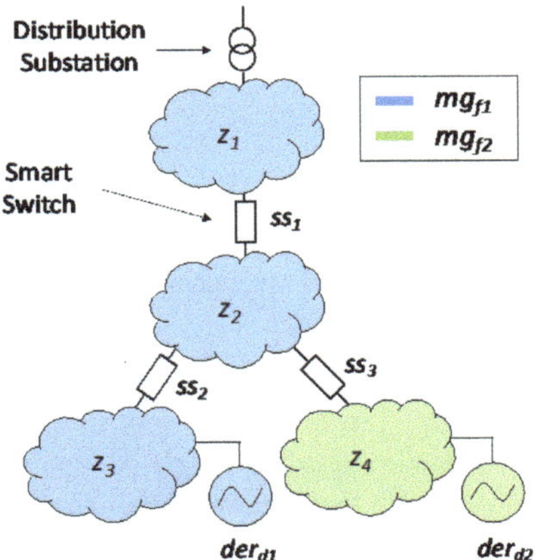

Figure 2. Example of a selected Network Solution.

The Network Solution (ns_{g1}) for the previous network example is described in Figure 3. The distribution network has four zones (z_1, z_2, z_3, and z_4), and two MGs (mg_{f1} and mg_{f2}) designed to improve the reliability of the network. These MGs are dimensioned considering the load within two groups of zones (gz_{c1} and gz_{c2}) and comprise two sets of DERs (der_{d1} and der_{d2}), respectively.

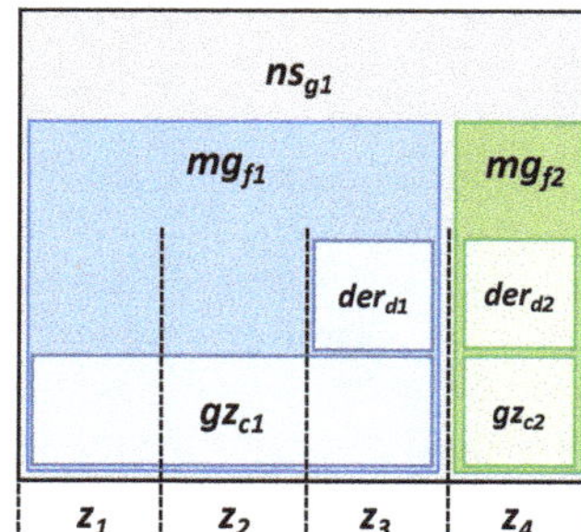

Figure 3. Nomenclature diagram for a selected Network Solution.

Considering the same network, under a conventional scenario, when a failure takes place in z_2, the nearest upstream switching element (ss_1) would isolate it from the upstream grid, leaving z_3 and z_4 without supply, while z_1 would continue being supplied from the upstream substation. However, with the Network Solution exemplified in Figure 2, the same fault in z_2, would only lead to the loss of supply of the affected zone, since der_{d1} would supply z_3, and der_{d2} would supply z_4, while z_1 would continue being supplied from the upstream substation. It should be noted that DERs would only be operated for this purpose while the failed zone is out of service, and only supplying those zones that cannot be supplied from the upstream substation due to the fault location. The rest of the time, these DERs may be used for other purposes like voltage control, energy production, or energy arbitrage in the case of batteries. It is important to note that if the contingency takes place in the same zone where the DERs are located, it is assumed that the corresponding micro-grid would lose its ability to be operated in an islanding manner. For instance, in the example of Figure 2, if the fault occurs in zone z_3, the load of this zone would not be supplied, counting as non-served energy, whereas zones z_1 and z_2 will be supplied by the upstream substation, and der_{d2} would supply z_4.

The methodology presented in this paper is observed from a DSO point of view. The utility can use this tool to plan the resources of its property used in case of failure, and which traditionally have consisted of diesel units. These DERs could be located in the point located further upstream of the indicated zone. In this way, the network will operate in normal conditions and congestion or voltage problems will be avoided. However, different strategies can be applied according to the wishes of the operator as described in the paper.

3. Terminology and Mathematical Formulation

The exact definitions of the concepts described in Section 2, as well as their mathematical formulation, are provided below in order to facilitate readers' understanding of the description of the methodology in Section 4.

- **Smart Switches (ss_a):** Tele-controlled switching devices able to isolate damaged parts of the network when a failure or contingency takes place. It is assumed that their actuation—status change to open and network reconfiguration—is fast enough to avoid worsening reliability indexes commonly counted for interruption durations longer than 3 min [27]. The set that includes all the smart switches located in the network is defined as SS, while A is the total number of elements—number of smart switches—as shown in Equation (1)

$$SS = \{ss_a\}_{a=1}^{a=A} \tag{1}$$

- **Zone (z_b)**: Network subset bounded by smart switches. The set that includes all the network zones is defined as Z, and B is the total number of zones. B can be obtained as the number of smart switches A plus one. The proposed notation is shown in Equation (2).

$$Z = \{z_b\}_{b=1}^{b=B=A+1} \tag{2}$$

- **Group of Zones (gz_c)**: Set of adjacent network zones potentially supplied by the same generation and storage installations under islanded operation. The set that includes all the possible groups of zones is defined as GZ, and the total number is C. This value depends on the network topology and the number/location of smart switches, and it is calculated as the number of different sets of connected zones as Equation (3) shows.

$$GZ = \{gz_c\ gz_c \subsetneq \mathrm{P(Z)}\ AND\ Zones(gz_c)\ are\ connected\}_{c=1}^{c=C} \tag{3}$$

where:

  - "*P (input element)*" is a function that provides the power-set of an input set, in this case, the Zones set. The power-set, in mathematics, is defined as the set of all the subsets of a set.
  - "*Zones (input element)*" is a function that provides the set of zones of which the input element is composed.

- **Distributed Energy Resources ($ders_d$)**: Tuple of generation facilities and storage installations as shown in Equation (4). In this paper, different solar PV installations (pv_i), diesel units (du_j), and batteries ($bess_k$) are considered. The indexes i, j, and k determines which elements of the equipment catalog are selected for each DERs design der_d. The PV installations and the diesel units are defined by their rated power (kW). The storage installations are defined by their rate capacity (kWh) and their ratio power/capacity (kW/kWh). In addition, the annualized CAPEX (Capital Expenditure) and OPEX (Operating Expenditures) are considered for all the installations. The set that includes all the possible distributed energy resources combinations, according to the equipment catalog, is defined as *DER*, and the total number is equal to D.

$$\mathrm{DER} = \{der_d \mid d = (i, j, k)\}_{d=1}^{d=D} \tag{4}$$

Moreover, the set that includes all the optimal DERs for the group of zones gz_c in terms of investment and non-served energy (reliability index) is defined as DER_OPT^c as shown in Equation (5). The total number of optimal DERs for the group of zones gz_c is equal to E. Thus, DER_OPT^c is just the optimal subset of DER for the group of zones gz_c.

$$DER\_OPT^c = \{der\_opt_e^c \mid der\_opt_e^c \subsetneq \mathrm{DER})\}_{e=1}^{e=E} \tag{5}$$

- **Micro-Grid (mg_f)**: Tuple composed by a group of zones gz_c, a tuple of optimal DERs $der\_opt^c{}_e$ feeding that group of zones, and the zone z_b, where the DERs are located as shown in Equation (6).The set that includes all the possible MG designs is defined as *MG*, and the total number is equal to F.

$$MG = \left\{mg_f\ mg_f = (\ gz_c,\ der_e^c,\ z_a\ )\ \mathrm{AND}\ z_a Ie\ \mathrm{Zones}\left(\mathrm{mg_f}\right)\right\}_{f=1}^{f=F} \tag{6}$$

- **Network Solution (ns_g)**: Subset of the micro-grids (*MG*) set, where all the zones are included but only once, as Equation (7) details. In other words, all the network is included in a network solution like the combination of different MGs. The set that includes all the possible network solutions is defined as NS, and the total number is equal to G.

$$NS = \left\{ns_g\ ns_g\ \subseteq MG \text{ AND } Z \subseteq \text{Zones}\left(ns_g\right) \text{ AND } \text{Zones}\left(ns_g\right)\backslash Z = o\right\}_{g=1}^{g=G} \tag{7}$$

Section 4 presents the developed methodology for obtaining the set of optimal Network Solutions to improve the continuity of supply under network contingencies while minimizing the associated investment of DER installations.

4. Methodology

This section details the methodology developed to obtain the aforementioned Network Solutions, understood as a combination of micro-grid system designs and locations. As shown in Figure 4, the construction of the set of Network Solutions starts with a network partition based on the location of the existing smart switches. Next, based on the grid connectivity layout, all possible Groups of Zones are identified. Subsequently, the DER installations for supplying each of the Groups of Zones—and therefore the MGs—are sized and located, finding the optimal combinations of technologies for each case. This optimization is carried out attending to a double perspective: minimization of investment costs and non-supplied energy. Since this is a multicriteria optimization, the sizing step makes it possible to find several optimal DER combinations for each MG, and therefore, different Network Solutions. Finally, network reliability is assessed for each Network Solution, selecting those ones that are optimal (non-dominated solutions in the Pareto front) in terms of reliability and investment.

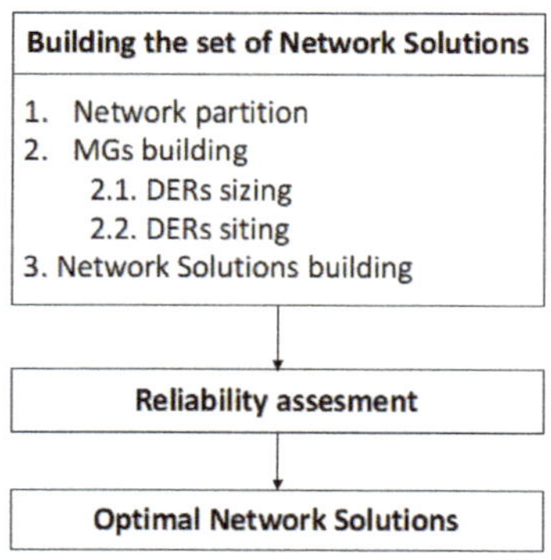

Figure 4. Methodology flowchart.

Broadly speaking, multicriteria optimization problems such as the one addressed herein can be tackled through two main approaches [28]. On the one hand, weights can be assigned to each of the functions to be optimized in order to obtain a single objective function which, once solved, yields a single optimal solution. This solution strongly depends on the weighting factors selected, which vary according to the planner preferences. On the other hand, a second alternative consists of evaluating all the individual objective functions separately and obtain a set of optimal, non-dominated, solutions that belong to the Pareto front. In this case, there is no longer a single solution, but a set of non-dominated ones, i.e., there is no other possible solution that performs better than these in all the separate objective functions at the same time. The methodology proposed in this paper follows this last approach.

Distribution network planning, and specifically the installation of DERs, entails important investments. Thus, it is really convenient for distribution companies to have a wide range of well-performing solutions so that they can choose the one that fits better their needs. The proposed methodology offers the advantage of obtaining all the non-dominated solutions without exploring all possible combinations. This is achieved thanks to a two-step optimization following a bottom-up approach. In the first stage, MGs are constructed sizing the DERs that minimize the MG non-served energy and investment cost. In a second step, an optimization that builds the Network Solutions with minimal non-served energy and investment is performed. Therefore, the total complexity of the whole design problem is significantly reduced. The computational burden of solving this problem is essentially determined by the number of smart switches in the feeder, as they increase the network reconfiguration possibilities. Nevertheless, the type of practical applications of the proposed planning solutions would be mostly installed in rural areas. In rural areas, the level of distribution network automation is typically low (normally between three and eight smart switches per feeder); consequently, the methodology described in this section is totally valid as demonstrated by the case study in Section 4.

In the following sub-sections, the different methodology steps, according to Figure 3, are described in depth, including the assumptions made and the developed mathematical formulation.

4.1. Network Partition

In highly automated distribution networks, as in the case addressed in this paper, when a fault occurs, a fast isolation of the faulted area and a quick network reconfiguration in MGs are essential to avoid affecting reliability indices. For this reason, the zone partition is made depending on the smart switches location. Thus, the total number of zones in which a network is divided is computed as the number of smart switches plus one. Taking as reference the example network structure represented in Figures 2 and 3, the SS and Z sets are calculated, see Equations (8) and (9).

$$SS = \{ss_1, ss_2, ss_3\} \tag{8}$$

$$Z = \{z_1, z_2, z_3, z_4\} \tag{9}$$

Before determining the location of DER, it is important to determine the zones they would be able to supply. For this purpose, all the possible Groups of Zones are determined. According to the definition in Equation (3), the set of all the Group of Zones GZ is determined by every group of interconnected zones that belongs to the power-set of Z. In the example, Equation (10) shows the resulting power-set of Z.

$$P(Z) = \{\{\ \}, \{z_1\}, \{z_2\}, \{z_3\}, \{z_4\}, \{z_1, z_2\}, \{z_1, z_3\}, \{z_1, z_4\}, \{z_2, z_3\}, \{z_2, z_4\}, \{z_3, z_4\}, \{z_1, z_2, z_3\}, \{z_1, z_2, z_4\}, \{z_2, z_3, z_4\}, \{z_1, z_3, z_4\}, \{z_1, z_2, z_3, z_4\}\} \tag{10}$$

The pseudocode shown in Table 1 summarizes the process used to obtain GZ from the power-set of Z, and its total number of elements C.

Table 1. Pseudocode for the Group of Zones GZ formation.

Pseudocode

$GZ = ø; c = 0$
for each "element" in $P(Z)$
    if the "element" is connected
        Add "element" to GZ
        c += 1

Accordingly, Equation (11) presents the GZ obtained for the example network through the pseudocode in Table 1.

$$GZ = \{\{z_1\}, \{z_2\}, \{z_3\}, \{z_4\}, \{z_1, z_2\}, \{z_2, z_3\}, \{z_2, z_4\}, \{z_1, z_2, z_3\}, \{z_1, z_2, z_4\}, \{z_2, z_3, z_4\}, \{z_1, z_2, z_3, z_4\}\} \tag{11}$$

4.2. Micro-Grid Building

Once all the possible Groups of Zones have been obtained, the next step is MG formation. As shown in Equation (6), a MG is defined through the tuple of three elements: a Group of Zones

gz_c, a tuple of optimal DERs $der\_opt^c{}_e$, and finally the zone z_a where the DER installations are located. These last two elements will be the object of study in this sub-section.

The process of obtaining the MGs is carried out in two stages. In the first one, for each Group of Zone gz_c, the optimal elements of the DER set "$der\_opt^c{}_e$" are determined. These are those that minimize investment costs and non-served energy. Secondly, the obtained $der\_opt^c{}_e$ is located in each one of all the zones z_a, which belongs to the considered Group of Zones gz_c. This process is described as pseudocode in Table 2 however, the sizing and siting procedures will be further explained in Sections 4.2.1 and 4.2.2 respectively.

Table 2. Pseudocode for the Micro-Grids MG formation.

Pseudocode	

$MG = \emptyset;\ DER\_OPT^c = \emptyset$
for each gz_c in GZ
    solutions = ø;
        for each der_d in DER
            $nse^c{}_d$= Non-Served Energy in gz_c supplied by der_d
            inv_d = Investment of der_d
            solutions.add ($nse^c{}_d$, inv_d)
        for each "element" in solutions
            if "element" belongs to Pareto front
                DER_OPT^c.add (der_d)
	DER sizing

for each $der\_opt^c{}_e$ in DER_OPT^c
    for each z_a in gz_c
        mg_f= (gz_c ,$der\_opt^c{}_e$, z_a)
        MG.add(mg_f)
	DER siting

4.2.1. DER Sizing

In this sub-section, the goal is to obtain for each one of the possible MGs, a set of optimal DERs DER_OPT^c that minimizes the micro-grid non-served energy (it should be noted that the micro-grid non-served energy calculated in this section will be an upper limit, since the actual value can only be determined when the locations of the DERs and the zone where the contingency or fault occurs are established, and how each of the zones is affected by the fault (Section 4.3)) and the investment costs at the same time. Therefore, the following process is repeated for each Group of Zone gz_c in order to explore all the MGs combinations.

The annualized investment (*A.Inv*) and operation and maintenance (*O&M*) cost of each DER installation are specified in an input equipment catalog (see Annex-A). Investment costs are annualized in order to compare facilities with different useful lives. The methodology used to obtain the annualized investment cost of each technology is not described since it is based on basic financial calculations and is outside the scope of the paper; however, Equation (12) aims to provide a simplified expression of the total annual DER cost calculation procedure. On the contrary, the MG non-served energy calculation associated with DER sizing and siting implies a more complex calculation and is described below.

$$Total\ investment = A.Inv_{PV} + A.Inv_{Die} + A.Inv_{Bat} + O\&M_{PV} + O\&M_{Die} + O\&M_{Bat} \quad (12)$$

It should be noted that all selected DER technologies (solar PV, diesel units, and batteries), specified in the input catalog—DER set—are considered in each MG. Thus, those DER combinations that minimize the investment and the MG non-served energy are considered optimal and part of the DER_OPT^c set, while the rest of other DER combinations are discarded. Figure 5 shows this process.

The red triangles represent the optimal subset of DER for the analyzed MG—DER_OPT^c—while the grey circles represent all the rest DER combinations included in the catalog—*DER* set.

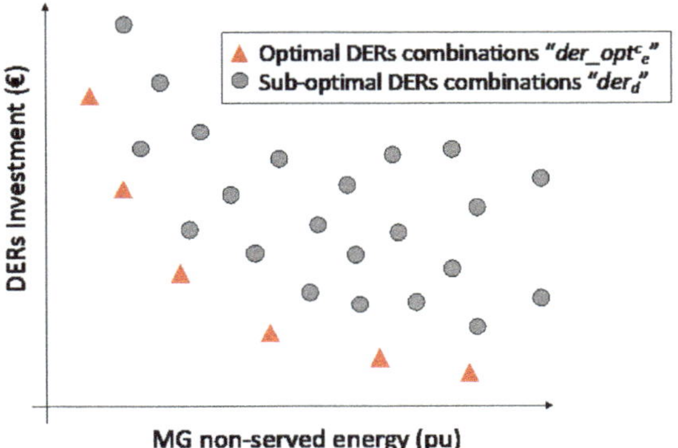

Figure 5. Pareto optimization for DER combinations for each analyzed MG.

To calculate the MGs' non-served energy, two algorithms are used. The first one is based on the installed DER capacity (kW), while the second one is based on the energy demanded by the affected customers during the contingency. The latter is needed due to the energy-constrained operation of batteries. The capacity (kWh) and the power (kW) are the two design parameters of batteries.

NSE Computing Based on Power-Limitation

As shown in Equation (13), the micro-grid non-served energy based on the power-limitation ($NSE_P(h)$), during a contingency that starts at a certain hour h, and with a pre-defined repair time h_{rep}, is calculated by adding the non-served energy during all hours ($NSE_1(h_1)$) in which the failure repair is taking place.

$$NSE_P(h) = \sum_{h_1=h}^{h_1=h+hrep} NSE_1(h_1) \tag{13}$$

It should be mentioned that only those hours with a generation deficit are counted according to Equation (14). In the deficit case, the difference ($Dif_P(h)$) between the micro-grid demand ($D(h)$) and the DER availability at that hour is computed. As previously mentioned, DERs are composed of photovoltaic generation, where ($P_{PV}(h)$) is the estimated PV production at hour h, diesel units, where (*Pdie*) is the rated capacity (kW), and batteries, where (*Pbat*) is the maximum power that can be delivered by the battery, as shown in Equation (15). This calculation is made for all of the hours of the year, since the demand and solar generation profiles vary hour by hour, and the hour when the fault occurs is not a priori known.

$$NSE_1(h) = \begin{cases} Dif_P(h) & Dif_P(h) \geq 0 \\ 0 & Dif_P(h) < 0 \end{cases} \tag{14}$$

$$Dif_P(h) = D(h) - P_{PV}(h) - P_{BAT} - P_{DIE} \tag{15}$$

NSE Computing Based on Energy-Limitation

As previously mentioned, the existence of storage units like batteries adds inter-temporal constraints that necessarily have to be considered under an energy-based approach. Resembling the previous power-based calculation, this micro-grid non-supplied energy approach ($NSE_E(h)$), only accounts for those hours with a generation deficit, see Equation (16).

$$NSE_E(h) = \begin{cases} Dif_E(h) & Dif_E(h) \geq 0 \\ 0 & Dif_E(h) < 0 \end{cases} \tag{16}$$

The difference ($Dif_E(h)$) between the energy to be supplied to the demand ($\sum_{h1=h}^{h1=h+hrep} D(h_1)$) during the repair time of the fault, and the generation that can be provided by the photovoltaic panels ($\sum_{h_1=h}^{h_1=h+h_{rep}} P_{PV}(h_1)$), the batteries ($E_{bat}$) and the diesel units ($P_{die} * h_{rep}$) is shown in Equation (17). It should be noted that the available energy storage in the battery is calculated considering the battery capacity (kWh) and the state of charge of the battery when the fault occurs ($S.C._{pu}$).

$$Dif_{E(h)} = \sum_{h1=h}^{h1=h+hrep} D(h1) - \sum_{h1=h}^{h1=h+hrep} P_{PV}(h1) - E_{BAT} * S.C._{pu} - P_{DIE} * h_{rep} \tag{17}$$

Combined Power and Energy-Based NSE Computing

The final micro-grid non-served energy (*NSE(h)*) from the hour when the fault occurs (*h*) and during the repair time of the fault (*hrep*), is calculated as the most restrictive of the two previous calculations, power-based ($NSE_P(h)$) or energy-based ($NSE_E(h)$), according to Equation (18).

$$NSE(h) = max(NSE_P(h),\ NSE_E(h)) \tag{18}$$

The micro-grid energy demand ($E(h)$) during the contingency time is obtained through the sum of the hourly demands in that period according to Equation (19).

$$E(h) = \sum_{h_1=h}^{h_1=h+h_{rep}} D(h_1) \tag{19}$$

To express the micro-grid non-served energy in unitary terms, the MG non-served energy coefficient is computed as the ratio between the micro-grid non-served energy (*NSE(h)*) and the total demanded energy ($E(h)$) as shown in Equation (20), assuming that the fault may equiprobably happen in any of the hours of the year.

$$NSE_{pu} = \frac{\sum_h NSE(h)}{\sum_h E(h)} \tag{20}$$

As mentioned above, those non-dominated combinations of the *DER* set that minimize the non-served energy and the investment will be selected as optimal to be part of the DER_OPT^c set. Only the set of optimal DER combinations will be considered in the algorithms presented in the following sections, while the rest of *DER* combinations will be discarded.

4.2.2. DER Siting

The location of DERs is intrinsically related to the reliability of the network. For instance, one DER combination, der_d, can result in very different reliability indexes with the same investment depending on the zone where it is located as shown in the Problem Statement section. For instance, zones with a high number of customers and a reduced network length, assuming that the power lines fault rate is constant per km, would be good candidates to place a set of DERs.

Therefore, the siting algorithm simulates locating the obtained optimal DER combinations, $der\_opt^c{}_e$, in all possible locations, understanding as location, each of the zones that are included in the considered Group of Zones. Thus, for each Group of Zones, gz_c, and optimal set of DER, $der\_opt^c{}_e$, there will be as many different micro-grids, mg_e, as zones are included in the corresponding Group of Zones. This algorithm is clarified in the second part of the pseudocode presented in Table 1.

4.3. Network Solution Building

After the DER siting and sizing process, the complete set of MGs is obtained. However, these MGs only include subsets of the original network; therefore, it is necessary to select different combinations of MGs that connected together form the whole distribution network. As defined in the terminology section, the set of MGs that comprises the whole network is defined as a Network Solution, ns_g, and the set that includes all possible Network Solutions is defined as NS. In addition, it should be noted that all zones must be included in each of the ns_g, but only once—no zone should be included in more than one MG within the same network solution. This process is carried out in two steps, as described below for our example network:

1. The Group of Zones are combined to compose the whole network. Table 3 shows all the possible network combinations for the example network shown in Figure 2.
2. These Groups of Zones that compose the network combinations are substituted with all the MGs that contain them to create the Network Solutions set, NS. It should be emphasized that more than one MG may contain the same group of zones.
3. In the case of our example network (see Figures 2 and 3), the nsg_1 is formed following the network combination number 5, where the Group of Zones $\{z_1, z_2, z_3\}$ and $\{z_4\}$ correspond to mg_{f1} and mg_{f2}, respectively.

Table 3. Network combinations for the example network.

Combination Number	Network Combination
1	$\{\{z_1\}, \{z_2\}, \{z_3\}, \{z_4\}\}$
2	$\{\{z_1\}, \{z_2, z_3\}, \{z_4\}\}$
3	$\{\{z_1\}, \{z_2, z_4\}, \{z_3\}\}$
4	$\{\{z_1, z_2\}, \{z_3\}, \{z_4\}\}$
5	$\{\{z_1, z_2, z_3\}, \{z_4\}\}$
6	$\{\{z_1, z_2, z_4\}, \{z_3\}\}$
7	$\{\{z_1, z_2, z_3, z_4\}\}$

4.4. Reliability Assesment

Once all the possible Network Solutions have been obtained, it is necessary to perform the reliability assessment of each one of them (ns_g included in the NS set). In the following, the method to evaluate the reliability of the system is described. In this case, the SAIDI index is selected to measure grid reliability levels. A Zone Interaction matrix "ZI^{ns_g}", is created for each Network Solution, seeking to identify how a contingency in a given zone affects the continuity of supply in another zone. Each element of this matrix is defined by the indices $b1$—rows—and $b2$—columns—where $b1$ represents the analyzed zone and $b2$ the zone where the contingency takes place. The element value can be interpreted as the per-unit non-served energy in z_{b1} under a contingency in z_{b2}. This value is equal to 0 when the z_{b1} is not affected by a contingency in z_{b2}; in the case of being equal to 1, the analyzed z_{b1} is affected by a contingency in z_{b2} and cannot be supplied from any DER, thus not supplying the entire zone demand. Finally, when the z_{b1} is affected by a contingency in z_{b2}, but there are DERs connected to the same MG that are able to supply totally or partially it, this element of the matrix can take a value between 0 and 1 equal to the MG non-served energy coefficient obtained in Equation (20) (see Section 4.2.1). It is important to mention that when a contingency affects one of the MG zones, but there are some zones in this MG that can still be supplied by the associated

DERs, the MG non-served energy coefficient will decrease proportionally to the installed power in the non-served MG zones.

It should be noted that under normal conditions (without contingencies) the operation of the entire network is radial. Only in the contingency case is the failure zone isolated, and the rest of the zones that cannot be supplied from the upstream substation (due to the fault location) would be supplied by the DERs allocated within their associated MGs.

For illustrative purposes, the example network used in the previous sections is analyzed. This network solution is composed of two MGs, the first one "mg_{f1}" is designed to supply z_1, z_2, and z_3 where the der_{d1} are located, the second one "mg_{f2}" is only formed by z_4 where der_{d2} are placed. Assuming the installed power in each zone is the same, and mg_{f1} and mg_{f2} have a 0.46 pu and 0.38 pu MG non-served energy coefficients, respectively (result of Equation (20)), the obtained zone interaction matrix for the ns_{g1} is the one shown in Equation (21).

$$ZI^{ns_g} = \begin{pmatrix} 1 & 0 & 0 & 0 \\ 0.19 & 1 & 0 & 0 \\ 0.19 & 0 & 1 & 0 \\ 0.38 & 0.38 & 0 & 1 \end{pmatrix} \quad (21)$$

Once the Zone Interaction matrix ZI^{ns_g} is calculated for each Network Solution, the reliability assessment for each Network Solution is carried out until complete the whole NS set. For the reliability assessment, it is necessary to consider some additional parameters, such as the number of supply points in each analyzed zone (N_b), the average repair time of the fault in the considered zone ($h\ rep_b$), and the annual failure rate of the network included in the considered zone (f_b) obtained as the sum of the product of the failure rate of overhead lines, expressed per km, times the overhead line lengths in the zone plus the product of the failure rate of underground lines, expressed per km, times the underground line lengths in the zone. Equation (22) shows the $SAIDI$ calculation for a generic Network Solution ns_g.

$$SAIDI^{ns_g} = \frac{\sum_{b2=1}^{b1=B} \sum_{b2=1}^{b2=B} f_{b2} * h\ rep_{b2} * N_{b1} * ZI_{b1,b2}^{ns_g}}{\sum_{b1=1}^{b1=B} N_{b1}} \quad (22)$$

4.5. Optimal Non-Dominated Network Solutions

Finally, a second multi-attribute optimization is carried out in which both the calculated reliability indices and the annualized total investment costs of all the Network Solutions included in the NS set are compared. Figure 6 shows an example of the results of such optimization. In this case, the optimal—non-dominated—network solutions are represented by triangles, while sub-optimal—dominated—NS are represented by circles.

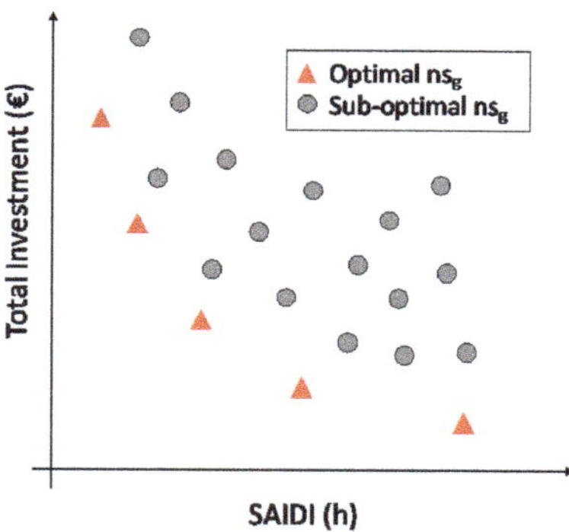

Figure 6. Pareto optimization for NS.

Once the Network Solutions have been selected, a specific location will be assigned for the DERs of each zone according to the established criteria, and a powerflow analysis will be carried out. Those Network Solutions that do not cause operational problems will be validated to be implemented.

5. Case Study

5.1. Description

In this section, the proposed methodology is applied to an actual distribution network. The analyzed feeder presents reliability problems. In addition, the feeder is located in a protected natural area where traditional network reinforcements are not an option due to the difficulty of obtaining permits. For this reason, the DSO in charge of supplying this area is exploring the use of non-conventional network solutions that minimize the impact on the environment.

The MV feeder analyzed presents three smart-switches and follows an identical structure as the example used for describing the proposed methodology, as shown Figure 7.

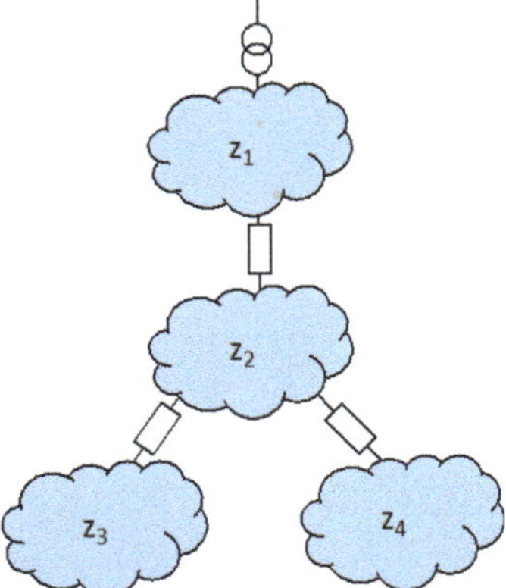

Figure 7. Network structure.

Table 4 shows the number of supply points located in each of the feeder zones, as well as the overhead and underground lengths of the power lines included in each zone.

Table 4. Network characteristics.

Zone	Number of Supply Points	Overhead Length (km)	Underground Length (km)
z_1	0	10.99	0.04
z_2	585	9.54	0.82
z_3	1386	9.94	2.07
z_4	771	6.04	0.15

Regarding failure rates, 12 outages/100 km-year was set for overhead lines and 6 outages/ 100 km-year for underground lines. Concerning the fault repair time, an average time of 6 h was considered. This data is consistent with the values posed in reference [29].

Regarding the catalog of DERs, Appendix A details the rated capacities, as well as investment and operation and maintenance costs of PV, batteries, and diesel units. It should be mentioned that the irradiation profile used to obtain PV generation is based on the geographical coordinates of the analyzed zone and obtained from [30].

5.2. Results

Applying the abovementioned methodology on the proposed network, 1,531 network solutions were obtained. Figure 8 shows a Pareto diagram in which the obtained reliability index (SAIDI) and the associated annualized investment in each network solution are displayed. Each of the dominated network solutions is represented by a circle, while the non-dominated solutions are represented by (red) triangles. Furthermore, in order to facilitate the selection of the solution that best suits the system needs, an additional parameter has been calculated. This parameter is, for each network solution, the standard deviation of the zonal SAIDIs with respect to the overall network SAIDI, measured in percentage. This parameter is represented by a color code in which the warmer the color the higher disparity between zonal reliability indicators. Therefore, at the same level of overall SAIDI and investment, solutions with similar reliability indices between zones (cooler colors) would be preferred ensuring more homogeneous solutions in terms of reliability.

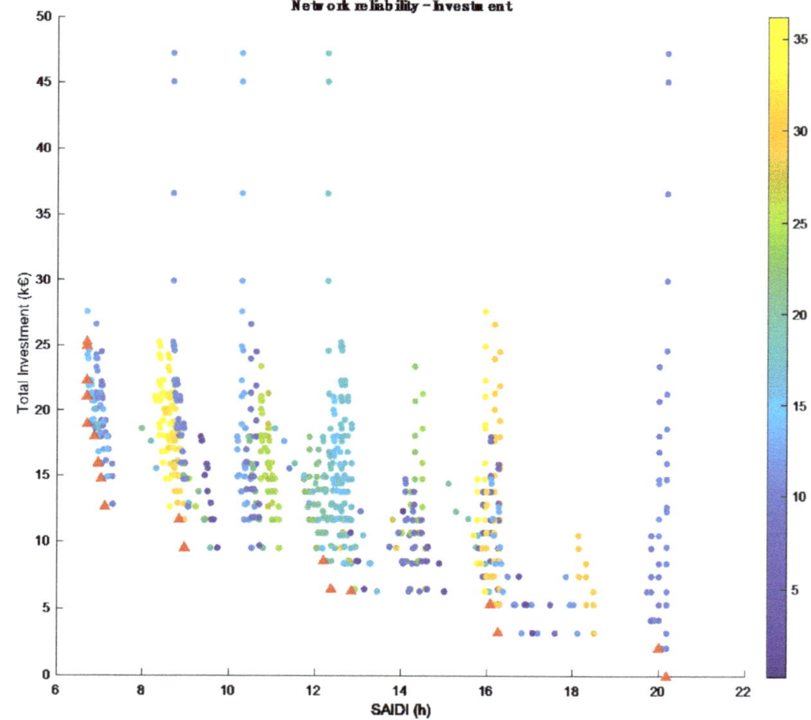

Figure 8. Network solutions for the case study network (colors represent the standard deviation of zonal SAIDI in percentage).

Table 5 presents all the optimal—non-dominated—solutions ordered by their annualized investment cost, the overall SAIDI in hours, and the SAIDI standard deviation between zones. It should be noted that the Base Case (current status of the network without DERs) belongs to the set of optimal non-dominated solutions since it does not imply any investment. On the other hand, it can be seen that as the DER investment increases, solutions with better reliability (lower SAIDI) are obtained. Network solutions, such as number 10, located in the Pareto front elbow, would bring important reductions in SAIDI with moderate investment. The set of 19 optimal network solutions achieve an improvement of up to 13.5 h with an annualized investment of up to 25,263 €.

Table 5. Optimal network solutions for the case study network.

ns_g	Annualized Investment (€)	SAIDI Overall Network (h)	SAIDI Reduction (%)	SAIDI Standard Deviation (%)	SAIDI z_1 (h)	SAIDI z_2 (h)	SAIDI z_3 (h)	SAIDI z_4 (h)
1	25,263	6.7	−67%	14%	7.8	7.1	4.4	0.0
2	24,945	6.7	−67%	14%	7.8	7.1	4.4	0.0
3	22,291	6.7	−67%	14%	7.8	7.1	4.4	0.0
4	21,052	6.7	−67%	14%	7.8	7.1	4.4	0.0
5	18,947	6.7	−67%	14%	7.8	7.1	4.4	0.0
6	17,980	6.9	−66%	16%	8.1	7.2	4.4	0.0
7	15,975	6.9	−66%	13%	8.0	7.3	4.8	0.0
8	15,875	7	−65%	17%	8.3	7.2	4.4	0.0
9	14,736	7	−65%	14%	8.1	7.3	4.8	0.0
10	12,632	7.1	−65%	15%	8.3	7.3	4.8	0.0
11	11,664	8.8	−56%	32%	8.3	15.0	5.1	0.0
12	9560	9	−55%	30%	8.3	15.0	5.6	0.0
13	8592	12.2	−40%	19%	14.7	15.0	5.6	0.0
14	6488	12.4	−39%	19%	15.1	15.0	5.6	0.0
15	6316	12.8	−37%	17%	8.3	15.0	19.4	0.0
16	5348	16.1	−20%	7%	14.7	15.0	19.4	0.0
17	3244	16.3	−19%	7%	15.1	15.0	19.4	0.0
18	2105	20	−1%	11%	22.4	15.0	19.4	0.0
Base Case	0	20.2	0%	11%	22.8	15.0	19.4	0.0

Another important issue to analyze is, for each non-dominated network solution, in which network zone DERs are located and which DER technologies are selected. Table 6 presents this information. For each network solution, it identifies in which zone the DERs are located (DERs *Zone*) and the zones they supply (z_1, z_2, z_3, and z_4). The size of the PV systems (P_{PV}), batteries (P_{BAT} and E_{BAT}) and diesel units (P_{DIE}) are also included. It can be seen that all the optimal solutions include DERs in zone number 3. This is an expected result since this is the area with the highest number of connected supply points and located farthest from the supply point. The same happens with zone number 4 —the zone with the second-highest number of supply points—for similar reasons.

Regarding the DER combinations selected, it is observed that all network solutions except one include diesel units, in some cases supported by batteries. These batteries have a high power (kW)–capacity (kWh) ratio, which leads to the conclusion that batteries are mainly used for peak shaving, avoiding the installation of an additional diesel unit with a low use rate. This effect can be observed by comparing the evolution of selected DERs starting from the Base Case and going through network solutions 18, 17, 16 and 15. It can be observed that as the reliability index improves, the selection of batteries (solutions 18 and 16) is alternated with that of diesel units (solutions 17 and 15). Therefore, a higher granularity in diesel unit sizes presumably would eliminate the need for batteries in the optimal network solutions.

On the other hand, it is observed that there is no optimal solution selecting PV facilities. This is mainly explained by two reasons. The first is the comparatively high PV investment cost. Nowadays, PV systems can be cost-effective when the produced energy is self-consumed and/or sold to the market during the expected life of the installation. However, in this paper, we are assuming that PV is only used to obtain benefits of its production during the small number of hours per year in which network outages take place. The second reason is that, unlike storage or diesel units, the PV production is limited to the sunlight hours. This effect is modeled through solar radiation hourly profiles for the specific network location. The use of annual profiles is recommended to capture seasonality. For example, if the network failure occurs at night or at the end of the day, PV will not be able to provide the necessary energy to supply the isolated customers and to improve the continuity of supply indices.

Table 6. Optimal network solutions for the case study network.

ns_g	DERs Zone	z_1	z_2	z_3	z_4	P_{PV} (kW)	P_{BAT} (kW)	E_{BAT} (kWh)	P_{DIE} (kW)
1	2	0	1	0	0	0	0	0	200
	3	0	0	1	0	0	0	0	400
	4	0	0	0	1	0	0	0	200
2	3	0	1	1	0	0	100	250	400
	4	0	0	0	1	0	0	0	200
3	3	0	1	1	0	0	100	100	400
	4	0	0	0	1	0	0	0	200
4	3	0	1	1	0	0	100	30	400
	4	0	0	0	1	0	0	0	200
5	3	0	1	1	0	0	0	0	400
	4	0	0	0	1	0	0	0	200
6	2	0	1	0	0	0	0	0	100
	3	0	0	1	0	0	100	30	200
	4	0	0	0	1	0	0	0	200
7	3	0	0	1	0	0	100	100	200
	4	0	1	0	1	0	0	0	200
8	2	0	1	0	0	0	0	0	100
	3	0	0	1	0	0	0	0	200
	4	0	0	0	1	0	0	0	200
9	3	0	0	1	0	0	100	30	200
	4	0	1	0	1	0	0	0	200
10	3	0	0	1	0	0	0	0	200
	4	0	1	0	1	0	0	0	200
11	3	0	0	1	0	0	0	0	200
	4	0	0	0	1	0	100	30	100
12	3	0	0	1	0	0	0	0	200
	4	0	0	0	1	0	0	0	100
13	3	0	0	1	0	0	100	30	100
	4	0	0	0	1	0	0	0	100
14	3	0	0	1	0	0	0	0	100
	4	0	0	0	1	0	0	0	100
15	3	0	0	1	0	0	0	0	200
16	3	0	0	1	0	0	100	30	100
17	3	0	0	1	0	0	0	0	100
18	3	0	0	1	0	0	100	30	0

Therefore, it can be concluded that for the network studied, and with current investment costs, the use of diesel units by themselves or in combination with batteries is the most cost-effective network solution to increase the reliability of the system. During network faults, the diesel units would cover the base demand and the batteries would cover the peaks of the isolated supply points.

A sensitivity analysis to the investment cost of the batteries was performed. To isolate the effect of this parameter, a new case was run for which battery systems were the only DER technology available. There are four reasons for this. Firstly, batteries are less environmentally pollutant than diesel units (at least locally). Additionally, batteries are already being used in the optimal solutions of 6. Secondly,

PV systems do not seem to be a good option to improve system reliability because their use is restricted to the sunlight hours and, therefore, they cannot be used when outages occur at night. Moreover, unlike batteries, in none of the 18 optimal network solutions the PV systems are chosen as a design option, being more cost-effective to invest in batteries. Third, diesel units are a mature technology from which no major price variations are expected. In this case, the only factor that could increase their cost would be a sharp rise in the cost of diesel (variable cost included in the O&M cost defined in the Annex). However, these are units that operate very few hours per year and are therefore not very sensitive to this factor. Finally, the cost of batteries nowadays presents a steady downwards trend. Therefore, the break-even point for batteries to become the preferred design option was evaluated.

Figure 9 shows the Pareto front for different percentages of battery cost reductions. It can be observed that to obtain a Pareto front similar to the one presented with current investment prices for PV and batteries and including diesel units, reductions in the cost of batteries by close to 80% would be needed.

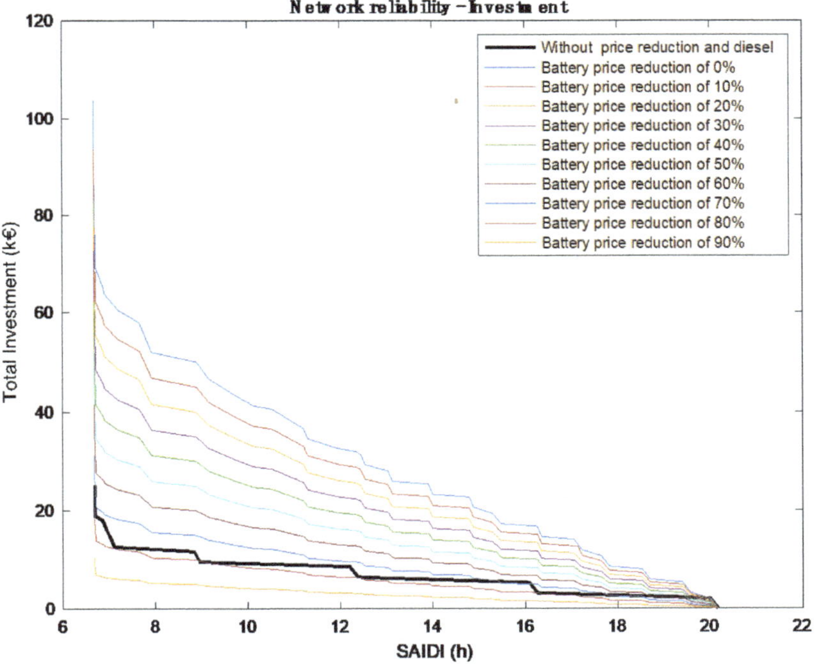

Figure 9. Pareto fronts for different battery price reductions.

Analyzing the results obtained, it can be concluded that diesel generators and batteries (as long as their initial state of charge is sufficient) are the preferred option, since, unlike PV systems, they are available at any time of the day and under any atmospheric conditions. This does not mean that PV systems cannot be used for this purpose, but they do have some features that make them difficult to apply. However, each case must be studied individually and new clean and hybrid solutions such as PV systems with batteries should be considered in future low-cost scenarios.

6. Conclusions

This paper proposes a novel methodology for determining the optimal location and sizing of MGs based on a multicriteria optimization in which both the DER investment cost and the network reliability levels are considered. Unlike other publications, this methodology allows the mix of DERs to be selected that best fits the needs of the network. Additionally, the result is a set of non-dominated solutions, allowing the DSO to select the one that best suits other possible non-modeled requirements (e.g., local emissions). This methodology was tested in a real case study, from which some relevant findings have been extracted, as discussed below.

The results show that the combination of diesel units and batteries seems to be the most cost-effective option to increase the reliability of the network through islanded operation. During the failure repair time, diesel units would be in charge of covering the base demand of islanded zones, whereas batteries would operate as peaking units. Moreover, the results show that selecting solar PV installations for this purpose is not cost-effective. The reason for this is twofold. On the one hand, PV presents a comparatively high investment cost. On the other hand, PV can only supply the demand during sunlight hours, thus being useless to tackle outages occurring at night. However, each case must be studied individually and adjusted to the specific climatic conditions of the network location. Finally, it has been shown that if battery investment costs were to drop by close to 80% with respect to current values, MG solutions based mainly on batteries instead of more highly polluting solutions based on diesel units would be the preferred option.

As observed in the case study, emission-free DERs are expensive and are only used in less than 1% of the hours to enhance the reliability of the distribution grid. For this reason, DSOs may explore the procurement of a service through which third parties, such as DER owners, would provide network support when a network outage takes place. The remuneration of this service would be based on the location of the DER in the network and its commitment to offering the service when requested. In this way, the DSO would reduce investments in network infrastructure and the owners of DERs could obtain additional benefits to the regular ones derived from their participation in the market or cost-savings thanks to behind the meter generation.

This work opens up future research lines to be explored. The main limitation of this study is the degree of automation of the network. The greater the degree of automation of the network, the greater the number of smart-switches, and the possibilities of reconfiguring the network would grow exponentially. For this reason, the proposed solution perfectly fits with the problems of rural networks, which are characterized by low automation levels. However, urban areas with higher automation levels may sharply increase the computational load. For this reason, this limitation opens, for example, a future research line in applying metaheuristic algorithms (e.g., genetics algorithms) to overcome this issue.

Author Contributions: Conceptualization, F.P.M., C.M.D., T.G.S.R. and R.C.A.; methodology, F.P.M., C.M.D. and T.G.S.R.; software, F.P.M. and C.M.D.; validation, F.P.M., C.M.D., T.G.S.R. and R.C.A.; writing—review and editing, F.P.M., C.M.D., T.G.S.R. and R.C.A. All authors have read and agreed to the published version of the manuscript.

Funding: This research has received funding from the Fundación Iberdrola España through the 'Ayudas a la investigación 2019' program.

Acknowledgments: The authors would like to thank i-DE for partially supporting this research. In particular, comments and suggestions for improving the practical application of the developed algorithms are acknowledged to Diego Haro Ginés, Roberto García Gómez, Fernando Palacios González, and Cristina Vilá Castro.

Conflicts of Interest: The authors declare no conflict of interest.

Appendix A DER Investment

The investment in DERs is based on a review of the state of the art of the technologies employed: PV systems, battery-based storage, and diesel units.

The price of the PV systems is composed of the cost of the panels, the cost of the inverter, and the cost of installation. The cost of the panels is based on [31]. For the rest of the costs like de O&M cost, a breakdown according to the reference has been considered. The useful life is based on [32]. Table A1. shows the data used.

Table A1. PV systems data.

Rated Capacity (kW)	Investment Cost (€/kW)
0	850
100	**O&M (€/kW-year)**
200	18.08
400	**Useful life (years)**
800	23

The price of the batteries is based on [33]. Other additional costs like de O&M costs have been considered according to the breakdown given in [34]. For the power-capacity ratio, the costs indicated in [32] have been taken into account. To obtain the total investment cost of batteries for different power-capacity ratios, the power and energy values are averaged. The useful life considers that very few charge/discharge cycles are expected per year in this application. Table A2 shows the data used.

Table A2. Batteries data.

Rated Power (kW)	Power (kW)—Capacity (kWh) Ratio	Investment Cost (€/kW)
0	3.33	314.26
100	1	**Investment cost (€/kWh)**
200	0.4	346.21
400	0.25	**O&M (€/kWh)**
800		0.356
		Useful life (years)
		15

The price of diesel units depends mainly on their rated capacity [35]. Apart from the price of the diesel unit itself, an installation cost of 60% was considered. For the fuel, a cost of 1.3 €/L and an expense of 0.3l/kWh has been assumed [36] to obtain the total O&M cost. A useful life of 20 years is considered, assuming that the unit will be used only a few hours per year. Table A3 shows the data used.

Table A3. Diesel units data.

Rated Capacity (kW)	Investment Cost (€/kW)	O&M (€/kW-year)
0	0	8
100	223.34	**O&M (€/kWh)**
200	125.76	0.3
400	121.26	**Useful life (years)**
		20

References

1. European Commission. *Digitalisation of the Energy Sector*; SETIS Magazine; European Commission: Brussels, Belgium, 2018.
2. Makris, P.; Efthymiopoulos, N.; Nikolopoulos, V.; Pomazanskyi, A.; Irmscher, B.; Stefanov, K.; Pancheva, K.; Varvarigos, E. Digitization Era for Electric Utilities: A Novel Business Model Through an Inter-Disciplinary S/W Platform and Open Research Challenges. *IEEE Access* **2018**, *6*, 22452–22463. [CrossRef]
3. Staschus, K. *ENTSO-E Memo 2011*; Secretariat of ENTSO-E: Brussels, Belgium, 2011.
4. Pavla Mandatova; Gunnar Lorenz. *Power Distribution in Europe: Facts & Figures*; Eurelectric: Brussels, Belgium, 2013.
5. Short, T.A. *Electric Power Distribution Handbook*; CRC Press: Boca Raton, FL, USA, 2014.
6. Electric Power Distribution Engineering. Available online: https://www.crcpress.com/Electric-Power-Distribution-Engineering-Third-Edition/Gonen/p/book/9781482207002 (accessed on 20 February 2019).

7. Pérez-Arriaga, I.J. *Regulation of the Power Sector*; Springer: New York, NY, USA, 2013; ISBN 978-1-4471-5033-6.
8. Chen, T.-H.; Huang, W.-T.; Gu, J.-C.; Pu, G.C.; Hsu, Y.F.; Guo, T.Y. Feasibility study of upgrading primary feeders from radial and open-loop to normally closed-loop arrangement. *IEEE Trans. Power Syst.* **2004**, *19*, 1308–1316. [CrossRef]
9. Lasseter, R.H.; Paigi, P. Microgrid: A conceptual solution. In Proceedings of the 2004 IEEE 35th Annual Power Electronics Specialists Conference (IEEE Cat. No.04CH37551), Aachen, Germany, 20–25 June 2004; Volume 6, pp. 4285–4290.
10. Hatziargyriou, N.; Asano, H.; Iravani, R.; Marnay, C. Microgrids. *IEEE Power Energy Mag.* **2007**, *5*, 78–94. [CrossRef]
11. Marnay, C.; Asano, H.; Papathanassiou, S.; Strbac, G. Policymaking for microgrids. *IEEE Power Energy Mag.* **2008**, *6*, 66–77. [CrossRef]
12. Pérez-Arriaga, J.I.; Batlle, C.; Gómez, T.; Chaves, J.P.; Rodilla, P.; Herrero, I.; Dueñas, P.; Ramírez, C.R.V.; Bharatkumar, A.; Burger, S.; et al. *Utility of the Future: An MIT Energy Initiative Response to an Industry in Transition*; Massachusetts Institute of Technology: Cambridge, MA, USA, 2016.
13. Types of Microgrids|Building Microgrid. Available online: https://building-microgrid.lbl.gov/types-microgrids (accessed on 10 April 2020).
14. Asmus, P.; Lawrence, M. Emerging Microgrid Business Models. Available online: /paper/Emerging-Microgrid-Business-Models-Asmus-Lawrence/67052852b5b104d9e7f61daaa1b7220d74cdb39d (accessed on 9 June 2020).
15. Rocabert, J.; Luna, A.; Blaabjerg, F.; Rodríguez, P. Control of Power Converters in AC Microgrids. *IEEE Trans. Power Electron.* **2012**, *27*, 4734–4749. [CrossRef]
16. Lopes, J.A.P.; Moreira, C.L.; Madureira, A.G. Defining control strategies for MicroGrids islanded operation. *IEEE Trans. Power Syst.* **2006**, *21*, 916–924. [CrossRef]
17. Lasseter, R.H. Smart Distribution: Coupled Microgrids. *Proc. IEEE* **2011**, *99*, 1074–1082. [CrossRef]
18. Patel, Y.P.; Patel, A.G. Placement of DG in Distribution System for loss reduction. In Proceedings of the 2012 IEEE Fifth Power India Conference, Delhi, India, 19–22 December 2012; pp. 1–6.
19. Anand, S.; Fernandes, B.G.; Guerrero, J. Distributed Control to Ensure Proportional Load Sharing and Improve Voltage Regulation in Low-Voltage DC Microgrids. *IEEE Trans. Power Electron.* **2013**, *28*, 1900–1913. [CrossRef]
20. Armendáriz, M.; Heleno, M.; Cardoso, G.; Mashayekh, S.; Stadler, M.; Nordström, L. Coordinated microgrid investment and planning process considering the system operator. *Appl. Energy* **2017**, *200*, 132–140. [CrossRef]
21. Ciller, P.; Ellman, D.; Vergara, C.; González-García, A.; Lee, S.J.; Drouin, C.; Brusnahan, M.; Borofsky, Y.; Mateo, C.; Amatya, R.; et al. Optimal Electrification Planning Incorporating On- and Off-Grid Technologies: The Reference Electrification Model (REM). *Proc. IEEE* **2019**, *107*, 1872–1905. [CrossRef]
22. Park, W.Y.; Shah, N.; Phadke, A. Enabling access to household refrigeration services through cost reductions from energy efficiency improvements. *Energy Effic.* **2019**, *12*, 1795–1819. [CrossRef]
23. DeForest, N.; MacDonald, J.S.; Black, D.R. Day ahead optimization of an electric vehicle fleet providing ancillary services in the Los Angeles Air Force Base vehicle-to-grid demonstration. *Appl. Energy* **2018**, *210*, 987–1001. [CrossRef]
24. Celli, G.; Ghiani, E.; Mocci, S.; Pilo, F. A multi-objective formulation for the optimal sizing and siting of embedded generation in distribution networks. In Proceedings of the 2003 IEEE Bologna Power Tech Conference Proceedings, Bologna, Italy, 23–26 June 2003; Volume 1, p. 8.
25. Ippolito, M.G.; Morana, G.; Sanseverino, E.R.; Vuinovich, F. Risk based optimization for strategical planning of electrical distribution systems with dispersed generation. In Proceedings of the 2003 IEEE Bologna Power Tech Conference Proceedings, Bologna, Italy, 23–26 June 2003; Volume 1, p. 7.
26. Popović, D.H.; Greatbanks, J.A.; Begović, M.; Pregelj, A. Placement of distributed generators and reclosers for distribution network security and reliability. *Int. J. Electr. Power Energy Syst.* **2005**, *27*, 398–408. [CrossRef]
27. Tabatabaei, N.M.; Ravadanegh, S.N.; Bizon, N. (Eds.) *Power Systems Resilience: Modeling, Analysis and Practice*; Springer International Publishing: Cham, Switzerland, 2019; ISBN 978-3-319-94441-8.
28. Ehrgott, M. *Multicriteria Optimization*, 2nd ed.; Springer-Verlag: Berlin/Heidelberg, Germany, 2005; ISBN 978-3-540-21398-7.
29. Kjolle, G.; Sand, K. RELRAD-an analytical approach for distribution system reliability assessment. *IEEE Trans. Power Deliv.* **1992**, *7*, 809–814. [CrossRef]

30. Renewables.ninja. Available online: https://www.renewables.ninja/ (accessed on 7 April 2020).
31. Magazine, pv Module Price Index. Available online: https://www.pv-magazine.com/features/investors/module-price-index/ (accessed on 10 April 2020).
32. McLaren, J.; Gagnon, P.; Anderson, K.; Elgqvist, E.; Fu, R.; Remo, T. *Battery Energy Storage Market: Commercial Scale, Lithium-ion Projects in the U.S.*; National Renewable Energy Lab. (NREL): Golden, CO, USA, 2016.
33. Worldwide - lithium ion battery pack costs 2019. Available online: https://www.statista.com/statistics/883118/global-lithium-ion-battery-pack-costs/ (accessed on 9 April 2020).
34. Wilson, M. *Lazard's Levelized Cost of Storage Analysis—Version 4.0*; Lazard: New Orleans, LA, USA, 2018.
35. What is the average price of the generator sets? Available online: https://grupoelectrogeno.net/precio-medio-grupos-electrogenos/ (accessed on 9 April 2020).
36. Precio en España de Ud de Grupo electrógeno. Generador de precios de la construcción. CYPE Ingenieros, S.A. Available online: http://www.generadordeprecios.info/obra_nueva/Instalaciones/Electricas/Generadores_de_energia_electrica/Grupo_electrogeno.html (accessed on 9 April 2020).

Article

Towards a Sustainability Assessment Model for Urban Public Space Renewable Energy Infrastructure

Kaan Ozgun

School of Design, Queensland University of Technology, Brisbane 4000, Australia; kaanozgun@gmail.com

Received: 22 May 2020; Accepted: 30 June 2020; Published: 3 July 2020

Abstract: As cities develop new interventions for climate change mitigation, incorporating renewable energy in urban public spaces becomes a common norm to address sustainability objectives. However, current built projects and assessment practices mainly uses a "techno-fixes" approach focusing on strategies that are related to the environmental benefits and neglecting other potential strategies instigating social and economic benefits of renewable energy. The purpose of this study is to present a potential sustainability assessment model introducing new strategies for public space renewable energy use where social and economic benefits of renewables become evident. Supplemented with theories and principles from ecology, the model's economic strategies refer whether the project considers meaningful part of the produced electricity for generating a local economy; environmental strategy comprises embodied energy, energy storage and self-maintenance; social strategy includes whether the project considers generating active and passive interaction using on-site electricity. Ballast Point Park in Sydney was used as a test bed to examine the model and sustainability of park's renewable energy use. The findings showed that environmental strategies were evident in the park, social strategies remained average and economic strategies with renewable energy were lacking. Interactions with on-site produced electricity was further claimed to be an imperative feature of any public space. Recommendations were made specific to operational and planning impacts of the integrated model.

Keywords: renewable electricity distribution for public space; sustainability assessment model; integrated assessment for public space; tripartite altruism; urban renewable energy; ecological infrastructures

1. Introduction

Energy autonomy is an emerging trend at city, neighborhood, community and building scales. Renewable energy, distributed generation, transition from alternative circuit to direct circuit infrastructure and resilient micro and smart grids, all indicate a transition to new energy environments [1]. Using renewable energy in built environment becomes a common solution for climate change mitigation. However, its current use into urban environments is problematic from a sustainability perspective. Despite some exceptions, most built interventions concentrate on improving cities' environmental sustainability by retrofitting urban environments with technological additions such as green walls and photovoltaic panels [2]. Rather than perceiving renewable energy applications as crude technological addition that simply address the environmental sustainability objective of a sustainable top-down developmental approach, scholars and practitioners need to treat renewable energy as "ecological infrastructures". For example, positioning the location of "energy–ecology–society relations in a 'commons' space" and for a focus on techniques and social arrangements can serve the aims of sustainability and equity [3]. Public space showcases for the renewable energy commons approach, a bridge that connects mainstream energy with the emerging alternative decentralized energy movements. Public space can also be designed as a real living demonstration of what can be achieved in renewable energy.

The success of sustainable development depends on several effective energy programs and their management strategies. This includes energy innovations; raising awareness through public and professional organizations; informing users about energy use and renewable types; increasing environmental knowledge through training and education; encouraging renewable energy use through finance, tax initiatives, and/or policies; and integrating assessment and auditing tools [4]. Because the concept of sustainability has evolved, scholars have recognized the importance of a multidisciplinary approach to the study and practice of sustainability. "The disciplines are essential for quality control" [5]. Each discipline creates its own quality control tools, one of which is the assessment of sustainability.

Tools that have been highly effective in pushing architectural industry to a better sustainability standard and have, by association, also influenced sustainable Landscape Architecture. The most appropriate ones concomitant to the scope of this study comprise 'LEED' (Leadership in Energy and Environmental Design) and 'SITES' (Sustainable Sites Initiative—a tool specifically for public open space assessment) in the USA, 'BREEAM' (Building Research Establishment Environmental Assessment Method) in the UK and 'Greenstar' in Australia.

SITES—which was jointly developed by the Lady Bird Johnson Wildflower Center at The University of Texas (Austin), the United States Botanic Gardens and the American Society of Landscape Architects—incorporates ecosystems services into its agenda aiming to foster sustainable land development and management practices [6]. While SITES aims to "create guidelines and performance benchmarks for sustainable design, construction and maintenance in Landscape Architecture projects" [6], it has not yet gained acceptance by the Landscape Architecture profession in Australia as a sustainability assessment tool for public spaces.

While renewable energy assessment is an important aspect of the SITES, their point rating scheme assesses public spaces based on a set of criteria focusing on environmental objectives such as making clean electricity available, promoting energy efficiency and reducing carbon footprint. For example, in their recently updated assessment guidelines (SITES v2), renewable energy is credited as "use renewable energy sources for landscape electricity needs" [7]. More specifically, if a project addresses fifty percent of annual outdoor site electricity, three points are scored. If a project generates one hundred percent of annual outdoor site electricity, four points are scored. Their description clearly states that renewable energy in a public space context currently does not count as a potential social sustainability initiator, but mainly considers the economic and environmental aspect of renewable energy and electricity production. While the relevant information in the guideline booklet includes a section about community renewable energy systems, its focus is on the produced electricity leasehold and management [7]. The current SITES v2 guidelines neglect the economic and social sustainability of on-site produced electricity and overlook renewable energy interactions from a public space context, thus misses the potential for enriching public space programs with new ways of public engagement.

Both theory and practice currently lack a sound knowledge base. Thus, the study formulates the following research question—which theories, principles and strategies can better contribute to the renewable energy-embedded public space and to what extent these principles can be used to form a potential assessment tool? With the advancement of renewable energy technologies and their increasing production capacities, renewable energy within a public space context can offer invaluable outcomes beyond the accepted environmental benefits. The purpose of this study is thus, to present a potential sustainability assessment model that realizes novel strategies instigating social and economic benefits of on-site produced electricity on top of the commonly accepted environmental benefits.

The optimal electricity distribution (OED) framework and its strategies were adapted and tested by assessing renewable energy projects designed for the Freshkills Park site in New York City [8]. A selection committee and multidisciplinary experts from the Land Art Generator Initiative (LAGI) design competition assessed and shortlisted twenty-five speculative projects [8]. This study considers LAGI as an authority in this emerging topic because LAGI organizes design competitions since 2010 and has gathered an archive of speculative projects over four different countries attracting experts, designers, and artists who work on the verge of science, art and technology. Therefore, LAGI's design proposals are likely to be indicative of some dominant contemporary approaches to the design of renewable energy-embedded public spaces. Part of the strategies presented in this study have been regularly used by the organizers in the last decade as criteria for judging schemes submitted to the LAGI design competitions. Both author's previous studies—as well as LAGI's frequently tested judging criteria—lay the groundwork for presented strategies in this study [8,9].

Because the SITES assessment framework is still in its development phase, the study's another objective is to enrich the scope of current assessment tools advocating creative processes and strategies that augment the sustainability of public space renewable energy. Contrary to the focus on the LAGI's speculative projects [8], this study adapts the OED framework and its strategies for assessing a built project site. Ballast Point Park (BPP) in Sydney is used as a test bed to demonstrate the model's performance and compatibility of its strategies.

To find out how each strategy is addressed in the design and how BPP design addresses sustainability from a public space renewable energy distribution perspective, the study acquires data from site observation, user survey and semi-structured interviews with consulted experts and designers. The following strategies are formulated coming out of author's earlier works and the LAGI's speculative project assessment criteria [8,9].

- Strategies under economic benefits involve selling 1/3 or meaningful portion of the produced on-site electricity to city grid for generating income to the public space and its local community;
- Strategies under environmental benefits include using 1/3 or meaningful portion of produced on-site electricity in the life time of renewable energy infrastructure for embodied energy, energy storage and maintenance of the renewable energy infrastructure and/or public space;
- Strategies under social benefits focus on the remaining portion of the onsite electricity to be used in creating activities whether the design considers social active and/or passive engagement with public space renewable energy.

Strategies are further substantiated with theories from ecology and described in the result section. The findings enable a blueprint for other sustainability practitioners who consider using renewable energy in their projects elsewhere. The study discusses the operations and planning impact of the potential assessment model and concludes with the limitations and future research. Recommendations are made how renewable energy can be effectively incorporated into public space. Local energy trading [10] and interaction with on-site produced electricity are further claimed to be an imperative feature for future public space designs. The need for an integrated approach when using renewables in public spaces is also discussed.

Next section first introduces the case then presents the methods of the study.

2. Material and Methods

This study used BPP as an "instrumental case" to comprehend an emerging paradigm, assessment of public space renewable energy infrastructure [11]. BPP was chosen as a case study because the park was recognized and awarded by the Australian Institute of Landscape Architecture (AILA) for its environmental sustainability innovations [12]. The park also represented the first Landscape Architecture project in Australia that incorporated renewable energy into its design. As presented in Figure 1, the 2.6 ha park was designed as a vegetated headland that retained its industrial footprint and opened to public in 2009 [13].

Mixed methods including semi-structured interviews with experts and designers, site observation and user survey were employed to investigate BPP. Five people in total were interviewed via Skype and in person. Three landscape architects from the leading firms involved in the design and planning of the park; one community expert and one project manager from a government authority responsible for the Sydney Harbor Foreshore parks engaged in project control and supervision.

Interviews were lasted between one to two hours. The model's strategies navigated the recorded interviews when analyzed using the NVivo software. The recordings were later transcribed as an intelligent verbatim [14]. The interview questions addressed the following topics; the essence of sustainability, triple-bottom-line (TBL), how renewable energy was integrated into the park design, and the original goals of overall design concerning social, economic, environmental and esthetic benefits.

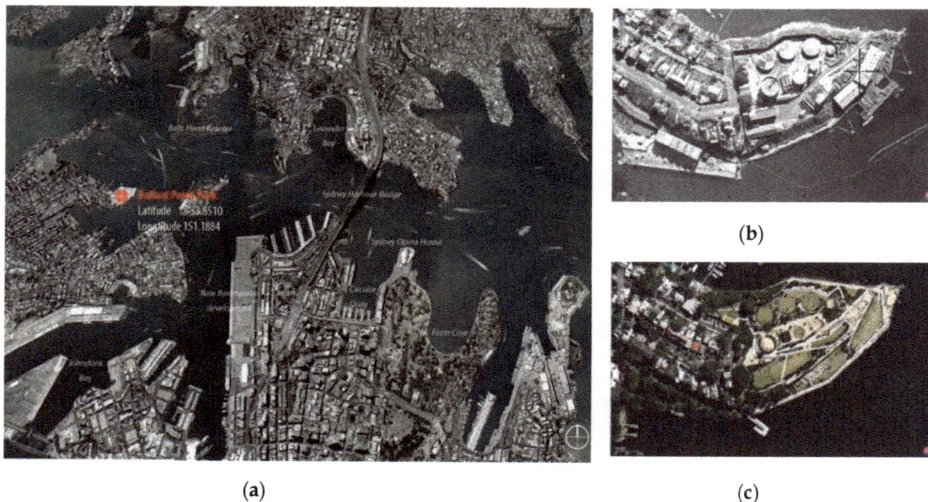

(**a**) (**b**) (**c**)

Figure 1. (**a**) Ballast Point Park (BPP) location map illustrates the Sydney Harbor context, map extracted from Google Earth; (**b**) 1943 BPP site; (**c**) 2010 BPP site, maps extracted from maps.six.ns.gov.au.

Site observation was used to understand the extend of activities concerning renewable energy infrastructure in the park. These included how renewable energy was incorporated into the design and how people interacted. The number of people using the specific location of the renewable energy infrastructure was also questioned against the use of other park functions. The observation was conducted during a two-week period in January 2014 including, weekdays, weekends and a public holiday when the summer started in the southern hemisphere. The site was divided into six zones and seven exposed spots were investigated, respectively to conduct the observations. By moving every twenty minutes from one spot to another "with rotating shifts of early morning 7:30–10:00 a.m.; morning 10:00 a.m.–12:00 p.m.; midday 12:00 p.m.–2:00 p.m., early afternoon 2 p.m.–4 p.m.; and late afternoon 4 p.m.–7 p.m.", park program was recorded and documented at each observation spot [14].

The type of renewable energy used in the design was vertical micro wind turbine, eight of which were integrated into a recycled steel construction in Figure 2 and erected at the park center as a renewable energy sculpture where the former tank 101 was standing.

Since wind was the primary renewable source in the project, wind intensity was measured using an anemometer from seven observations zones to understand if turbine locations associated with the site wind potentials [14].

To comprehend the social aspect of the model, thirty-four users at various locations were approached randomly during the site observation. A small survey was conducted asking two questions addressing if they reside near the vicinity and noticed the park's capacity for renewable electricity production. The outcome of the survey informed a better understanding of park's local or regional user group against their awareness about the park's renewable energy aspect [14].

Micro wind turbines technical specification notes and one interviewee's energy calculations were used to inform the turbines' life cycle and energy payback time which helped to recognize environmental strategies of the model.

Figure 2. Renewable energy sculpture incorporated eight micro wind turbines was made from the former tank 101 recycled steel structures. Image by the author.

3. Results

Copious amount of sustainability assessment tools are used for different human environments, ones specific to landscape architecture profession are inadequate. While the effectiveness and leadership of assessment tools are key to energy future, current tools including the 'SITE', disregards a comprehensive sustainability assessment of public space on-site renewable electricity.

Next section presents theories and principals from ecology substantiating the strategies of potential assessment model, its components—and finally, its performance and outcomes after tested against the case study.

3.1. Theories behind the Potential Assessment Model: The Link between Energy, Information and Public Space

To provide food, water and shelter and to improve the quality of our life, energy is required. Energy comes in many forms, and the science of thermodynamics assesses energy transfers and transformations in processes, systems and devices. Two laws in principle rules the nature of energy. According to the first law of thermodynamics, energy cannot be extinguished or created and only its state can be altered and kept. The second law of thermodynamics can be defined within these states. Respectively, when the state of useful energy, the work capacity, a.k.a., 'exergy' dissipates, the state of disorder, a.k.a., 'entropy' arises [15].

Howard T. Odum, a renowned 20th century ecologist, introduced the ecosystem concepts of energy hierarchy and energy flow to the human environment. His early ideas on energy and nature stem from a provisional hypothesis—'tripartite altruism' which was later developed to 'emergy' concept, a.k.a., 'the quality of energy' [16]. Emergy refers to the history of the system and the time and processes that have contributed to its current state [17]. The letters 'em' of emergy represents 'energy memory' and relate directly to the now more common term 'embodied energy'.

Odum's energy-nature equation so-called 'tripartite altruism' postulates that organisms in nature uses their energy based on 1/3 ratio comprising higher order, lower order and, maintenance and storage purposes. He presented this theory using rabbits as an example illustrated in Figure 3. According to this theory, "They eat grass to live, grow and reproduce. Their manure fertilizes the grass that feed[s] them and they 'sacrifice' weak rabbits to predators to keep the population fit and in balance" [18].

Figure 3. 'tripartite altruism' represented in rabbit's lifecycle.

Such formula can be useful when the public space is purposed to be designed toward an ecological sophistication. Particularly, Tripartite altruism's higher order, lower order and storage and maintenance strategy is simply about distributing the available energy on the system equally between three types of needs.

This study claims that active and passive interaction with renewable energy is an imperative feature of public spaces to establish a give-and-take connection between society and energy and to reach an evocative and assessable mode of sustainability. Similarly, the maximum empower principle also known as the Odum's fourth law states that during the self-organization process, systems advance interactions, processes and parts that amplify efficiency and production [19–21]. Odum's fifth law further asserts that "system processes maximize power by interacting abundant energy forms with ones of small quantity, but a larger amplification ability" [19]. Thus, design processes and outcomes targeting an ecological sophistication in public space can be used as a tool to educate the public about a sustainable energy lifestyle and to increase general environmental awareness, so maximizes energy efficiency and production in the broader community, leading to long-term benefits.

Focusing on the social aspect of public space where people mingle through communication, interaction and learning from their surroundings and from each other, Miller [22] argues "[p]ublic spaces do not exist as static physical entities, but are constellations of ideas, actions and environments". Therefore, public space can shift society into a sustainable energy lifestyle when used as an educational and information platform. When conceptualized as a self-organizing state, public space can become, in Odum's terms, "a platform to create useful information for community", in doing so, endorse a larger commitment to sustainable energy across multiple social realms.

3.2. *The Potential Assessment Model: Components*

Theories explained in the previous section substantiate a set of strategies for the assessment model to assess the sustainability of renewable energy infrastructure in BPP. While author's previous study focused on the OED framework's 1/3 ratio as a rigid constraint for scoring, this study slightly modifies wording for a real built case and proposes a flexible, but meaningful distribution of produced renewable electricity. In other words, a project can fulfil electricity distribution not equally in a quantitative sense, but still distribute electricity between social, economic and environmental strategies while the outcome can be more synergistic and transcends a quantitatively balanced approach.

Accordingly,

- Economic strategy refers to using one third or meaningful part of the on-site produced electricity to be sold to city grid for generating income to the public space and its local community;
- Environmental strategy suggests whether the design considers embodied energy, energy storage and maintenance of the renewable energy infrastructure and/or public space. That is, one third or meaningful part of the produced total electricity throughout the life time of renewable energy infrastructure needs to pay back its maintenance cost and embodied energy considering its depreciation value;

- The model's final one third or meaningful part of the produced electricity focuses on creating social strategies using on-site renewable energy. That is whether the design considers any active and/or passive engagement using public space renewable energy.

More precisely, the economic strategy is quantified by using a scale from zero to four points. Since each case study site would have different energy distribution needs, appropriateness of the distribution strategy would score the highest point. The project scores zero point if no generated electricity within public space is used for local energy trading. The project scores one point if designers are aware of and have the intention to implement the model's economic strategy however project fails to do so for several reasons as some of them are exemplified and presented in this case study. If the strategy is implemented, but unmeasured and unmonitored the project scores one and a half points. A project scores two points if total production capacity is used for local energy trading [10]. A project scores three points if environmental or social strategy is considered in addition to economic engagement. A project scores three and a half points both environmental and social strategies are considered in addition to the economic strategy. A project scores four points if full generated electricity amount is equally and/or meaningfully distributed between three assessment strategies.

To comply with the environmental strategy, the project shall consider embodied energy, energy storage and contributes to the maintenance of the infrastructure and/or the public space. A project scores zero point, either designers are not aware or consider any above factors. A project scores one point if an intention for addressing all three factors of environmental strategy is clear however project fails to achieve this. If an intention for addressing two of the three factors is obvious, the score would be 2/3 point. If an intention for addressing at least one of the three factors is apparent, the score would be 1/3 point. A project scores two points, if successfully and fully addresses one of the three factors. A project scores three points, if any two factors are fully addressed. A project scores four points, if all three factors are considered simultaneously.

The model's social strategy assesses whether public space renewable energy generates both active and passive engagement with public. Educational, informative and place making activities can involve active and/or passive engagement with on-site renewable energy. Here active and passive engagement is twofold.

The first one is about using on-site renewable electricity through either production or consumption activity within public space. An example of active renewable electricity production is Piezoelectric generator. Active renewable electricity consumption activities comprise mobile phone charge, car battery charge, using electricity for street performance while passive renewable energy consumption may include digital information screens where users consume electricity in an indirect manner.

The second one refers to 'energy quality' concept introduced earlier in the theory section which is literally about 'quality design' or quality of the renewable energy infrastructure. More specifically, how the design communicates with public, what diverse uses intentionally and unintentionally present, observable or hidden. Active engagement with renewable energy may include a meta sustainability narrative the design is trying to communicate regardless of functioning as a renewable electricity generator. Interpretative signage of renewable energy may be incorporated into the design to communicate with public at a symbolic level. Whereas, passive engagement involves spontaneous public space activities, for example the design is used as a backdrop for market stall installation, photography and performances. Or the design triggers questions in public about its function, evokes several other meanings and promote multifunctionality.

Respectively, a project scores zero point, if renewable energy design does not generate any engagement with public whatsoever. A project scores one point if any intention creating either active or passive interaction is apparent, but the project is failed. A project scores two points if any intention creating both active and passive interaction is obvious, but the project is failed. A project scores three points, if either active or passive engagement is considered. A project scores four points, if both active and passive engagements using on-site renewable energy are considered. Points in each segment would be divided into two–one covers using on-site renewable electricity and other one concerns with parameters related to design quality of the renewable energy infrastructure.

Table 1 in the following section presents the assessment scoring in detail while incorporating information from this prescribed text.

Table 1. Ballast Point Park renewable energy infrastructure sustainability assessment scores.

		Economic Assessment Score (4)	Environmental Assessment Score (4)	Social Assessment Score (4)	Top Assessment Score (12)
Sustainability Strategies					
Economic	-Local energy trading [10]	1	n/a	n/a	1
Environmental	-Embodied energy	n/a	2	n/a	2.7
	-Energy storage	n/a	1/3	n/a	
	-Maintenance of Infrastructure and/or public space	n/a	1/3	n/a	
Social	-Passive interaction modes with renewable infrastructure	n/a	n/a	2	4
Strategy	-Active interaction modes with renewable infrastructure	n/a	n/a	2	
			TOTAL		7.7

Economic Sustainability Strategy—Local Energy Trading *
(0) No local electricity trading
(1) Designer's intention exists, however project fails
(2) All generated electricity used for local trading
(3) Part of the generated electricity used for social and environmental strategies **
(4) All generated electricity distributed equally/meaningfully between each strategy

Environmental Sustainability Strategy—Embodied Energy/Energy Storage/Self Maintenance ***
(0) None
(1) Designer's intention exists; however, project fails ****
(2) If project successfully and fully addresses one of the three factors.
(3) If any two factors are fully addressed.
(4) A project scored four points, if all three factors are considered simultaneously.

Social Sustainability Strategy—Active/Passive Engagement with on-site renewable energy ^ ~
(0) None
(1) Designer's intention exists for Active or Passive engagement through direct electricity consumption ^^ or Production ^^^, however project fails to do so.
(2) Designer's intention exists for Active and Passive engagement through direct electricity consumption or production; however, project fails to do so.
(3) If project fulfils active or passive engagement using on-site renewable electricity.
(4) If project fulfils both active and passive engagements using on-site renewable electricity.

* If the strategy is implemented, but unmeasured and unmonitored, the project scores one and a half points; ** While social and environmental strategy scores (1) point, each strategy scores (1/2) point; *** Self maintenance factor is about the maintenance of energy infrastructure or/and public space, this includes primary electricity demand for permanent functions in public space such as lighting, heating, energy storage, bikeshare, electricity for pump, irrigation, etc.; **** Each three factors will get 1/3 of the score if considered in the design, but fails; ^ Educational, informative and place making activities can cover active and/or passive engagement with on-site renewable energy; ~ Points in each segment would be divided into two–one covers using on-site renewable electricity and other one concerns with parameters related to the design quality of renewable energy infrastructure. ^^ Active electricity consumption for public space users such as mobile phone charge, wireless use, car battery charge, etc. whereas passive electricity consumption for example digital information screens ^^^ For example, Piezoelectric generator produces on-site electricity.

3.3. *The Proposed Assessment Model: Performance and Outcomes*

Table 1 illustrates economic, environmental and social strategies against assessment scores for the BPP renewable energy infrastructure. The highest score for renewable energy infrastructure assessment is calculated as twelve points. BPP scores a total of seven point seven from twelve points which approximates a success rate of sixty-four percent.

3.3.1. Ballast Point Park Renewable Energy Infrastructure Economic Sustainability Assessment

- Renewable energy's economic benefit was an incentive to include the concept in the park; however, the integrated wind turbines did not function and did not generate any electricity [14,23–25].

While BPP was a well-regarded project that fulfilled many sustainability aspects, specific to local economy of the park, little information was found. Minor activities in the park such as wedding ceremonies support park's local economy and a recent study has revealed a value increase over fifty percent on surrounding properties since the park opened to public in 2009 [14,26].

Economic benefit from on-site generated electricity was initially planned to balance out the operational cost of the park, however micro wind turbines installed did not operate as anticipated [14,23–25]. Annual wind data were not incorporated into the process of decision making since no calculations were done by renewable energy consultant. Wind turbines' technical specification notes were used as the only primary source to predict possible yield. However, to anticipate the wind turbines' possible location, local wind measurements were needed which were never realized [23]. Following upon this gap, also to better understand reasons behind the dysfunctional wind turbines, wind measurements were taken by using an anemometer at seven different observation locations during site trips as presented in Figure 4a.

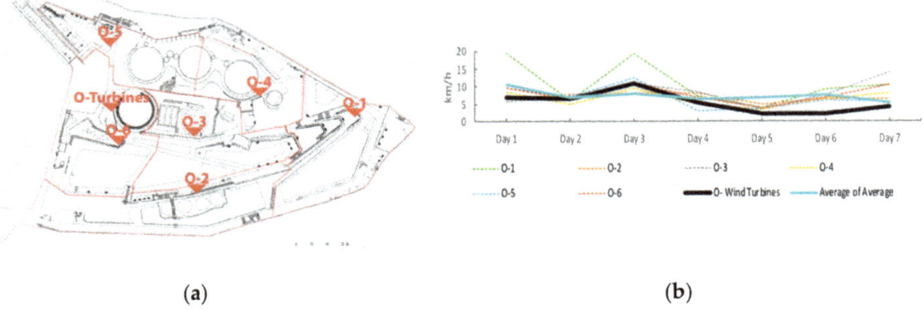

(a) (b)

Figure 4. (**a**) BPP location map showing observation spots; (**b**) graph showing the daily averaged wind values for each spot.

During regular visits for two weeks in January 2014, no spinning movement was monitored on any single wind turbine despite seeing the park at some extremely windy days.

Figure 4b presents a graph showing daily averaged wind values for each spot, respectively while Table 2 demonstrates each zone's average wind values during one week period. Each value in Table 2 represents daily average wind speed calculated from 7:30 a.m. to 7 p.m. by moving every twenty minutes from one observation location to another. Black thick line in Figure 4b represents the turbines' average wind speed at their exact location where wind values repeatedly stayed below the other measured average wind speeds. While local wind measurements are crucial to determine the turbines' exact location, the economy of generated electricity requires yearly wind data since every wind turbines including very successful ones show strong seasonality in its output. Therefore, data presented in Table 2 is indicative for a general understanding of site wind potentials and so possible ideal locations of the turbines.

According to the model's economic sustainability assessment, BPP scored one from four points. In this study, local electricity production's economic value was an essential strategy for sustainable renewable energy-embedded public spaces and BPP has failed to do so. BPP has accomplished critical success in its own right and has deservedly won many awards for addressing numerous environmental sustainability criteria, but in the context and for the purpose of this study,—when assessed against the more thorough renewable energy economy potential which the proposed model shows here—it fell short.

More specifically, the model indicated an equation using 1/3 or meaningful amount of the produced electricity for local energy trading which could bring income to the community. However, at the time, the technology was new and underdeveloped. Generated electricity was not efficient for providing enough economic incentive [23–25]. Such shortcoming was also discussed later as one limitation of the model. Presumably, past technologies would not able to address this strategy fully, therefore for the time being, designer intentions are invaluable to understand whether this strategy originally included in the project's agenda.

Table 2. BPP daily local and average wind values.

Date	O—1	O—2	O—3	O—4	O—5	O—6	O—Wind Turbines	Average of Average
Day 1	19.7	7.1	5.6	8.5	9.6	9.8	7.1	10.9
Day 2	6	7.6	6.9	4.8	7.4	6.5	6.4	7.1
Day 3	19.3	8.4	10.8	9.7	12.7	11.2	11	8.2
Day 4	7.4	8	8.6	4.9	3	7.1	5.2	6.6
Day 5	3.9	4.9	3.5	3.4	3.9	3.8	2.1	7.0
Day 6	9.5	6.7	7.8	6.7	6.3	6.8	2.1	7.9
Day 7	10.4	6.7	14.1	8.1	5.9	10.4	4.3	5.5
Average	10.9	7.1	8.2	6.6	7.0	7.9	5.5	7.6

O—N: observation location.

3.3.2. Ballast Point Park Renewable Energy Infrastructure Environmental Sustainability Assessment

According to the model's environmental assessment, BPP shall consider embodied energy, energy storage and the maintenance of the infrastructure and/or park successfully.

- The park's 'cradle-to-cradle' (CTC) design philosophy favorites embodied energy consideration of the renewable energy infrastructure.
- Wind turbines and batteries are part of the renewable energy infrastructure, however none of them functioned and delivered the anticipated energy yield.
- Public space maintenance using on-site renewable electricity was discussed by interviewees however never become a reality.

What was interesting about embodied energy aspect of the BPP renewable energy infrastructure was the 101 oil tank and its recycled steel frames repurposed as a renewable energy sculpture on the basis of park's post-industrial history. However, no life cycle analysis done; designers applied the 'cradle-to-cradle' (CTC) philosophy to the park's overall design at multiple scales indicating a focus on environmental sustainability for design decisions. According to the model, embodied energy prevails important implications for renewable energy devices. Embodied energy of a renewable energy infrastructure refers to 'energy pay back times' (EPT); in other words, the energy and revenue spent on a renewable energy device until constructed and functioning [27–29].

Commenting on EPT, one interviewee presented his preliminary calculations referring to twenty-nine years pay-back time when eight micro wind turbines spin at 10 m/s (36 km/h–Force 5 fresh breeze). The wind turbines catalog used in the project specified each turbine generating 800 W at 14 m/s [23]. Considering the latest developments in renewable energy technology, EPT has significantly dropped nowadays in comparison to eight years ago. Thus, despite the limitations of past technological shortcomings, designers were aware of the topic and considered embodied energy. In the case of BPP, the CTC design philosophy was applied to the design of energy sculpture. When assessed by the proposed model, the CTC approach taken by designers addressed embodied energy, which scored two from four points and partly fulfilled the model's environmental strategy. Designers used pre-existing structures and materials from the site during the construction phase, thus reduced the art piece's embodied energy cost.

Concerning wind turbines and energy storage, data from site observation and interviews revealed that neither batteries nor turbines functioned. Turbine moving segments have rusted over time because the sprayed sea salt prevented turbines spinning [25]. As originally calculated by the distributer, wind turbines' power would have sufficed park's energy need. However, to acquire full performance from the installed turbines, several criteria needed to be addressed. For example, determining the most productive wind harvesting location was a crucial strategy to be fulfilled which was not implemented. Therefore, the project scored 1/3 point partially addressing this strategy.

What stood out from an environmental assessment of the relation between maintenance and renewable energy infrastructure, designers intended to use generated electricity from the renewables for reducing park maintenance costs. However, because any technical and expert knowledge was lacking, the renewable energy devices did not function as intended, and did not meet the goal of supplying electricity to the park for daily use [23–25]. Thus, the park did not generate electricity for itself or the community, and thereby neglected both economic and partially environmental (self-maintenance) aspect of on-site electricity production. From this finding, the project achieved a score of 1/3 point. The environmental sum total was then calculated as 2.7 from 4 points.

3.3.3. Ballast Point Park Renewable Energy Infrastructure Social Sustainability Assessment

- No active and/or passive engagement with on-site renewable electricity was recorded on the site. This has also never become the designers' intention [23,25].
- Active engagement was generated through 'quality design'; powerful symbolic meaning from a former fossil fuel tank to a renewable energy sculpture; interpretive signage.
- Passive engagement was generated with renewable infrastructure through spontaneous use.
- Designer's esthetical concerns, community will and site wind exposure influenced the decision for using wind turbines in the design.

Throughout regular site visits over two weeks in January 2014, no spinning movement was monitored on any single wind turbine, owing to this, no active or passive engagement with on-site renewable energy was recorded. Moreover, designers did not intend creating activities in public space using on-site generated electricity [23,25]. However, transforming a former fossil fuel tank to a renewable energy sculpture conveyed a powerful symbolic meaning which was the main design intention. The principal designer emphasized his view on the societal impact of the designed renewable energy sculpture with his own words "the big fossil fuel tank turned into the biggest wind turbine thing on Sydney Harbor. There is poetry there" [14,25]. Such narrative was also reflected in using interpretive signages across the park as seen in Figure 5. In doing so, the project scored one point through addressing the 'quality design' strategy–active engagement side of renewable energy.

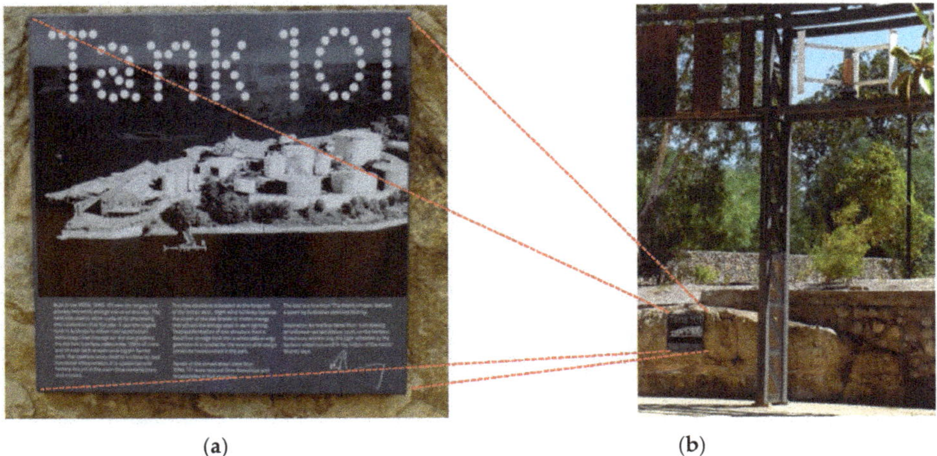

Figure 5. (**a**) The interpretive signage of renewable energy sculpture; (**b**) The renewable energy sculpture. Images by the author.

The assessment of renewable energy passive interaction targeted design multifunctionality. Several monitored and recorded activities as well as anecdotes from public users provided ample data to score another point for assessment. For example, the renewable energy sculpture was used for wedding photography in Figure 6a and its structure was pinned with geo-tags at several locations in Figure 6b. Anecdotally, the sculpture together with some other park divisions were used as meeting points for parkour performers.

(a) (b)

Figure 6. (**a**) Renewable energy sculpture used as a background for wedding photography; (**b**) geo-tags were pinned to the renewable energy sculpture. Images by the author.

During site visits, the survey presented in Figure 7 was completed with thirty-four participants. Only 24 percent of the park users were aware of the park's capacity to produce electricity, and these park users were living in the vicinity.

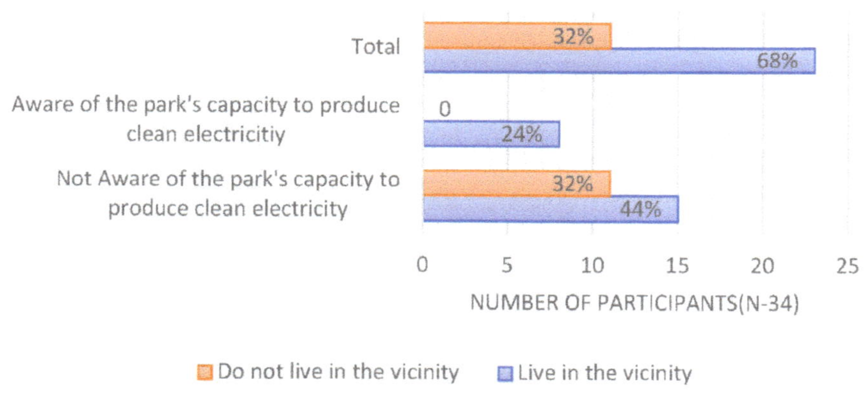

Figure 7. Survey results revealing user's response to the park's renewable energy.

This finding was further elaborated with interview data for example using on-site produced electricity to encourage public space activities was never brought up in any community consultation process. However, during these meetings some initial ideas were seeded for example using the wind as a choice for renewable source and retaining the former oil tank on the site. As discussed with the principal designer, the idea of generating on-site electricity originated from the community and its stories about flying kites on the site because of strong wind exposure. The idea of keeping the former tank was originally suggested by community to be used as a landmark for captain's sense of direction while roaming in the harbor. These ideas together with designers' esthetical concerns further formed the final design which was developed in a design charette and the former tank turned into a renewable energy sculpture [25].

According to the assessment model, simply relying on the renewable energy's environmental and economic aspect was not sufficient to create genuinely sustainable renewable energy embedded public space. An important factor in accomplishing social outcomes of renewable energy use was its public acceptance [30,31]. For example, photovoltaic (PV) panels, despite their controversial esthetic qualities, needed to be visible to users to increase their social acceptance [32]. These esthetic concerns were reiterated in the case study that designers were hesitant to integrate PV panels into the BPP design [23]. Moreover, because most park users did not know the park had the capacity to produce electricity, the socially sustainable knowledge and acceptance of renewable energy was limited [14]. Therefore, some issues emerging from these findings were related specifically to the designers' priority of focusing on the esthetic and technological aspect of renewable energy, rather than on the potential socioeconomic relations stemming from its local production.

From a social sustainability viewpoint, BPP's renewable energy scored two points through active and passive engagement. Several aspects of which referred to 'quality design'—by definition design quality of the energy sculpture and its well-thought narrative, but not from on-site produced electricity as clearly presented in this section.

4. Discussion

Current sustainability assessment tools specific to the Landscape Architecture profession devalue public space renewable energy use by neglecting the potential social, economic and environmental engagements of local electricity production that could otherwise increase the sustainability of the public space. One common issue that emerges from the BPP case study findings as well as the reviewed literature is that although a broad range of sustainability assessment tools have been used for different human environments, ones specific to the Landscape Architecture profession are limited.

Therefore, for the purpose of this study, the SITES initiative is selected as a typical example which is a legitimate well-regarded sustainability assessment tool in the profession. As discussed, the sustainable SITES initiative assesses the sustainability of public spaces with a point-based system [7]. However, SITES' assessment approach to renewable energy-embedded public open spaces is one-sided, often misses the social and economic relations around electricity production and, thus, only ticking the environmental sustainability box. This may be because SITES' focus is only on the renewable energy devices, rather than on the activity of electricity production. The SITES' limitation to assess the renewable energy side of a public space led the development of a potential future assessment tool as presented through the BPP case in Sydney.

Having a device-based focus on renewable energy is symptomatic of a top-down sustainability development approach. A possible explanation for this may be ingrained in the definition of 'techno-fix', which holds that present market conditions generally satisfy demand with quick solutions to increase consumerism [33]. Respectively, urban renewable energy is widely framed as a technological addendum, an afterthought, by definition a retrofit to public space designs. Currently renewable energy infrastructures embedded in urban environments have a relatively small electricity production capacity. In addition, energy storage utilities are still expensive, and distributed energy infrastructure systems in cities around the world are not widespread. For the time being, existing industry practices may be reluctant to make use of producing clean electricity in cities at maximum potential. However, such parameters indicate only present day constraints and it is foreseeable that these constraints will be overcome in the very near future. Therefore, a balanced quantity between the proposed strategies may gain more credit if not now, but later.

The presented model promotes social, economic and environmental engagement using on-site renewable energy. For this reason, it requires negotiation among many stakeholders, including policy makers, local government, neighborhood residents, public space users, designers and experts involved in the project. It makes the design process more transparent and gives each stakeholder equal opportunity for management and/or involvement. It is likely, therefore, that designers can be more cognizant of the sustainability of renewable energy. This in turn, can reduce the propensity for superficial 'green wash' outcomes and increase the importance of social and economic factors involved in electricity production.

The model can be scaled for household and neighborhood uses. For example, researchers can identify how much clean electricity a household generates; sells to the city grid; uses for its own needs including maintenance, storage, ordering, installation, transportation, as well as its own embodied energy. The electricity's specific consumption also becomes the focus of some recent studies that investigates integrated solutions for households' peak energy demand [34]. The proposed model can guide a multidisciplinary research team in specifying the social, economic and environmental engagement strategies according to a household's energy needs. Simply put, a balance is needed between these three aspects. The model can also be used for long-term sustainable development to evaluate the efficacy of renewable energy investments for impoverished communities. It may also benefit energy independence studies for energy poverty alleviation in remote settlements [35].

The model's economic engagement strategy not only fosters a surplus to community but can indicate a green rating for the neighborhood. For example, some emerging distributed energy neighborhoods have energy information centers. Their main roles are marketing and advertising the engagement with renewable energy and promote sustainable energy education. These neighborhoods are widespread in some European countries such as Austria, Germany, Denmark, Sweden and the Netherlands [36–40]. Public spaces designs in line with the thinking promoted through the proposed model here can showcase its neighborhood electricity use as a public space amenity. A distributed energy market economy is reliant on smart grid systems, which require active interaction with its stakeholders [41]. Therefore, each public space design can support such interaction addressing site-specific social, environmental and economic data while integrating renewable energy technology to support a local energy economy, monitoring electricity production and consumption, and informing residents about their energy use, with a view to helping them to be more energy responsive. Urban administrative policies can then use such data to channel subsidies for energy responsive neighborhoods and encourage the transition to sustainable energy.

Another local energy trading application is so-called transactive energy which uses the existing infrastructure for renewable energy distribution. A real world example was tested through Monash Micro grid case study in Melbourne, Australia where a complete hardware and software infrastructure are staged for micro-grid energy management [10]. Furthermore, some alternative administrative tools are evolving to provide on-site energy services. For example, the sustainable energy utility (SEU) offers on-site energy services to empower private energy markets and facilitate self-finance for communities [42,43]. However, the presented model sheds new light on the community energy distribution using public space as a point of departure, it requires side information, policies and administrative innovation similar to the SEU for better integration within human environments.

The proposed model's environmental assessment strategy formed a significant connection between embodied energy, energy storage and maintenance of the infrastructure and/or park. The BPP designers seemed to be more aware of such connection, because environmental, social and economic strategies using renewable energy scored highest to lowest, respectively. In the previous study of the author, twenty-five speculative designs submitted to the LAGI design competition were assessed against very similar strategies whereas the findings revealed an entirely different outcome—the shortlisted projects' renewable energy based economic, environmental, and social strategies scored highest to lowest average, respectively [8]. This outcome perhaps suggests environmental interventions in general prevail dominant trend for the mainstream urban sustainability practices which is clearly the case as presented in this study. However, LAGI's philosophy on renewable energy and public space is highly responsive, their approach to the sustainability esthetics, as noted by Elizabeth Meyer, needs to be revisited. Meyer states that the esthetics of sustainable design need to be considered within two contexts—'appearance' and 'function'—to have a significant cultural and societal impact. She argues that 'function', also referred to as 'fitness' or 'performance', cannot be experienced through representation, but through direct engagement [44].

In this study, direct engagement with renewable energy—recognized as an important attribute of public spaces—was presented as active and passive interaction with renewable energy. 'Active interaction' referred to the direct consumption of produced electricity; for example, wireless services and energy charging points. 'Passive interaction', whereas, referred to the indirect consumption of the produced electricity, such as video artworks and interactive energy signage. Another dimension of active and passive interaction with renewable energy was 'quality design'. In other words, how designer intended to communicate with public about the renewable energy design in the first place. In BPP, active interaction was depicted through a successful meta-narrative, interpretive signages and use of renewables across the public space. Whereas, passive interaction involved spontaneous uses of the design concerning multifunctionality and so obliquely related to the renewable energy feature; fo example, using the structure as a backdrop for a street performance or photography exhibition.

In the light of the introduced theory, this study argued that such enhancements in our interactions with renewable energy in public space correlated with the observed tendencies of self-organized mature ecosystems. For example, Odum's fifth law of thermodynamics states that, "system processes maximize power by interacting abundant energy forms with ones of small quantity, but a larger amplification ability" [19]. Therefore, the more ecologically sustainable public space was the one that more effectively responded to the fifth law by engaging with renewable energy at the highest level of interaction, that was, through both active and passive means. The central premise here was that, the greater the number of active and passive interactions between renewable energy and public spaces, the greater the likelihood that renewable energy would influence society's sustainable energy lifestyle.

The presented model and its informed social strategies concern with public space engagement and place-making through renewable energy and its use in and around the public space. This use can be elaborated through site specificity, designer's approach and community wishes. Parallel urban trends are evident in the fields of media architecture, urban informatics and interactive media and public art, which can be used with generated on-site clean electricity for advancing the message of renewable energy.

5. Conclusions

This study investigated the link between renewable energy and public space and developed a potential integrated model to assess the sustainability of public space renewable energy infrastructure. The model was tested through assessing a renowned public space awarded for its environmental sustainability innovations. Since incorporating renewable energy use in urban public space is an emerging field, this study is built upon author's previous work on assessing sustainability of speculative designs in which findings revealed an imbalance for distributing onsite generated electricity from a public space context.

The devised model, strategies and related ecological theories are the key contribution of this study. Sustainability practitioners have not developed a full understanding of public space renewable energy use beyond it being a mere design retrofit. The presented strategies in Section 3.2 would be useful for designers to better incorporate renewable energy into their designs. Such strategies would also benefit public space assessment bodies such as the SITES initiative. Importantly, the link between three strategies in this study was argued to become indispensable for an ecologically sophisticated public space design.

Another crucial contribution is the acknowledgement of the importance of using generated electricity immediately within public spaces through active and passive interactions. An ecologically sophisticated public space design should be energy responsive in the sense that it increases the social, economic and environmental sustainability of its locale. At the same time, such a public space may generate useful information for the community, which can then be used as a basis for sustainable energy transition and social change. This claim stems from the "maximum power principle", which is considered as Odum's fourth law of thermodynamics. Odum's fifth law also posits that active and passive interactions are imperative for on-site renewable energy, and, respectively, the more these interactions occur in public spaces, the greater the chances the public space would help society transitioning to a sustainable energy life style.

Finally, the study presented the need for an integrated approach to renewable energy and public space, to achieve not only meaningful and measurable sustainability, but also to communicate the reciprocal relations between society and energy.

5.1. Limitations

The model's major limitation is the devised strategies and their appropriateness with the renewable energy project. While these strategies explicitly engage with renewable electricity at multi-level; site-specific features, designer's esthetical concerns, decisions on renewable energy technology types and many other unforeseen factors may influence full exploitation of renewable energy as presented through the BPP. From an overall sustainability perspective, BPP is successful on its own right. However, when the renewable aspect of the project is scrutinized, a different outcome becomes evident.

Another limitation of the study is time factor. Because the project examined was built in 2009, at the time renewable energy technologies were less efficient and economically infeasible considering its highly expensive manufacturing cost against lower production capacity. Therefore, in the past a design fully responsive to the proposed model' strategies in an urban condition was not practical and was perhaps impossible. This is why the project used the BPP as a test bed, merely to demonstrate the model's function rather than focusing on the project flaws.

5.2. Future Research

A natural progression of this study is to increase the built project case study sampling. In turn, this will increase the assessment model's generalizability. This study simply presents the first project from a wide range of potential sites elsewhere. Initially, the model and its production-oriented approach will be best applied to public open spaces, where local electricity production is used in neighborhoods to create a more sustainable community.

In the future, opportunities may arise to improvise other interactions with generated electricity. Until that time though, the balanced and/or meaningful distribution of renewable electricity generated in public space should be served as a blueprint as long as site factors permit. Such thinking presented here would enable designers engage with the project's renewable energy at the outset rather than considering it an afterthought. Conceptualizing the public space interactions and renewable energy simultaneously while targeting an equal and/or meaningful balance between social, economic and environmental strategies would draw an effective path for transitioning to a sustainable energy future.

Rapid technological advancements on renewable energy coupled with increasing rate of efficiency and decreasing costs would accelerate transforming our understanding of urban renewable energy from a technological addition into a community production activity. This shift in thinking should be first embraced by design and assessment practices to create a sustainable energy future. Cities around the world are gradually taking steps toward building new social, economic and political frameworks and infrastructures such as resilient micro and smart grids, virtual renewable energy utilities, sustainable energy utilities and distributed energy neighborhoods. The spatial and social adaptation of these technologies in cities is as critically important as is their technical capability.

Cities are in crisis. Their increasing energy demands require radical sustainable design solutions at the supra-national to regional and regional to local scales. Given today's concerns with climate change and energy future, there is a need for energy autonomous cities. While we are at the verge of twenty-first century, information and energy technology has become the central tenants. Energy future and information technology are quite interlinked. Such link was noticed by many scholars from the late nineteen and early twenty century to present day. Departing from such link, this study presented a potential sustainable path to an ideal energy future by introducing a potential energy assessment model for public space renewable energy infrastructure building on the renowned ecologist Howard

T. Odum's early theories. To generate local energy economies and create smart learning spaces in addition to the social and ecological function of public space, this study conceptualized public space as an innovation and transformation hub for social and environmental change while considering advancements in renewable energy technologies and their increasing production rates. A transition from fossil fuel to renewable energy is a significant start on the road to future energy autonomy but requires systemic bottom-up and top-down policies and interventions to create a truly sustainable society. The outcome of this study is one such intervention that can inform and advance this critical transition process.

Funding: This research did receive a fee of waive from Energies Journal, many ideas presented here was seeded in my PhD which was funded by Australian Postgraduate Award (APA).

Acknowledgments: I would like to thank Ian Weir, Debra Cushing and Laurie Buys for their support during my PhD process. I would like to express my deep gratitude to the interviewees and Mcgregor + Coxall team for their time and support. I would also like to thank three reviewers for their constructive feedback helping me to improve this manuscript.

Conflicts of Interest: The author declares no conflict of interest.

References

1. Droege, P. *100 Per Cent Renewable: Energy Autonomy in Action*; Earthscan: London, UK, 2009.
2. Moldan, B.; Janoušková, S.; Hák, T. How to understand and measure environmental sustainability: Indicators and targets. *Ecol. Indic.* **2012**, *17*, 4–13. [CrossRef]
3. Byrne, J.; Martinez, C.; Ruggero, C. Relocating Energy in the Social Commons: Ideas for a Sustainable Energy Utility. *Bull. Sci. Technol. Soc.* **2009**, *29*, 81–94. [CrossRef]
4. Dincer, I. Renewable energy and sustainable development: A crucial review. *Renew. Sustain. Energy Rev.* **2000**, *4*, 157–175. [CrossRef]
5. Cairns, J. Sustainability and specialisation. *Ethics Sci. Environ. Politics ESEP* **2004**, *4*, 33–38. [CrossRef]
6. SITES, T.S.S.I. *Guidelines and Performance Benchmarks*; The American Society of Landscape Architects: Washington, DC, USA; The United States Botanic Garden: Washington, DC, USA; The Lady Bird Johnson Wildflower Center at the University of Texas: Austin, TX, USA, 2009.
7. SITES, T.S.S.I. *SITES v2 Rating System For Sustainable Land Design and Development*; The American Society of Landscape Architects: Washington, DC, USA; The United States Botanic Garden: Washington, DC, USA; The Lady Bird Johnson Wildflower Center at the University of Texas: Austin, TX, USA, 2014.
8. Ozgun, K.; Weir, I.; Cushing, D. Optimal Electricity Distribution Framework for Public Space: Assessing Renewable Energy Proposals for Freshkills Park, New York City. *Sustainability* **2015**, *7*, 3753–3773. [CrossRef]
9. Monoian, E.; Ferry, R. LAGI 2012 Design Guidelines. In *Regenerative Infrastructures: Freshkills Park, NYC: Land Art Generator Initiative*; Klein, C., Ed.; Prestel: Munich, Germany; London, UK; New York, NY, USA, 2013; p. 30.
10. Khorasany, M.; Azuatalam, D.; Glasgow, R.; Liebman, A.; Razzaghi, R. Transactive Energy Market for Energy Management in Microgrids: The Monash Microgrid Case Study. *Energies* **2020**, *13*, 2010. [CrossRef]
11. Silverman, D. *Doing Qualitative Research*, 4th ed.; SAGE: London, UK; Thousand Oaks, CA, USA, 2013; p. 470.
12. AILA. Ballast Point Park, Birchgrove, Sydney. Available online: http://www.aila.org.au/projects/nsw/ballast-point/overview.htm (accessed on 1 December 2013).
13. O'Neill, R.; Consultant, Sydney, Australia; Ozgun, K.; Researcher, Sydney, Australia. Personal Communication, 2014.
14. Ozgun, K.; Cushing, D.; Buys, L. Renewable energy distribution in public spaces: Analyzing the case of Ballast Point Park in Sydney, using a triple bottom line approach. *Jola J. Landsc. Archit.* **2015**, *10*, 18–31. [CrossRef]
15. Rosen, M.A.; Dincer, I. *EXERGY: Energy, Environment and Sustainable Development*; Elsevier: Amsterdam, The Netherlands, 2007.
16. Odum, H.T. Self-Organization, Transformity, and Information. *Science* **1988**, *242*, 1132. [CrossRef]

17. Bastianoni, S.; Marchettini, N. Emergy/exergy ratio as a measure of the level of organization of systems. *Ecol. Model.* **1997**, *99*, 33–40. [CrossRef]
18. Holmgren, D. *Permaculture: Principles & Pathways beyond Sustainability*; Holmgren Design Services: Hepburn, Australia, 2002; p. 286.
19. Tilley, D.R.; Howard, T. Odum's contribution to the laws of energy. *Ecol. Model.* **2004**, *178*, 121–125. [CrossRef]
20. Odum, H.T. *Environmental Accounting*; Wiley: Hoboken, NJ, USA, 1996.
21. Odum, H.T.; Odum, E.C. *A Prosperous Way Down: Principles and Policies*; Univ Pr of Colorado: Boulder, CO, USA, 2008.
22. Miller, K.F. *Designs on the Public: The Private Lives of New York's Public Spaces*; University of Minnesota Press: Minneapolis, MN, USA, 2007.
23. McDermott, L.; Landscape Architect, Brisbane, Australia; Ozgun, K.; Researcher, Brisbane, Australia. Personal Communication, 2014.
24. Kennedy, T.; Project Manager, Sydney, Australia; Ozgun, K.; Researcher, Sydney, Australia. Personal Communication, 2014.
25. Coxall, P.; Principal Landscape Architect, Sydney, Australia; Ozgun, K.; Researcher, Sydney, Australia. Personal Communication, 2014.
26. Kilbane, S.; Toland, A.; Pham, K. *Ballast Point Park*; Landscape Architecture Foundation: Washington, DC, USA, 2017. [CrossRef]
27. Alsema, E.A.; Fthenakis, V.M. Photovoltaics Energy Payback Times, Greenhouse Gas Emissions and External Costs: 2004–early 2005 Status. *Prog. Photovolt.* **2006**, *14*, 275–280. [CrossRef]
28. Hammond, G.P. Energy and sustainability in a complex world: Reflections on the ideas of Howard, T. Odum. *Int. J. Energy Res.* **2007**, *31*, 1105–1130. [CrossRef]
29. Roberts, F. Energy accounting of alternative energy sources. *Appl. Energy* **1980**, *6*, 1–20. [CrossRef]
30. Assefa, G.; Frostell, B. Social sustainability and social acceptance in technology assessment: A case study of energy technologies. *Technol. Soc.* **2007**, *29*, 63–78. [CrossRef]
31. Rogers, J.; Convery, I.; Simmons, E.; Weatherall, A. Local renewables for Local Places? Attitudes to Renewable Energy and the Role of Communities in Place-based Renewable Energy Development. In *Making Sense of Place: Multidisciplinary Perspectives*; Convery, I., Corsane, G., Davis, P., Eds.; Boydell & Brewer: Woodbridge, UK, 2012.
32. Thayer, R.L. Gray World, Green Heart (1994). In *Theory in Landscape Architecture: A Reader*; Swaffield, S.R., Ed.; University of Pennsylvania Press: Philadelphia, PA, USA, 2002.
33. Huesemann, M.; Huesemann, J. *Techno-Fix: Why Technology Won't Save Us Or the Environment*; New Society Publishers: New York, NY, USA, 2011.
34. Buys, L.; Vine, D.; Ledwich, G.; Bell, J.; Mengersen, K.; Morris, P.; Lewis, J. A Framework for Understanding and Generating Integrated Solutions for Residential Peak Energy Demand. *PLoS ONE* **2015**, *10*, e0121195. [CrossRef]
35. González-Eguino, M. Energy poverty: An overview. *Renew. Sustain. Energy Rev.* **2015**, *47*, 377–385. [CrossRef]
36. Dóci, G.; Vasileiadou, E.; Petersen, A.C. Exploring the transition potential of renewable energy communities. *Futures* **2015**, *66*, 85–95. [CrossRef]
37. Hauber, J.; Ruppert-Winkel, C. Moving towards Energy Self-Sufficiency Based on Renewables: Comparative Case Studies on the Emergence of Regional Processes of Socio-Technical Change in Germany. *Sustainability* **2012**, *4*, 491–530. [CrossRef]
38. Scheer, H. *Energy Autonomy: The Economic, Social and Technological Case for Renewable Energy*; EARTHSCAN: London, UK; Sterling, VA, USA, 2007.
39. Van Timmeren, A.; Zwetsloot, J.; Brezet, H.; Silvester, S. Sustainable Urban Regeneration Based on Energy Balance. *Sustainability* **2012**, *4*, 1488–1509. [CrossRef]
40. Wächter, P.; Ornetzeder, M.; Rohracher, H.; Schreuer, A.; Knoflacher, M. Towards a Sustainable Spatial Organization of the Energy System: Backcasting Experiences from Austria. *Sustainability* **2012**, *4*, 193–209. [CrossRef]
41. Ilic, D.; Da Silva, P.G.; Karnouskos, S.; Griesemer, M. An energy market for trading electricity in smart grid neighbourhoods. In Proceedings of the Digital Ecosystems Technologies (DEST), 2012 6th IEEE International Conference on, Campione d'Italia, Italy, 18–20 June 2012; pp. 1–6.

42. Byrne, J.; Taminiau, J. A review of sustainable energy utility and energy service utility concepts and applications: Realizing ecological and social sustainability with a community utility. *Wiley Interdiscip. Rev. Energy Environ.* **2015**. [CrossRef]
43. Houck, J.; Rickerson, W. The sustainable energy utility (SEU) model for energy service delivery. *Bull. Sci. Technol. Soc.* **2009**, *29*, 95–107. [CrossRef]
44. Meyer, E.K. Sustaining beauty. The performance of appearance: A manifesto in three parts. *J. Landsc. Archit.* **2008**, *3*, 6–23. [CrossRef]

Article

Adaptable Source-Grid Planning for High Penetration of Renewable Energy Integrated System

Ming Tang [1], Jian Wang [2,*] and Xiaohua Wang [2]

1. College of Electrical Engineering, Zhejiang University, Hangzhou 310027, China; tangmingtzju@zju.edu.cn
2. Tsinghua Sichuan Energy Internet Research Institute, Chengdu 610213, China; wangxiaohuawxh@tsinghua-eiri.org

* Correspondence: wjqhscnyy@163.com

Received: 4 June 2020; Accepted: 26 June 2020; Published: 28 June 2020

Abstract: To adapt to the growing scale of renewable energy and improve the consume ability of the power system, it is necessary to design a highly adaptable planning scheme for high penetration of the renewable energy integrated system. Thus, this paper firstly gives the conception of system adaptability and designs an adaptability index system, which considers the supply and demand balance, operation state, and network structure of the high penetrated renewable energy integrated system. It can help to comprehensively evaluate the system ability towards uncertain shocks. Then, a two-stage source-grid coordinative expansion planning model is presented. The adaptability indexes of supply and demand balance are used as objection of the source planning stage, the adaptability indexed of the operation state and network structure are used to guide the grid planning stage. The model is further solved based on the coordination between the source and grid planning stage. Finally, the case study verifies that the obtained optimal plan has good adaptability to the impact of renewable energy on the power supply capacity and security operation.

Keywords: high penetration; renewable energy; adaptability planning; source-grid coordination

---

1. Introduction

1.1. Motivation

The pursuit of a sustainable low-carbon society has made renewable energy, represented by wind power and photovoltaics, popularized and applied on a large scale. With the rapid development of related technologies and the breakthrough of key issues, the penetration of renewable energy increases each year [1]. The newly installed capacity of renewable energy units in the world reached 181 GW by 2018, in which wind and photovoltaic power accounted for 83%, while the renewable energy unit supplied more than 26% of load demand [2]. The obvious influence of natural resources and seasonal climate cause the renewable energy unit to have strong volatility and randomness [3]. Variable renewable energy sources, such as solar photovoltaic and wind power, not only change the patterns of electricity utilization on the demand side, but also bring challenges for the balancing and security of power supply on the source side [4]. Considering the scale effect, the uncertain characteristics of multi-spatiotemporal differential distribution of renewable energy are coupled in the operation features of the power system. That seriously impacts the safety and stability of the power system and brings great challenges to the planning and operation. The limitation of network structure and shortage of flexibility resources will further restrict the adaptability of the power system for renewable energy integration. Thus, a source-grid coordinative planning scheme adaptable for the high penetration system is crucial to improve the system adaptability to renewable energy [5].

1.2. Previous Work

To reduce the impact of renewable energy uncertainties on the power system and maximize its advantages of cleanliness, high efficiency, and environment friendly, existing research analyzes the adaptability of high penetration of the renewable energy integrated system [6]. Flexibility is an important theory that is currently widely studied and used to solve the uncertain impact of high penetration of renewable energy. This theory mainly analyzes the ability of flexibility resources in the system to copy with the demand under uncertain operating environments [7]. The characteristics of flexibility such as directionality, time scale, and the key points of flexible assessment were summarized in [8]. A quantitative assessment index of flexibility based on interval evaluation was established in [9]. Based on the flexibility assessment, the flexibility balance mechanism was explored in [10] and various types of flexibility coordination planning models were established. Although flexibility assessment is an important part of the scope of adaptability analysis, it pays more attention to the adaptability of supply and demand balance and lacks consideration of the state of the power system. The impact of high penetration of renewable energy is not only reflected in the supply and demand balance, but also needs to consider the safety, stability, economic efficiency of system operation, and also the affordability of the network structure. The system adaptability characterizes the ability to achieve optimal development in the mutual influence and coordination between itself and external factors, it can more fully reflect the consumption ability for renewable energy integration [11]. Hence, considering the impact of high penetration of renewable energy on other aspects of the power system, [12] analyzed the adaptability of the power system to integrated photovoltaic power plants from the perspective of voltage stability and power flow safety and proposed adaptive expansion measures for reactive power improvement. The influence of the centralized wind farm on the evolution of the self-organized state of the power system is discussed in [13] based on the entropy theory. [14] further considered the probabilistic operating characteristics of high penetration of the renewable energy integrated system and proposed an adaptive evaluation index system from the prospects of safety, economy, and stability. As far as current research is concerned, the system adaptability to the renewable energy integrated system mainly focuses on the adaptability of existing power systems, which belongs to the category of post evaluation and rarely works on the relevant application of adaptability evaluation to the source and grid expansion planning of the power system [15].

1.3. Current Contribution

Based on the above, this paper proposed an adaptable source-grid coordinated expansion plan for high penetration of the renewable energy integrated system with the help of the adaptability evaluation. Firstly, an adaptability index system for high penetration of the renewable energy integrated system towards source-grid expansion planning is presented in this paper, which covers three aspects: adaptability of the supply and demand balance, adaptability of the operating state, and adaptability of the network structure. Among them, the state evaluation model for supply and demand balance is established considering the directivity of supply and demand regulation, diversity of time scale, and complexity of network constraint, then the adaptability index comprising both the insufficient rate and insufficiency of supply and demand is presented. Quantifying the fluctuation and variable operation state of high penetration of the renewable energy integrated system, the weighted entropy index of the expected line load rate that takes into account the safety and economic operation and the weighted entropy index of the line power fluctuation rate to evaluate the stability operation are proposed, then a comprehensive index is designed based on joint weighted entropy to evaluate the rationality of the system operation state. Quantifying the electrical characteristics of the network structure, the weighted entropy index that considers the safety of bus and branch is proposed, then a joint weighted entropy index of the network structure is established to reflect the robustness of the overall structure. Furthermore, based on the proposed index system and accounting for the system safety constraints, power flow constraints, equipment operation constraints, and renewable energy output volatility, an adaptable source-grid coordinative two-stage expansion planning model is proposed, which combines the source planning and the gird planning. While preserving the power balance and regulation capabilities, the multiple attributes of the operation state and network structure are optimized. The matching and adaptability between the source planning scheme and grid planning scheme are enhanced, improving consumption ability of the system to high penetration of renewable energy. The simulation and analysis carried out in the Garver-18 bus test system verify the practicability and effectiveness of the proposed method.

1.4. Structure

The rest of this paper is organized as follows. Section 2 presents the system adaptability index system. Section 3 introduces the adaptive source-grid coordinative expansion planning model and its solution method. The case study is illustrated in Sections 4 and 5 concludes this paper.

2. Evaluation of the System Adaptability to High Penetration of Renewable Energy

Renewable energy is obviously restricted by natural resources and its essential feature is strong uncertainty with multi-temporal distribution. Its large-scale integration will have a strong impact on all aspects of the power system, as shown in Figure 1. It can be seen that under a new background, fast and sufficient regulation resources are the key to ensuring the balance between supply and demand [16].This paper defines the adaptability of the high penetration of the renewable energy integrated system as: under the uncertain operating environment of high penetration of renewable energy integration, power system calls for fast regulation resources to respond to the changes in demand and relies on a robust topology to resist uncertain shocks. On the basis of the balance between the supply and demand under various time scales, the ability to maintain the safe, efficient, and stable operation state of the power system is always maintained. Combined with the above definition, we can see that the evaluation perspective of the adaptability of high penetration of the renewable energy integrated system includes three parts: multi-time scale power supply and demand balance, high-quality and reliable level of operation state, and robustness of network topology.

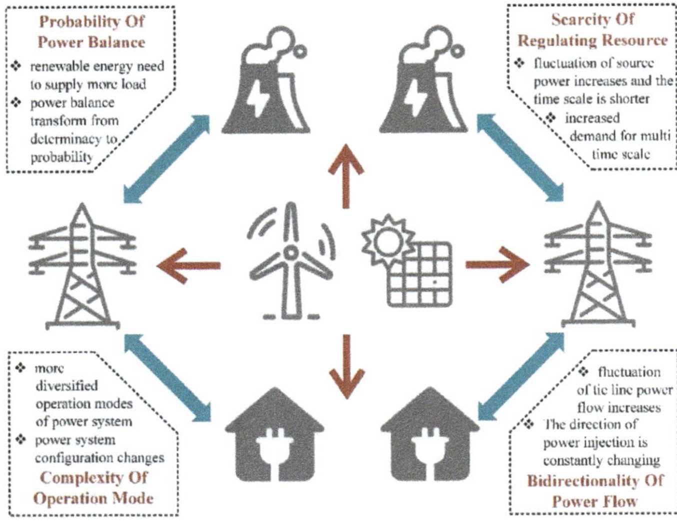

Figure 1. Typical characteristics of high penetration of the renewable energy integrated system.

2.1. Adaptability Index System

The adaptability index system should take into account the typical characteristics of high penetration of renewable energy, which is suitable for the new background with multi-spatiotemporal uncertain characteristics and then effectively characterizes the adaptability to the uncertain impact of renewable energy from multiple perspectives [17]. At the same time, the design of the index system should be oriented to be a guidance criterion for source-grid planning. Here, the proposed index system for the adaptability of high penetration of the renewable energy integrated system is given in Figure 2.

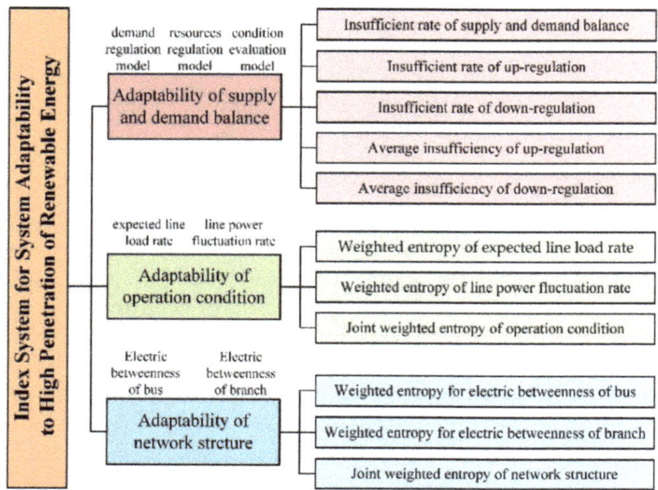

Figure 2. Evaluation index system for system adaptability to high penetration of renewable energy.

2.2. *Adaptability of Supply and Demand Balance*

Adaptability of supply and demand balance means the ability to maintain multi-time scale balance of supply and demand, where the system responds to the regulation demand caused by the fluctuating output of renewable energy by fast regulation resources. In this section, we firstly introduce the basic characteristics, and the quantitative evaluation model of supply and demand balance considering network constraints is proposed. Then, we give the indexes in detail to evaluate the probability and average degree of supply and demand shortage.

2.2.1. Basic Characteristics

(1) Directivity of Supply and Demand Regulation

The regulation demand of traditional power systems mainly comes from the power shortage caused by faults. Specifically, when the power system lacks power supply, spare capacity is required to fill the power gap in time to ensure the supply and demand balance. So, the fault regulation demand has a single directionality, i.e., up-regulation. However, under the high penetration of renewable energy integration, renewable energy sources have become the main source of the power supply and will bear most of the load demand. Due to the strong fluctuation and intermittency of renewable energy, its output power will continuously change with time and the direction of change is uncertain. In this situation, the system needs to be able to quickly adjust the resources to match it, respond to changes in the renewable energy, form a complementary relationship of upward and down-regulation, and ultimately ensure reliable power supply and efficient consumption of renewable energy. Therefore, the high penetration of the renewable energy integrated system has the regulation demand coming from the renewable energy power change and its directionality has the upward and downward bidirectional characteristics.

(2) Diversity of Time Scale

When there is a generator and transmission line failure, it will cause up-regulation demand. However, with the improvement of the safety and reliability of the power system, the probability of its occurrence is very low and the frequency of the up-regulation demand is the same as the frequency of the failure. At the same time, the response time of its regulation is usually required to be within 10–30 min, which is a medium-long regulation time scale. However, because the output power of renewable energy varies with natural resources, the fluctuation is relatively frequent and there are extreme cases of steep drop and steep increase. Therefore, the frequency of upward and down-regulation requirements resulting from changes in the output of renewable energy is relatively higher. Similarly, the response time of the regulation resources is shorter, which belongs to the short-term regulation time scale. Comparing with the temporal characteristics of traditional reserve regulation, the regulation demand caused by renewable energy has a higher frequency and shorter time scale.

(3) Complexity of Network Constraint

The actual regulation capability of the power system is determined by the capacity of the system configuration regulation resource and network structure, the capacity of the regulation resource configuration determines the upper limit of the system regulation capability. In the operation of the system, due to the limitation of network transmission, the actual regulation capability of the power system is often less than the regulation capacity. Therefore, in the analysis of the power supply and demand balance relationship between source and demand, it is not possible to simply use the sum relationship to determine whether the available regulation resources can match the demand. On the one hand, the quantitative relationship can theoretically reflect whether the upper limit of regulation capability can meet demand; on the other hand, when further analyzing the balance between supply and demand, power flow constraint should be added to consider the support of the network.

2.2.2. State Evaluation Model

(1) Demand Regulation Model

In high penetration of the renewable energy integrated system, the fluctuation of renewable energy output power is the main reason for the regulation demand. Compared with changes in load power, the fluctuation of renewable energy power is larger and more frequent and there is no corresponding rule to follow. This paper mainly considers the regulation demand caused by the fluctuation of renewable energy, the relevant regulation demand is formulated as [18]:

$$\begin{cases} P_{\mathrm{D}}^{\mathrm{up}}(t,\tau) = \sum\limits_{P_{i,\mathrm{REG}} \in \Omega_{\mathrm{REG}}} \left|P_{i,\mathrm{REG}}(t+\tau) - P_{i,\mathrm{REG}}(t)\right| & P_{i,\mathrm{REG}}(t) > P_{i,\mathrm{REG}}(t+\tau) \\ P_{\mathrm{D}}^{\mathrm{down}}(t,\tau) = \sum\limits_{P_{i,\mathrm{REG}} \in \Omega_{\mathrm{REG}}} \left|P_{i,\mathrm{REG}}(t+\tau) - P_{i,\mathrm{REG}}(t)\right| & P_{i,\mathrm{REG}}(t) < P_{i,\mathrm{REG}}(t+\tau) \end{cases} \tag{1}$$

where ${P_{\mathrm{D}}}^{\mathrm{up}}(t, \tau)$ and ${P_{\mathrm{D}}}^{\mathrm{down}}(t, \tau)$ represent the up- and down-regulation demand respectively; $P_{i,\mathrm{REG}}(t)$ and $P_{i,\mathrm{REG}}(t + \tau)$ represent the output power of the renewable energy unit i at the time t and $t + \tau$; Ω_{REG} is the set for the renewable energy unit.

It can be seen from (1) that when the output power of renewable energy decreases at $t + \tau$, up-regulation demand will be generated. If the fast regulation resources can match this demand, the power supply reliability can be maintained continuously. Otherwise, there will be a load shedding. Similarly, when the output power of renewable energy increases at $t + \tau$, there will be a down-regulation demand. If the fast regulation of resources cannot be matched, there will be a curtailment of renewable energy.

(2) Resources Regulation Model

The regulation resources include various types of generators with rapid regulation capabilities, such as rapid regulation units, energy storage devices, etc., and various regulation methods, such as demand-side response, rapid load shedding. However, as far as the technical level and practical application of the above-mentioned various measures are concerned, energy storage devices are expensive to build and have limited capacity and are mostly used in distribution networks and below. Demand-side responses are similar. Two types of regulation measures, such as rapid load shedding and curtailment of wind and photovoltaic, are contrary to the original intention of reliable power supply and efficient consumption of renewable energy and are often used as the last level of security control. Therefore, at the transmission grid level, large-capacity fast regulating units are currently the most important regulation resource. For such resources, its ability to regulate can be expressed as [18]:

$$\begin{cases} P_{\mathrm{S}}^{\mathrm{up}}(t,\tau) = \sum\limits_{P_{i,\mathrm{G}} \in \Omega_{\mathrm{G}}} \min\left[R_{i,\mathrm{G}}^{\mathrm{up}} \cdot \tau, P_{i,\mathrm{G}}^{\max} - P_{i,\mathrm{G}}(t)\right] \\ P_{\mathrm{S}}^{\mathrm{down}}(t,\tau) = \sum\limits_{P_{i,\mathrm{G}} \in \Omega_{\mathrm{G}}} \min\left[R_{i,\mathrm{G}}^{\mathrm{down}} \cdot \tau, P_{i,\mathrm{G}}(t) - P_{i,\mathrm{G}}^{\min}\right] \end{cases}, \tag{2}$$

where $P_{\mathrm{S}}{}^{\mathrm{up}}(t, \tau)$ and $P_{\mathrm{S}}{}^{\mathrm{down}}(t, \tau)$ represent the up- and down-regulation capability and the subscript S refers to "Supply"; Ω_{G} is the set of fast regulating units; $P_{i,\mathrm{G}}(t)$ is the output power of the fast regulating units i at time t; $P_{i,\mathrm{G}}{}^{\max}$ and $P_{i,\mathrm{G}}{}^{\min}$ are the upper and lower output of the fast regulating generator units respectively; $R_{i,g}{}^{\mathrm{up}}$ and $R_{i,g}{}^{\mathrm{down}}$ respectively represent the maximum up and down ramp rate of the fast regulation units.

As can be seen from (2), that the regulation capability of the fast regulation unit is determined by the current operating state and regulation capability of the unit. Taking the up-regulation capability as an example, the actual regulation capability is the smaller of the maximum climbing regulation rate, the upper limit of the output power, and the difference between the actual output power.

(3) Condition Evaluation Model for Supply and Demand Balance

$$\begin{cases} P_{\mathrm{M}}^{\mathrm{up}}(t,\tau) = P_{\mathrm{S}}^{\mathrm{up}}(t,\tau) - P_{\mathrm{D}}^{\mathrm{up}}(t,\tau) \\ P_{\mathrm{M}}^{\mathrm{down}}(t,\tau) = P_{\mathrm{S}}^{\mathrm{down}}(t,\tau) - P_{\mathrm{D}}^{\mathrm{down}}(t,\tau) \end{cases}, \tag{3}$$

$$\text{s.t. } -\boldsymbol{B\theta}(t,\tau) + \boldsymbol{P}_{\mathrm{g}}(t,\tau) + \boldsymbol{P}_{\mathrm{G}}(t,\tau) + \boldsymbol{P}_{\mathrm{REG}}(t,\tau) = \boldsymbol{P}_{\mathrm{L}}(t,\tau), \tag{4}$$

where $P_{\mathrm{M}}{}^{\mathrm{up}}(t, \tau)$ and $P_{\mathrm{M}}{}^{\mathrm{down}}(t, \tau)$ represent the upper and lower regulation resource margins and the subscript M refers to "Margin"; $\boldsymbol{B}$ is bus admittance matrix; $\boldsymbol{\theta}$, $\boldsymbol{P}_{\mathrm{g}}$, $\boldsymbol{P}_{\mathrm{G}}$, $\boldsymbol{P}_{\mathrm{REG}}$, $\boldsymbol{P}_{\mathrm{L}}$ are voltage angel vector, the output vector of the conventional generator unit integrated bus, the output vector of the fast regulation generator unit integrated bus, the output vector of the renewable energy unit integrated bus and the power vector of the load bus.

The presented state evaluation model of the supply and demand balance consists of two parts. (3) can be used to judge whether the upper limit of system regulation resources is sufficient or insufficient, so as to indicate whether the regulation resources allocated by the system can meet the demand of regulation. (4) requires that the operation calls of each unit meet the network transmission constraints. Therefore, (3) and (4) can be used to comprehensively reflect the satisfaction relationship between the actual available regulation capacity and the demand. To sum up, when $P_{\mathrm{M}}{}^{\mathrm{up}}(t, \tau)$ or $P_{\mathrm{M}}{}^{\mathrm{down}}(t, \tau)$ are greater than 0, it indicates that the system can maintain the balance of power supply and demand at this time; on the contrary, there is a lack of fast regulation in the system, which will lead to a certain degree of forced load shedding or curtailment of renewable energy [18].

2.2.3. Evaluation Indexes

Based on the state evaluation model for supply and demand balance, further construct the adaptability indexes of supply and demand balance, which are described below.

(1) Insufficient Rate of Supply and Demand Balance (IRSD)

IRSD is introduced to characterize the ratio of the time when the power supply and demand balance is insufficient to the total operation time, i.e., the probability of the insufficient supply and demand balance [18]:

$$R_{\text{ins}} = \frac{T_{\text{ins}}}{T_{\text{total}}}, T_{\text{ins}} = \sum_{t \in T_{\text{total}}} t_{P_{\text{M}}(t,\tau)<0}, \tag{5}$$

where T_{ins} and T_{total} represent the time of insufficient supply and demand balance and the total running time; $t_{\text{PM} < 0}$ represents the moment when the supply and demand balance is insufficient.

(2) Insufficient Rate of Up-Regulation (IRUR)

IRUR is introduced to characterize the ratio of the time between insufficient supply and demand balance due to lack of up-regulation resources and the total time of up-regulation demand, i.e., the probability of the supply and demand imbalance and load shedding due to lack of up-regulation resources [18]:

$$R_{\text{ins}}^{\text{up}} = \frac{T_{\text{ins}}^{\text{up}}}{T_{\text{total}}^{\text{up}}}, T_{\text{ins}}^{\text{up}} = \sum_{t \in T_{\text{total}}^{\text{up}}} t_{P_{\text{M}}^{\text{up}}(t,\tau)<0}, \tag{6}$$

where ${T_{\text{ins}}}^{\text{up}}$ and ${T_{\text{total}}}^{\text{up}}$ are the time when the up-regulation resources are insufficient and the total operation time when the up-regulation demand occurs respectively, ${t_{\text{PM}}}^{\text{up}} < 0$ is the time when the up-regulation is insufficient.

(3) Insufficient Rate of Down-Regulation (IRDR)

IRDR is introduced to characterize the ratio of the time when the supply and demand balance is insufficient due to the lack of down-regulation resources and the total time when the down-regulation demand occurs, i.e., the probability of the supply and demand imbalance and renewable energy curtailment due to the lack of down-regulation resources [18]:

$$R_{\text{ins}}^{\text{down}} = \frac{T_{\text{ins}}^{\text{down}}}{T_{\text{total}}^{\text{down}}}, T_{\text{ins}}^{\text{down}} = \sum_{t \in T_{\text{total}}^{\text{down}}} t_{P_{\text{M}}^{\text{down}}(t,\tau)<0}, \tag{7}$$

where ${T_{\text{ins}}}^{\text{down}}$ and ${T_{\text{total}}}^{\text{down}}$ represent the time when the down-regulation resource is insufficient and the total operation time when the down-regulation demand occurs; ${t_{\text{PM}}}^{\text{down}} < 0$ represents the time when the down-regulation is insufficient.

(4) Average Insufficiency of Up-Regulation (AIUR)

AIUR is introduced to characterize the average shortage of up-regulation resources per minute, due to the mismatch of up-regulation resources and requirements [18]:

$$E_{\text{ins}}^{\text{up}} = \frac{\sum_{t \in T_{\text{ins}}^{\text{up}}} \left| P_{\text{M}}^{\text{up}}(t,\tau) \right|}{T_{\text{ins}}^{\text{up}}}, \tag{8}$$

where the numerator represents the total insufficiency of up-regulation resource.

(5) Average Insufficiency of Down-Regulation (AIDR)

AIDR is introduced to characterize the average shortage of down-regulation resources per minute, due to the mismatch of down-regulation resources and requirements [18]:

$$E_{\text{ins}}^{\text{down}} = \frac{\sum_{t \in T_{\text{ins}}^{\text{down}}} \left| P_{\text{M}}^{\text{down}}(t, \tau) \right|}{T_{\text{ins}}^{\text{down}}}, \tag{9}$$

where the numerator represents the total insufficiency of down-regulation resource.

2.3. *Adaptability of Operation State*

The adaptability of operation state refers to the ability to stabilize power disturbances, resist uncertain shocks, and maintain a safe, efficient, and stable operating state by relying on its own network structure. Aiming at the uncertain operating environment of high penetration of the renewable energy integrated system, the expected load rate and line power fluctuation rate of the line are proposed and the actual operating state of a single line is characterized from both the load level and the degree of fluctuation. Furthermore, based on the weighted entropy and joint weighted entropy theory, the operating state characteristics of each single line are integrated to obtain the overall operating state of the power system, so as to achieve a comprehensive assessment of the safety, efficiency, and stability of the actual operating state of the power system from the spatiotemporal dimension.

2.3.1. Basic Characteristics

(1) Operational Security

System operation security refers to the safety degree of the overall state, which can be quantitatively characterized by measuring the safety distance between the current operating state and the fault state. Self-organized critical theory is a typical evaluation method based on the above ideas [19]. The self-organized critical state represents the marginal state of the system operation. In this state, any disturbance may cause system failure or even large-scale cascading failure. Since the traditional operation mode of the power system is relatively fixed, the self-organized critical theory mainly starts from the spatial dimension and evaluates the safety of the system by analyzing the concentration and order of the load levels between lines under a certain operation mode. However, due to the superposition of the fluctuations of the renewable energy output, the load level of lines shows spatiotemporal differences. In addition, the traditional self-organized critical theory emphasizes that the lower the load rate, the safer it is and ignores the economics of operation. Therefore, in high penetration of the renewable energy integrated system, when the operating state of each line is concentrated and orderly at a reasonable load level recognized by the operator, the actual operating state of the power system is safer.

(2) Operational Efficiency

The operational efficiency refers to the economic utilization level of power system transmission equipment and reflects the reasonable degree of the power system investment, planning, and operation. If the utilization efficiency is too low, it will cause a large number of transmission equipment to be idle and redundant, indicating that the grid investment is too large and there is a lot of waste of resources. The high utilization efficiency of the power system improves the economics of power system investment and operation, but this situation often has a limited safety margin and there is a risk of safe operation. Existing operational efficiency indexes just reflect the average utilization level of transmission equipment and cannot reflect the penetration of lines with high or low utilization efficiency and the impact on overall efficiency. In addition, for the operation state with reasonable utilization efficiency, the operational efficiency of the system cannot be simply characterized by the load level. Therefore, in an uncertain operating environment where high penetration of renewable energy is integrated, operation state with good efficiency and adaptability means that each line is always operating within a reasonable range as determined by the planner. The utilization level of each line is similar, so that the overall operation state is efficient and reasonable.

(3) Operational Stability

The stability of the operating state can be understood as the adaptability of the operation state to power fluctuations such as renewable energy or the ability to maintain the operation state in a dynamic environment. The uncertainty of the integrated renewable energy unit is coupled with the overall operation state and there will be a phenomenon where the power flow distribution is continuously shifted and the system operating state changes rapidly. At this time, if the ability to maintain the operation state is poor, it will result in a large difference in the safety and efficiency characteristics of the system state under the flow section at different times. This will not only lead to poor safety, low utilization efficiency, or heavy overloading, but also the evaluation index cannot cover the above extreme scenarios, which ultimately leads to the ambiguity of the index and reduces the accuracy of the overall assessment. In the face of various power fluctuations and shocks coupled in the operation, the operating state of the various lines has a low degree of change and the fluctuation range of the transmission power is small. The power disturbance is suppressed to the greatest extent and eventually it will not cause a huge change in the overall operating state. Therefore, when the load level of all lines is concentrated and orderly within the reasonable operating range recognized by the operating personnel, it indicates that the safety and security of the current operating state is at a good level. In other words, showing good adaptability to the uncertainty of the complex spatiotemporal distribution of renewable energy.

2.3.2. State Evaluation Model

(1) Expected Line Load Rate

The high frequency and large range of output power changes of renewable energy, the superimposition of spatiotemporal characteristic lead to the continuous transfer of the state. For each line, the transmission power fluctuates and changes, which can be understood as a time series. The line undertakes the corresponding power transmission task at different times according to the power flow distribution. Therefore, this paper first defines the expected line load rate to describe the average load level of the transmission power in each scenario [18]:

$$F_i = \frac{\int_{-\infty}^{+\infty} \left| P_i \varphi(P_i) \right| \mathrm{d}P_i}{S_{i,\max}}, \tag{10}$$

where P_i and $\varphi(P_i)$ represent the actual transmission power of line i and the probability density function respectively; the numerator represents the expected value of the transmission power; $S_{i,\max}$ represents the rated transmission power capacity of the line.

(2) Line Power Fluctuation Rate

Fluctuation is a typical characteristic of high penetration of renewable energy integrated systems compared with traditional grids. The line power fluctuation rate is used to characterize the deviation of the actual transmission power of the line at different times from the average load level in each scenario. This index represents the fluctuation range of the actual transmission power from the average load level and reflects the dispersion degree of the transmission power [18]:

$$B_i = \frac{\sqrt{E(P_i^2) - E(P_i)^2}}{\min\left\{ \left| S_{i,\max} - E(P_i) \right|, \left| E(P_i) \right| \right\}}, \tag{11}$$

$$E(P_i) = \int_{-\infty}^{+\infty} P_i \varphi(P_i) \, \mathrm{d}P_i, \tag{12}$$

$$E(P_i^2) = \int_{-\infty}^{+\infty} P_i^2 \varphi(P_i) \, \mathrm{d}P_i, \tag{13}$$

where $E(\cdot)$ is the expected value of the relevant parameters; the numerator is the fluctuation value of the transmission power; the denominator is the allowable power fluctuation range, i.e., the power fluctuation capacity.

2.3.3. Evaluation Indexes

The above evaluation model can characterize the safety margin, efficiency level, and stability of each line, the actual operating state should be an overall evaluation of all lines, and the operating state of each line will constitute the overall operating state. Therefore, based on the weighted entropy [20] and joint weighted entropy theory [21], this paper proposes adaptability evaluation indexes for the operation state of overall system.

(1) Weighted Entropy of Expected Line Load Rate (WEELLR)

The operating state is regarded as the system as a whole and each load rate interval is regarded as a system event. The ratio of the number of lines in each interval to the total number of lines is regarded as the probability of occurrence of events in this interval. The weights of each interval are subjectively determined according to planners, and set the optimal operating interval weight to the minimum. The weighted entropy theory can be used to establish WEELLR to evaluate the concentration and rationality of the operating state distribution.

The load rate interval is expressed as a set $M = [M_1, M_2, \ldots, M_5]$, where interval M_1 indicates that the load rate interval is (0, 20%] of the rated capacity and so on, M_5 indicates (80%, 100%]. N_{Mi} is the number of lines whose expected load rate belongs to the interval. Then, the probability that the line is in the ith expected load rate interval can be expressed as [20]:

$$p(F_{M_i}) = \frac{N_{M_i}}{\sum\limits_{i=1}^{5} N_{M_i}}. \tag{14}$$

The weight of the expected load rate interval of each line $\omega(F_{Mi})$ can be determined according to the planner's comprehensive consideration of the grid efficiency and safety margin. For example, the rational ranking of the load rate interval is: $M_3 > M_2 > M_1 > M_4 > M_5$, where the ">" symbol indicates that the left side of the symbol is more reasonable than the right-side object. The above sorting method shows that the planner thinks that when the line load rate in (80%, 100%] of the rated capacity is the most reasonable and its weight is the smallest. Therefore, WEELLR is expressed as [20]:

$$H_F = -\sum_{i=1}^{5} \omega(F_{M_i}) p(F_{M_i}) \ln p(F_{M_i}). \tag{15}$$

This index reflects the concentration and rationality of the distribution of the average load levels between lines from the spatial dimension. The smaller the index value, the more the expected load rate of the line is concentrated in the range that the operator believes is more reasonable. The overall average state has both safety and efficiency and has good adaptability.

(2) Weighted Entropy of Line Power Fluctuation Rate (WELPFR)

Due to fluctuations of renewable energy output, the operating state will shift from the average state. When the offset is large, it will cause the safety and efficiency of the running state to not maintain the average state. Therefore, WELPFR is introduced to quantify the degree of this deviation and to evaluate the stability of the system operating state.

The operating state is still regarded as the system as a whole and each power fluctuation rate interval is regarded as a system event. The ratio of the number of lines existing in each interval to the total number of lines is regarded as the probability of occurrence of events in this interval. Because the smaller the line power fluctuation, the more stable the overall system state, so the weight of the operating interval with the smallest power fluctuation rate is set to the minimum. The weighted entropy theory can be used to establish WELPFR to evaluate the stability of the grid operating state distribution. The power fluctuation rate interval is expressed as a set V = [V_1, V_2, … , V_{10}], where interval V_1 represents the power fluctuation rate value interval is (0, 10%] and so on, V_{10} indicates (90%, 100%]. N_{Vj} is the number of lines whose line power fluctuation rate belongs to the interval V_j. Then, the probability that the line is in the jth power fluctuation rate interval can be expressed as [20]:

$$p(B_{V_j}) = \frac{N_{V_j}}{\sum\limits_{j=1}^{10} N_{V_j}}. \tag{16}$$

The smaller the power fluctuation of each line is, the more stable the system is. Therefore, the smaller the weight of the power fluctuation rate interval, the higher the rationality and the average value of the line power fluctuation rate of each interval can be directly used as the weight of the interval. Let k denote the line whose line power fluctuation rate is in the interval V_j, the weights is calculated as follows:

$$\omega(B_{V_j}) = \frac{1}{N_{V_j}} \sum_{k \in V_j} B_k. \tag{17}$$

Thus, WELPFR can be expressed as:

$$H_B = -\sum_{j=1}^{10} \omega(B_{V_j}) p(B_{V_j}) \ln p(B_{V_j}). \tag{18}$$

This index considers the fluctuation of the transmission power of all lines and can reflect the stability operating state from the time and space dimensions. The smaller the index value, the more the line power fluctuation range is smaller, power system has better adaptability and smoothing ability to the renewable energy power fluctuation, and shows high inertia operating characteristics.

(3) Joint Weighted Entropy of Operation State (JWEOS)

WEELLR evaluates the safety and efficiency operating state, WELPFR evaluates the stability of the operating state. The above two indexes and three attributes jointly determine the adaptability and rationality of the system operating state. However, in practical applications, it is often more desirable to use an index to directly assess the adaptability and rationality of the operation state. Therefore, this paper combines the two indexes and further propose JWEOS to comprehensively evaluate the safety, efficiency, and stability of operation state.

The operating state of high penetration of the renewable energy integrated system can be characterized by the expected line load rate and line power fluctuation rate. If the grid operating state is regarded as the whole system, the expected line load rate and line power fluctuation rate can be regarded as the two types of events that constitute the overall system, the concentration and rationality jointly determine the adaptability operating state. Because, under the considered operation scenario and corresponding operation mode, the expected load rate and power fluctuation rate distribution are obtained through simulation. Therefore, JWEOS is formulated as [21]:

$$H_{FB} = -\sum_{i=1}^{5} \sum_{j=1}^{10} \omega(F_{M_i}) \omega(B_{V_j}) p(F_{M_i}) p(B_{V_j}) \ln\left[p(F_{M_i}) p(B_{V_j})\right]. \tag{19}$$

The smaller the index value, the more lines in the system are operating within the load level interval that the planner thinks reasonable and the more line with smaller power fluctuation range. Combined with the definition of adaptability of operation state, it can be seen that the system considers both safety and efficiency in this operating state, with strong robustness and a comprehensive high level of adaptability.

2.4. Adaptability of Network Structure

The adaptability of network structure means that in an uncertain operating environment where high penetration of renewable energy is connected to the grid, the power system will ultimately improve the ability components and the overall structure to withstand uncertain impacts by optimizing the network topology and reducing weak links. According to the complex network theory and vulnerability theory, bus and branch electric betweenness indexes are used to characterize the importance and robustness of a single component [22]. Based on the uniformity theory, weighted entropy and joint weighted entropy theory, the impact of buses and branches on the topology is integrated. From a single element to the overall power system, the weak links are identified and the ability of the overall network structure to withstand uncertain shocks.

2.4.1. Basic Characteristics

In the power system analysis model, buses and branches are the most basic components that make up the network structure. The rationality of the connection relationship between buses and branches will directly determine the ability of the network structure to resist the impact of uncertain factors. In the context of high penetration of the renewable energy integrated system, if there are weak buses or branches, it is very easy to cause large-scale cascading failures from renewable energy power shocks. Therefore, evaluating the robustness of buses and branches is the key to understanding the ability of the overall structure to withstand uncertainties.

(1) Bus Robustness

On the one hand, bus robustness is determined by the load level and the integrated generator capacity, on the other hand, by the spatial position of bus in the power system topology. Bus that bears important loads or bus with important generators often has a high importance level and needs to pay special attention to ensure reliable power supply of the critical load. Bus with more complicated contact relationships in the system also tends to have a higher importance level. Once this type of bus withdraws due to various uncertain shocks or reduces its contribution to the operation, it has a wide range of coverage and is often easy to cause larger failures. Therefore, bus with higher criticality tends to become weak links, it is of great significance to improve the robustness of the network structure to identify the key bus and reduce its weakness through reasonable planning.

(2) Branch Robustness

The transmission line assumes the key role of connecting various types of generator and load buses, which is the basic platform to ensure the reliable power supply. On the one hand, branch robustness is determined by the importance of buses connected by the transmission corridor, on the other hand, by the transmission capacity undertaken by the transmission corridor in the system power supply and demand balance. The transmission lines with more complicated contact relationships and contact buses belonging to important buses often have higher importance levels. Once such lines exit the system due to various uncertain impacts, it is easy to cause large-scale chain failures and large-scale power outages. Therefore, identifying the weak links of the power transmission line and on this basis, reducing the number of weak links and reducing the degree of weakness is the key to improving the robustness of network structure.

2.4.2. State Evaluation Model

Complex network theory is an abstract description of complex systems composed of different nonlinear basic units. Its core idea is to abstract complex systems into complex networks composed of buses and branches and use graph theory, statistical theory, etc. The classic index of the method to quantify the network structure characteristics of complex systems. Among them, buses of the complex network represent the basic units that make up the system and branches reflect the association between the basic units. In the analysis of complex networks in power systems, the electric betweenness is the most commonly used. This index is based on the network intermediary index and combined with the improvement of the actual characteristics of the power flow distribution. It is divided into electric betweenness of buses and branches. Then, clarify the network's weak links and their weaknesses. It means that any power generation and load bus pair in the power system is selected, the power generation and load power of the bus pair are used as power weights and then the unit current is added between bus pairs and Kirchhoff's law solves the current in the entire network After the traversal considers all bus pairs, the current through a bus or branch is weighted and accumulated and the electric betweenness of bus or branch can be obtained. The index can reflect the power capacity of the different power generation and load bus pairs by adjusting the weights and can accurately characterize the occupancy of each bus and branch element based on Kirchhoff's theorem.

(1) Electric Betweenness of Bus

Electric betweenness of bus characterizes the occupancy of all bus components by each power transmission path of the grid, i.e., the contribution of different bus in the power transmission. The larger the electric betweenness of bus, the greater the contribution and importance of bus in the power transmission of the whole network and the more easily and disturbed the system power is. By obtaining the electric betweenness values of all buses in the power system, the weak ranking of buses can be obtained. The electric betweenness of bus i can be calculated as [22]:

$$D_i = \sum_{a \in \mathbf{Ge}, b \in \mathbf{Lo}} \sqrt{\omega_a \omega_b} B_{e,ab}(i), \tag{20}$$

where a, b respectively represent the number of generator and load bus; **Ge** and **Lo** represent the set of generator and load bus; ω_a, ω_b represent the weights of generator and load bus, their values are denoted as the actual power; $B_{e,ab}(i)$ is the current distribution of bus i, after add a unit current element between pairs of generator load bus (a, b), which can be calculated as [22]:

$$B_{e,ab}(i) = \begin{cases} \frac{1}{2} \sum_j \left| I_{ab}(i,j) \right| & i \neq a,b \\ 1 & i = a,b \end{cases}, \tag{21}$$

where $I_{ab}(i,j)$ is the current of branches (i,j) after adding the unit current element between the generator load bus pair (a, b) and bus j represents the remaining buses directly connected to bus i. When bus i is one of the generator load bus pairs, the current value passing through bus is regarded as the unit current.

(2) Electric Betweenness of Branch

Electric betweenness of branch characterizes the occupancy of all power system components by each power transmission line, i.e., the contribution of each branch in the power transmission of the entire network. The greater the electric betweenness of branch, the greater the contribution and importance of branch in the power transmission. It also indicates that branch is more susceptible to system power disturbances and has less robustness. The easier it is to become the weak link in the overall grid. The electric betweenness of branch m can be calculated as [22]:

$$L_m = \sum_{a \in \mathbf{Ge}, b \in \mathbf{Lo}} \sqrt{\omega_a \omega_b} \left| I_{ab}(m) \right|, \tag{22}$$

where $I_{ab}(m)$ is the current of branch m after the unit current element is added between the generator load bus pair (a, b).

2.4.3. Evaluation Indexes

(1) Weighted Entropy for Electric Betweenness of Bus (WEEBBus)

Although the electric betweenness of bus can clarify the order of the weakness of each bus within the power system, it cannot compare the robustness of different network structures from the perspective of the overall bus. Therefore, this paper proposes WEEBBus to clarify the difference in robustness of different network structures. The network topology is regarded as the system, the electric betweenness interval of each bus is regarded as event. The ratio of the number of buses in each interval to the total number of buses is regarded as the probability of occurrence of events in this interval. Since the smaller the electric betweenness of bus, the more robust the component and the overall structure, the interval weight with the smallest electric betweenness is set to the minimum. Thus, the weighted entropy theory can be used to establish WEEBBus and the robustness of network structure can be evaluated from the perspective of bus.

Represent the interval for electric betweenness of bus as a set O = [$O_1, O_2, \ldots, O_5$], N_{Oi} is the number of bus whose electric betweenness is in the interval O_i. Then, the probability that the electric betweenness of bus belongs to the ith interval O_i can be expressed as [20]:

$$p(D_{O_i}) = \frac{N_{O_i}}{\sum\limits_{i=1}^{5} N_{O_i}}. \tag{23}$$

The smaller the electric betweenness of bus, the less the weak links of the network, the better the robustness of the structure. Therefore, the smaller the electric betweenness value is, the smaller the weight value of interval is. Here, let k denote bus whose electric betweenness belongs the interval O_i, the average value of the electric betweenness for each interval is used as the weight value of the interval [20]:

$$\omega(D_{O_i}) = \frac{1}{N_{O_i}} \sum_{k \in O_i} D_k. \tag{24}$$

Thus, WEEBBus can be formulated as [20]:

$$H_D = -\sum_{i=1}^{5} \omega(D_{O_i}) p(D_{O_i}) \ln p(D_{O_i}). \tag{25}$$

The smaller the index value, the more buses in the power system are concentrated in a smaller value interval. This shows that from the perspective of buses, the overall structure is balanced, robust, and has few weak links and is not easily affected by power disturbances. Network structure has higher adaptability and stronger ability to withstand uncertain shocks.

(2) Weighted Entropy for Electric Betweenness of Branch (WEEBBra)

When the electric betweenness of all branches is concentrated in a relatively small range, it indicates that there is no weak link in the branch of the network structure. The topological structure is regarded as the system as a whole, the electric betweenness interval of each branch is regarded as a system event and the ratio of the number of branches in each interval to the total number of branches is regarded as the probability of occurrence of events in this interval. The interval weight with the smallest electric betweenness is set to the minimum and so on. In this way, the weighted entropy theory can be used to establish WEEBBra and the robustness of the network structure can be evaluated from the perspective of the branch.

Represent the interval for electric betweenness of branch as a set $Z = [Z_1, Z_2, \ldots, Z_5]$, N_{Zj} is the number of branches whose electric betweenness belongs to interval Z_j. Then, the probability that the electric betweenness of bus belongs to the jth interval Z_j can be expressed as [20]:

$$p(L_{Z_j}) = \frac{N_{Z_j}}{\sum\limits_{j=1}^{5} N_{Z_j}}. \tag{26}$$

On this basis, determine the weight $\omega(L_{Zi})$ of the interval for electric betweenness of each branch. Similarly, the smaller the electric betweenness is, the smaller the weight setting. Let k denote the branch whose electric betweenness belong the interval Z_j, the average value of electric betweenness of the branch for each interval is used as the weight of the interval [20]:

$$\omega(L_{Z_j}) = \frac{1}{N_{Z_j}} \sum_{k \in Z_j} L_k. \tag{27}$$

Thus, WEEBBra can be formulated as [20]:

$$H_{\mathrm{L}} = -\sum_{j=1}^{5} \omega(L_{Z_j}) p(L_{Z_j}) \ln p(L_{Z_j}). \tag{28}$$

The smaller the index, the more branches are concentrated in the smaller value interval of electric betweenness. This shows that from the perspective of branch circuit, the overall structure is balanced, robust, and not easily affected by power disturbances. Network structure has higher adaptability and stronger ability to withstand uncertain shocks.

(3) Joint Weighted Entropy of Network Structure (JWENS)

From the above analysis, we can see that buses and branches are the most basic components of the power network and their respective robustness will directly affect the ability of the overall structure to withstand uncertain shocks. Although the above two weighted entropy indexes can describe the robustness of all buses or all branches separately, they cannot describe the robustness of the overall structure. Therefore, this paper further proposes JWENS. Based on the overall consideration of all the basic components, the robustness topology is comprehensively evaluated and then the adaptability of network structure to uncertain shocks is reflected.

The most basic components topology includes two types of buses and branches. If the power system topology is regarded as the system as a whole, then buses and branches can be regarded as two types of events that constitute the system as a whole. The rationality together determines the adaptability of the network structure. For a certain network structure, the connection relationship between its internal buses and branches is fixed and the distribution of electric betweenness of buses and branches is also independent and determined. Therefore, the joint weighted entropy model of the decoupling calculation is used to establish JWENS as follows [21]:

$$H_{\mathrm{DL}} = -\sum_{i=1}^{5} \sum_{j=1}^{5} \omega(D_{\mathrm{O}_i}) \omega(L_{Z_j}) p(D_{\mathrm{O}_i}) p(L_{Z_j}) \ln\left[p(D_{\mathrm{O}_i}) p(L_{Z_j})\right]. \tag{29}$$

The smaller the index value, the more buses and branches have smaller electric betweenness. In other words, the fewer weak links in system, the smaller the impact of power disturbance, the better the overall robustness, and the higher the adaptability to uncertain shocks.

3. Adaptable Source-Grid Expansion Planning Model

From the source side, the upper limit of the system regulation capability is determined by the capacity and regulation rate of the power supply regulation resources. However, due to the limitation of network transmission, the actual regulation capacity of the system is often less than the regulation capacity of the source. In the process of power supply and demand balance, although it does not provide or consume regulation resources in essence, the power system plays a supporting role in the process of supply and demand balance. Therefore, the establishment of a grid structure that is compatible with the process of supply and demand distribution of source and load power can ensure the full utilization and efficient transmission of regulation resources and maximizes the release of system regulation capabilities.

From the grid side, the transmission network not only needs to support the basic task of power supply and demand balance, but also needs to maintain excellent operation state and resist renewable energy during the above process. The impact of energy power fluctuations on the operating state ensures the safety, stability, and efficiency of the operating state. On the one hand, the state depends on the reasonable degree of network structure, on the other hand, it is also greatly affected by the power supply and system operation mode. Therefore, the establishment of a network structure that is compatible with power supply transformation and operation schemes is also the key to ensuring an efficient and reliable actual operating state and improving the adaptability of network structure.

Based on the above analysis, it is known that designing mutually source-grid coordinated planning schemes is the key to comprehensively improving the adaptability of high penetration of renewable energy integrated systems. Therefore, this paper proposes a two-stage expansion planning method. In the first stage, the current supply and demand balance of the system is evaluated. If the supply and demand balance adaptability meet the threshold preset by the planner, then the power supply does not need to be upgraded and directly entered into the second stage and the power system is adapted through the adaptable grid planning model planning and reconstruction to optimize the structure and operation state. If the supply and demand balance adaptability index does not meet the threshold set by the planner, the first stage adopts the adaptable source planning model to upgrade the technical parameters of the existing units. Then, the optima source plan is utilized as input in the second stage planning. On this basis, the adaptable grid planning model is applied to obtain the optimal grid plan. The details are as follows.

3.1. *Obective Funuction*

3.1.1. Adaptable Source Expansion Planning Model

The objection function is to minimize the comprehensive cost of source transformation, which includes transformation cost for the quick regulation unit, curtailment cost for renewable energy power, and load shedding cost, as shown below:

$$\min F = C_{\mathrm{G}} + C_{\mathrm{WR}} + C_{\mathrm{WL}}, \tag{30}$$

$$C_{\mathrm{G}} = (k_1 + k_2) \sum_{i \in \Omega_{\mathrm{G}}} z_i \cdot (c_{1,i} x_i + c_{2,i} y_i), \tag{31}$$

$$k = \frac{r(1+r)^n}{(1+r)^n - 1}, \tag{32}$$

$$C_{\mathrm{WR}} = D \sum_{s \in \mathbf{S}} p(s) \sum_{t \in T_{\mathrm{ins}}^{\mathrm{down}}} c_{\mathrm{REG}} \cdot \left| P_{\mathrm{M}}^{\mathrm{down}}(t,\tau) \cdot \tau \right|, \tag{33}$$

$$C_{\mathrm{WL}} = D \sum_{s \in \mathbf{S}} p(s) \sum_{t \in T_{\mathrm{ins}}^{\mathrm{up}}} c_{\mathrm{LOAD}} \cdot \left| P_{\mathrm{M}}^{\mathrm{up}}(t,\tau) \cdot \tau \right|, \tag{34}$$

where C_G is the cost of regulation capacity transformation, C_{WR} is the annual cost of renewable energy generation, C_{WL} is the annual load shedding cost; k_1, k_2 are the fund recovery factor and the project fixed operating rate, which are used to allocate the one-time investment cost to each year of life span, r is the discount rate, n is the economically applicable year of project; Ω_G is set of fast regulating units; z_i is the sign for transformation, x_i is the transformation capacity of unit, y_i is the transformation quantity of regulation rate; $c_{1,i}$ is the unit cost for capacity transformation, $c_{2,i}$ is unit cost for transformation of regulation rate; D is the number of days of operation per year, S is the set for typical scenarios of renewable energy outputs, $p(s)$ is the probability of typical scenarios s, c_{REG} is the unit cost of renewable energy curtailment, and ${T_{ins}}^{down}$ is the time of renewable energy curtailment; $|{P_M}^{down}(t,\tau)\cdot\tau|$ is the curtailment power of the renewable energy, c_{LOAD} is the cost of load shedding, ${T_{ins}}^{up}$ is the time of load shedding, $|{P_M}^{up}(t,\tau)\cdot\tau|$ is the power of load shedding.

3.1.2. Adaptable Grid Expansion Planning Model

Considering the economy of grid planning, adaptability of operation state, and adaptability of network structure, the objective function is to minimize the cost of grid expansion, joint weighted entropy index of operation state, and joint weighted entropy index of network structure. Due to the dimensional differences between the various objectives, the objection function is expressed as a multi-objective function in vector form:

$$\min F = \min(C_N, H_{FB}, H_{DL}), \tag{35}$$

$$C_N = (k_1 + k_2)\sum_{i\in\Omega_l} c_i n_i l_i, \tag{36}$$

$$H_{FB} = -\sum_{i=1}^{5}\sum_{j=1}^{10} \omega(F_{M_i})\omega(B_{V_j})p(F_{M_i})p(B_{V_j})\ln\left[p(F_{M_i})p(B_{V_j})\right], \tag{37}$$

$$H_{DL} = -\sum_{i=1}^{5}\sum_{j=1}^{5} \omega(D_{O_i})\omega(L_{Z_j})p(D_{O_i})p(L_{Z_j})\ln\left[p(D_{O_i})p(L_{Z_j})\right], \tag{38}$$

where C_N is the cost of grid expansion; H_{FB} is JWEOS; H_{DL} is JWENS; Ω_l is the set of expanded lines, c_i is the unit investment of reconstructed line, n_i is the circuit number of expanded lines, l_i is the total length of the reconstructed line.

3.2. *Constraints*

3.2.1. Constraints for Source Expansion Planning

The constraints of adaptable source expansion planning model include power balance constraints (39), unit output constraints (40) and (41), unit ramp constraints (42), power flow constraints (43), and insufficient rate index of supply and demand balance constraints (44). The specific formulas are as follows.

$$\sum_{i\in\Omega_g} P_{i,g}(s,t) + \sum_{i\in\Omega_G} P_{i,G}(s,t) + \sum_{i\in\Omega_{REG}} P_{i,REG}(s,t) = \sum P_L(s,t), \tag{39}$$

$$P_{i,G}^{min} \le P_{i,G}(s,t) \le P_{i,G}^{max}, i \in \Omega_G, \tag{40}$$

$$P_{i,g}^{min} \le P_{i,g}(s,t) \le P_{i,g}^{max}, i \in \Omega_g, \tag{41}$$

$$-R_{i,G}^{down}\cdot\tau \le P_{i,G}(s,t) - P_{i,G}(s,t-\tau) \le R_{i,G}^{up}\cdot\tau, i \in \Omega_G, \tag{42}$$

$$-\boldsymbol{B\theta}(s,t) + \boldsymbol{P}_g(s,t) + \boldsymbol{P}_G(s,t) + \boldsymbol{P}_{REG}(s,t) = \boldsymbol{P}_L(s,t), \tag{43}$$

$$R_{ins} \le \alpha, \tag{44}$$

where $\Sigma P_L(s,t)$ is the total load at time slot t in scenario s; $P_{i,g}(s, t)$, $P_{i,G}(s,t)$, and $P_{i,REG}(s,t)$ are the actual output power of the conventional unit, the fast regulating unit, and the renewable energy unit at time slot t in scenario s; Ω_g, Ω_G, Ω_{REG} are the sets of the corresponding units; $P_{i,G}{}^{max}$ and $P_{i,G}{}^{min}$ are the upper and lower limits for the output power of fast regulating unit; $P_{i,g}{}^{max}$ and $P_{i,g}{}^{min}$ are the upper and lower limits for the output power of the conventional unit; $R_{i,G}{}^{up}$ and $R_{i,G}{}^{down}$ are the maximum ramp up and down rate of the fast-regulation unit under the time scale τ; B is the admittance matrix; $\boldsymbol{\theta}$ is the voltage phase angle vector; P_i is the transmission power of line i; $S_{i,max}$ is the rated transmission capacity of line i; R_{ins} is the insufficient rate index of supply and demand balance; α is the set threshold value.

3.2.2. Constraints for Grid Expansion Planning

The constraints of adaptable grid expansion planning model include the integer constraint for the circuit number of new lines, the rated conditions, the N−1 conditions, and the safe operation constraints in each scenario. The specific formulas are as follows.

$$n_i^{\min} \le n_i \le n_i^{\max},\ i \in \Omega_l\ n_i \in \mathbf{Z}, \tag{45}$$

$$\begin{cases} -\boldsymbol{B\theta} + \boldsymbol{P}_g + \boldsymbol{P}_G + \boldsymbol{P}_{REG} = \boldsymbol{P}_L \\ |P_i| \le S_{i,\max}, i \in \Omega_L \\ P_{i,g}^{\min} \le P_{i,g} \le P_{i,g}^{\max}, i \in \Omega_g \\ P_{i,G}^{\min} \le P_{i,G} \le P_{i,G}^{\max}, i \in \Omega_G \end{cases}, \tag{46}$$

$$\begin{cases} -\boldsymbol{B}^{N-1}\theta^{N-1} + \boldsymbol{P}_g^{N-1} + \boldsymbol{P}_G^{N-1} + \boldsymbol{P}_{REG}^{N-1} = \boldsymbol{P}_L^{N-1} \\ \left|P_i^{N-1}\right| \le S_{i,\max}, i \in \Omega_L \\ P_{i,g}^{\min} \le P_{i,g}^{N-1} \le P_{i,g}^{\max}, i \in \Omega_g \\ P_{i,G}^{\min} \le P_{i,G}^{N-1} \le P_{i,G}^{\max}, i \in \Omega_G \end{cases}, \tag{47}$$

$$\begin{cases} -\boldsymbol{B\theta}(s,t) + \boldsymbol{P}_g(s,t) + \boldsymbol{P}_G(s,t) + \boldsymbol{P}_{REG}(s,t) = \boldsymbol{P}_L(s,t) \\ \left|P_i(s,t)\right| \le S_{i,\max}, i \in \Omega_L \\ P_{i,g}^{\min} \le P_{i,g}(s,t) \le P_{i,g}^{\max}, i \in \Omega_g \\ P_{i,G}^{\min} \le P_{i,G}(s,t) \le P_{i,G}^{\max}, i \in \Omega_G \\ -R_{i,G}^{down} \cdot \tau \le P_{i,G}(s,t) - P_{i,G}(s,t-\tau) \le R_{i,G}^{up} \cdot \tau, i \in \Omega_G \end{cases}, \tag{48}$$

where $n_i{}^{min}$, $n_i{}^{max}$ are the upper and lower limits of the circuit number of lines; the superscript "N−1" represents the variable in the case of N−1 condition.

3.3. Model Solution

The proposed adaptable source-grid expansion planning model consists of two-stage planning. The source expansion planning model is a single objective mixed integer linear programming problem [23], which can be easily solved by a commercial solver, the Yalmip toolkit with CPLEX solver is utilized in this paper. The gird expansion planning model is a multi-objective mixed integer nonlinear programming problem, the multi-objective particle swarm optimization algorithm is utilized to search the pareto non-dominated solution set, the data envelopment analysis and analytic hierarchy process (DEAHP) decision method is used to comprehensively evaluate the optimal solution set to obtain the final optimal plan [24]. The flowchart of the detail process is shown in Figure 3.

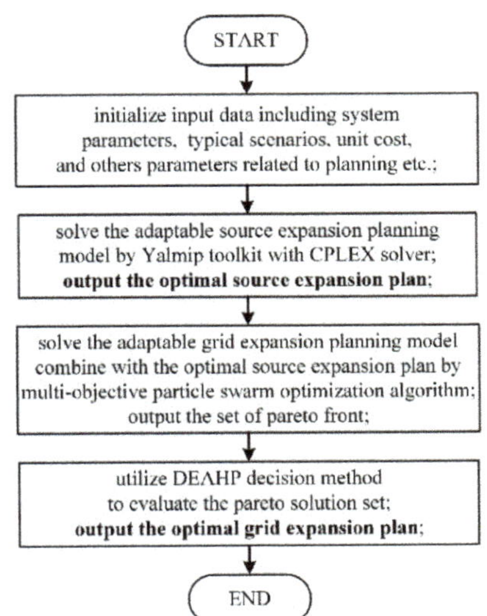

Figure 3. Flow chart of grid-generator coordination planning.

4. Results and Discussion

4.1. Basic Parameters

The Garver-18 bus power transmission system is used as a test system [25]. The network structure is shown in Figure 4. The remaining parameters are set as follows: a centralized wind farm is integrated at bus 16 and a centralized photovoltaic power plant is integrated at bus 14. The installed capacity of renewable energy unit is 20% of the conventional unit. The typical output scenario is given in Figures 5 and 6; the system runs for 360 days, totaling 8640 h; the threshold of insufficient rate index of supply and demand balance is set as 10%; the investment cost of transmission line construction is set to 800,000 yuan/km; the discount rate is set to 10%; the service life of the project is set to 15 years; the fixed operation rate of the project is set as 5%.

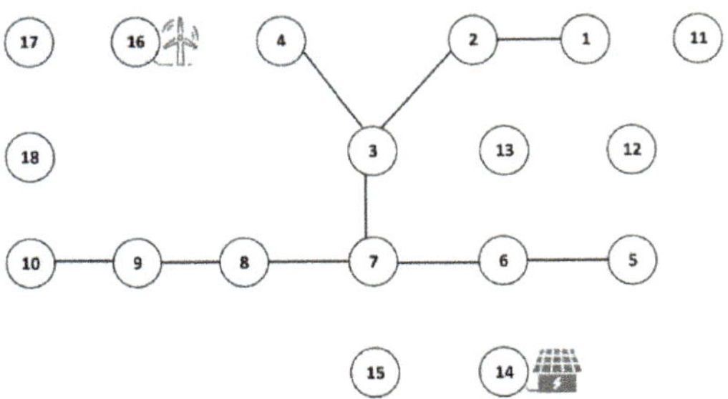

Figure 4. Garver-18 testing system.

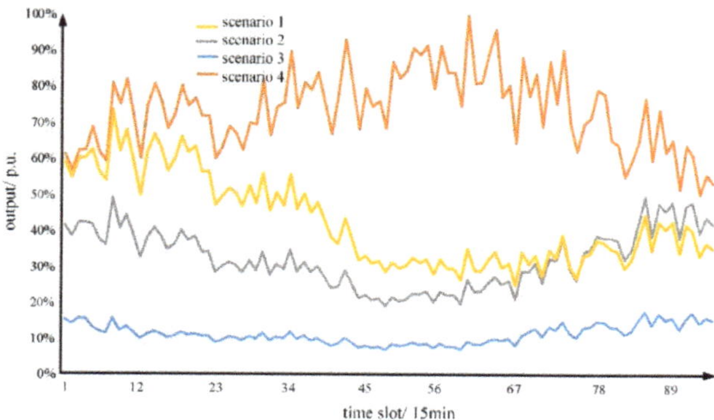

Figure 5. Typical scenarios of a wind farm.

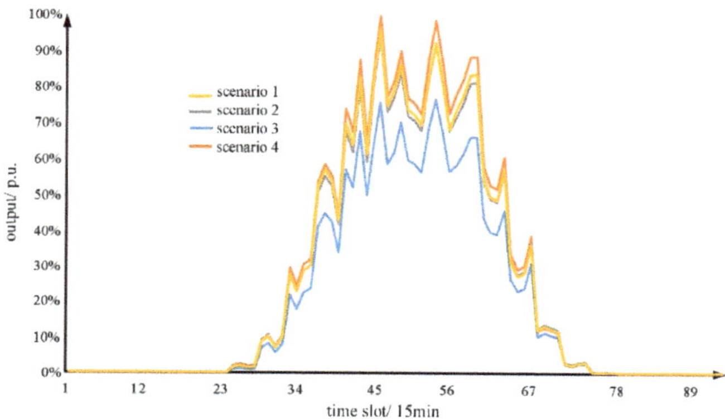

Figure 6. Typical scenarios of a photovoltaic power plant.

4.2. Model Validation

By solving the proposed adaptable source-grid two-stage expansion planning model, the optimal source-grid planning scheme is given in Table 1, and the economic optimal plan is also given in the last column as comparison.

Table 1. Optimal adaptable source-grid planning scheme.

Integrated Bus of Conventional Unit	Expanded Capacity of Ramp Rate/MW·15 min^{-1}	Expanded Circuit Number of Lines	Branch of Expanded Lines	
			Comprehensive Optimal Plan	Economic Optimal Plan
2 5 10	7.6 6.4 6.1	1	1-2, 3-7, 4-7, 4-16, 6-13, 7-13, 7-15, 9-16, 10-18, 11-12, 11-13, 16-17;	1-11, 4-16, 5-12, 6-13, 6-14, 7-8, 7-15, 9-16, 10-18, 11-12, 16-17;
11 14	5.5 5.8	2	1-11, 5-12, 6-14, 7-8, 14-15;	6-7, 7-13, 8-9, 9-10, 14-15, 17-18
16	5.9	3	8-9, 17-18,	\
18	7.6	4	9-10	\

4.2.1. Optimal Adaptable Source-Grid Plan

From the optimal source planning results in Table 1, it can be seen that in order to make the system regulation ability meet the threshold condition of insufficient supply and demand balance rate of less than 10%, the optimal scheme has been modified and improved the regulation rate parameters of each conventional generator set, but the unit has not been expanded. This is because the power capacity resources of the system are relatively abundant and the sum of the theoretical output of conventional power generation resources and renewable energy resources far exceeds the load power supply requirements of the system. At the same time, even if there is an extreme situation where the output of renewable energy is zero, the upper limit of the capacity allocation of conventional power generation resources can meet the load power. The main contradiction that causes the supply and demand balance to remain unmaintained is that the system cannot compensate for the regulation demand brought about by changes in renewable energy power in a relatively short time scale. Therefore, the planning model optimizes the regulation capability parameters of the conventional generator set, so that the regulation capability of the system can be improved. It can follow the magnitude and rate of the response to the regulation demand and truly releases the capacity advantage of the system. Therefore, under the new background of renewable energy grid connection, the main demand for power system regulation has changed from capacity demand to demand for rapid regulation capability.

The objective value of optimal grid planning scheme in Table 1 is given in Table 2. As can be seen from the results that although the economically optimal scheme can guarantee various safety constraints operation and meet the power supply requirements of the load, its power system status and structural adaptability indexes are weaker than the comprehensive optimal scheme. This can be understood as that although the comprehensive optimal scheme sacrifices part of the economy, it effectively considers the operating state and adaptability of the structure in the planning process and has a more reasonable operating state and a more robust grid structure. It can better adapt to the development trend of large-scale grid connection of renewable energy. At the same time, the comprehensive optimal plan and the power supply construction and operation plan have a better matching degree. The following is a further analysis of the specific differences between the operating state and the network structure.

Table 2. Objective value for optimal grid planning scheme.

Items	Comprehensive Optimal Plan	Economic Optimal Plan
cost of grid expansion/Ten thousand yuan	52,583.84	34,668.75
joint weighted entropy index of operation state	1.02	2.75
joint weighted entropy index of network structure	9493.70	10,668.53

4.2.2. Comparison for Adaptability of Supply and Demand Balance

The adaptability index of supply and demand balance is shown in Table 3. As we can see, the technical parameters have been improved, which has significantly improved the balance between supply and demand in the system. Before the transformation, the insufficient supply and demand balance rate of each typical scenario exceeded 20%. In the scenario where the demand for renewable energy regulation is relatively frequent, this index reaches a maximum of 46.87%. At the same time, the indexes of the upward and downward balance of supply and demand balance before the transformation are both large, indicating that there are frequent cases of urgent load or the curtailment of renewable energy. After the power supply adaptability evaluation and transformation, IRSD of each typical scenario are reduced to below 10%. Further analysis shows that due to the threshold constraint, all scenarios need to meet the supply and demand balance shortage rate of less than 10%. Therefore, the transformation result needs to ensure that the scene with the most stringent regulation requirements satisfies this condition and also makes the index far less than 10% in the remaining scenes, which is about 5%. Secondly, IRUR decreased by an average of 86.09% compared with that before the transformation, indicating that the improvement of the unit's regulation capability reduced the urgent load phenomenon caused by the insufficient power up-regulation capability and improved the reliability of power supply. Similarly, IRDR decreased by an average of 80.99%, indicating that the improvement of the unit's regulation capability reduced the phenomenon of renewable energy curtailment due to insufficient power down-regulation and increased the level of renewable energy consumption.

Table 3. Comparison for adaptability index of supply and demand balance.

Items	Scenario	IRSD	IRUR	IRDR
Before expansion	1	46.87%	48.94%	45.83%
	2	30.21%	36.17%	25.00%
	3	22.92%	21.28%	25.00%
	4	46.87%	56.25%	38.30%
After expansion	1	9.37%	4.26%	14.58%
	2	5.21%	4.26%	6.25%
	3	3.12%	4.26%	2.08%
	4	6.25%	8.51%	4.17%

4.2.3. Comparison for Adaptability of Operation State

Table 4 shows the specific distribution of the expected load rate in the two optimal plans. The first column is the load rate interval, which is used to evaluate the reasonable degree of line load level. In this example, the rational ranking of each interval is set as: (40–60%) > (20–40%) > (0–20%) > (60–80%) > (80–100%). It can be seen from the results that the comprehensive optimal scheme does not have lines running in the last two intervals of the rational order of load levels and the economically optimal scheme has lines distributed in each interval and the penetration of lines above 80% load rate is as high as 19%. It can be seen that the economic optimal plan has a situation where the load level of individual lines is too high and the concentration of its overall operating state is significantly weaker than the comprehensive optimal plan. From the self-organized critical theory, it can be seen that the safety level of the grid state of the economically optimal solution is poor. On the contrary, the comprehensive optimal scheme not only makes the grid state more orderly, but also ensures the efficiency of grid operation to the greatest extent. A total of 54% of the lines in the comprehensive optimal plan operate in the most reasonable load rate range preset by planners, while the data for the economically optimal plan is only 25%. This shows that the running state of the comprehensive optimal scheme is more reasonable, taking into account the running efficiency and safety and is more in line with the running ideal state preset by the planning operator. Although the absolute value of efficiency is higher in the economically optimal solution, it is actually because some heavy-load lines increase the overall value and its true efficiency is significantly worse than the comprehensive optimal solution. It can be seen that the comprehensive optimal scheme has higher operating quality.

Table 4. Distribution of expected line load rate.

Interval of Load Rate	Comprehensive Optimal Plan		Economic Optimal Plan	
	Number of Lines	Penetration	Number of Lines	Penetration
0–20%	11	27%	10	31%
20–40%	8	20%	3	9%
40–60%	22	54%	8	25%
60–80%	0	0%	5	16%
80–100%	0	0%	6	19%

Table 5 shows the specific distribution of the power fluctuation rate of each line of the two schemes. Lines whose actual maximum load rate is less than 10% are not considered, because the absolute load level of such lines is too low and small changes in power will cause a large power fluctuation rate index, but the absolute magnitude of the change in load level is still small. It can be seen from the table that the fluctuation rate of 86% line in the comprehensive optimal scheme is below 20% of its allowable fluctuation interval, of which the line power fluctuation rate is less than 10% and the penetration of lines is 71%. The two data corresponding to the optimal economic plan correspond to 71% and 21%, respectively. In summary, the operating state of the optimal scheme is more stable, which can better resist the impact of renewable energy power fluctuations on the system state under uncertain operating environments and always maintain the safety and efficiency of the operating state during actual operation. At the same time, it can be found that the two schemes still have a small amount of line power fluctuation rate, which is because the line mainly undertakes the task of renewable energy transmission. Large fluctuations in the output of renewable energy have led to large fluctuations in the power of these lines, but they have not had a significant impact on the overall operating state.

Table 5. Distribution of line power fluctuation rate.

Interval of Fluctuation Rate	Comprehensive Optimal Plan		Economic Optimal Plan	
	Number of Lines	Penetration	Number of Lines	Penetration
0–10%	24	71%	6	21%
10–20%	5	15%	14	50%
20–30%	3	9%	6	21%
30–40%	0	0%	0	0%
40–50%	2	6%	1	4%
50–60%	0	0%	0	0%
60–70%	0	0%	1	4%
70–80%	0	0%	0	0%
80–90%	0	0%	0	0%
90–100%	0	0%	0	0%

4.2.4. Comparison for Adaptability of Network Structure

Tables 6 and 7 show the distribution of the interval of the electric betweenness of buses and branches of the two schemes. The first column in the table is the interval of the electric dielectric value, which is the objective standard used to evaluate the robustness degree of branch and bus. The higher the ratio of the number of components in the smaller area is, the more reasonable and robust network structure is. The results show that 88% of the total number of branches in the comprehensive optimal scheme are distributed in the range of (0–80) low electric medium and 81% in the economic optimal scheme. In the same way, 72% of the total buses of the comprehensive optimal scheme are distributed in the (0–80) low electric medium range, while the data of the economic optimal scheme are only 56%. In conclusion, in the comprehensive optimal scheme, the electrical mediums of branches and buses are more centralized and orderly distributed in the range of low severity, the rationality of its network structure is obviously superior to the economic optimal scheme and its resistance and adaptability to uncertain impact are stronger.

Table 6. Distribution of electric betweenness of bus.

Interval of Electric Betweenness	Comprehensive Optimal Plan		Economic Optimal Plan	
	Number of Buses	Penetration	Number of Buses	Penetration
0–40	7	39%	7	39%
40–80	6	33%	3	17%
80–120	1	6%	4	22%
120–160	4	22%	4	22%
160–200	0	0%	0	0%

Table 7. Distribution of electric betweenness of branch.

Interval of Electric Betweenness	Comprehensive Optimal Plan		Economic Optimal Plan	
	Number of Lines	Penetration	Number of Lines	Penetration
0–40	20	49%	16	50%
40–80	16	39%	10	31%
80–120	1	2%	4	13%
120–160	4	10%	2	6%
160–200	0	0%	0	0%

5. Conclusions

From the perspective of adapting to the security integration of renewable energy, this paper firstly proposes the adaptability indexes system and applies it to the two-stage adaptable expansion planning of the transmission grid. The following conclusions can be draw:

(1) The adaptability of supply and demand balance not only evaluates the power balance of the system, but also focuses on the adequacy of the system regulation capacity and rate, which can reflect the level of renewable energy consumption and the risk of being urgently loaded; the adaptability of operation state consider the characteristics of power flow distribution change caused by renewable energy output fluctuation and the security, efficiency and stability of the actual operation of system; the adaptability of network structure evaluation the balance and robustness of system based on the order and severity of the electric betweennesses.
(2) The weighted entropy index can evaluate the adaptability state and network structure based on the given objective criteria. Not only can it achieve the ranking of relative advantages and disadvantages between the programs, but also it can refine the specific differences of program states and structures based on the objective standards. The objective evaluation criteria can be set by planner according to different considerations in different development periods, which improves the flexibility and applicability of the indexes. The application of joint weighted entropy can combine the physical meaning of each index to achieve dimensionality reduction and integration of indexes, so as to comprehensively measure the security of the state and structure with a single value. It is more suitable for the analysis safety adaptability under multi-dimensional influence factors and the multi-objective planning.
(3) The expansion planning of the high renewable energy penetrated system is vital and decisive for the actual consumption capacity in the operation stage, both the planning of the power plant in the source side and transmission line in the grid side are interrelated. For the improvement of safe integration of renewable energy in the planning stage, it is suggested to comprehensively consider the adaptability of supply and demand balance, operation state, and network structure in source-grid planning issues.

Author Contributions: Conceptualization, M.T.; methodology, J.W.; validation, J.W. and X.W.; writing—original draft preparation, J.W.; writing—review and editing, M.T. and X.W. All authors have read and agreed to the published version of the manuscript.

Funding: This work is founded by the National Key R&D Program of China (2018YFB0905200)—Research and application demonstration on complementary combined power generation technology for distributed photovoltaic and cascade hydropower.

Conflicts of Interest: The authors declare no conflict of interest.

Nomenclature

$P_D{}^{up}$, $P_D{}^{down}$	up- and down-regulation demand (MW)
$P_{i,REG}$	output power of the renewable energy unit i (MW)
Ω_{REG}	the set for renewable energy units
$P_S{}^{up}$, $P_S{}^{down}$	up- and down-regulation capability (MW)
Ω_G	the set of fast regulating units
$P_{i,G}$	output power of the fast regulating units i at time t (MW)
$P_{i,G}{}^{max}$, $P_{i,G}{}^{min}$	upper and lower output of the fast regulating generator units (MW)
$R_{i,g}{}^{up}$, $R_{i,g}{}^{down}$	maximum up and down ramp rate of the fast regulation units (MW/15min)
$P_M{}^{up}$, $P_M{}^{down}$	upper and lower regulation resource margins (MW)
$T_{ins}{}^{up}$	the time when the up-regulation resources are insufficient
$T_{total}{}^{up}$	the total operation time when the up-regulation demand occurs
$T_{ins}{}^{down}$	the time when the down-regulation resource is insufficient
$T_{total}{}^{down}$	the total operation time when the down-regulation demand occurs
P_i	the actual transmission power of line i
$S_{i,max}$	the rated transmission power capacity of the line
ω_a, ω_b	the weights of generator and load bus
$B_{e,ab}(i)$	the current distribution of bus i
$I_{ab}(i, j)$	the current of branches (i, j)
C_G	the cost of regulation capacity transformation
C_{WR}	the annual cost of renewable energy generation
C_{WL}	the annual load shedding cost
C_N	the cost of grid expansion
$\Sigma P_L(s,t)$	the total load at time slot t in scenario s
IRSD	insufficient rate of supply and demand balance
IRUR	insufficient rate of up-regulation
IRDR	insufficient rate of down-regulation
AIUR	average insufficiency of up-regulation
AIDR	average insufficiency of down-regulation
WEELLR	weighted entropy of expected line load rate
WELPFR	weighted entropy of line power fluctuation rate
JWEOS	joint weighted entropy of operation state
WEEBBus	weighted entropy for electric betweenness of bus
WEEBBra	weighted entropy for electric betweenness of branch
JWENS	Joint weighted entropy of network structure

References

1. Eitan, A.; Rosen, G.; Herman, L.; Fishhendler, I. Renewable energy entrepreneurs: A conceptual framework. *Energies* **2020**, *13*, 2554. [CrossRef]
2. Murdock, H.E.; Gibb, D.; André, T. *Renewables 2019 Global Status Report*; REN21: Paris, France, 2019.
3. Kiviluoma, J.; Holttinen, H.; Weir, D.; Scharff, R.; Soder, L.; Menemenlis, N.; Cutululis, N.A.; Danti Lopez, I.; Lannoye, E.; Estanqueiro, A.; et al. Variability in large-scale wind power generation. *Wind Energy* **2016**, *19*, 1649–1665. [CrossRef]
4. Zsiborács, H.; Baranyai, N.H.; Vincze, A.; Zentkó, L.; Birkner, Z.; Máté, K.; Pintér, G. Intermittent Renewable Energy Sources: The Role of Energy Storage in the European Power System of 2040. *Electronics* **2019**, *8*, 729. [CrossRef]
5. Lu, Z.; Huang, H.; Shan, B.; Wang, Y.; Du, S.; Li, J. Morphological evolution model and power forecasting prospect of future electric power systems with high proportion of renewable energy. *Autom. Electr. Power Syst.* **2017**, *41*, 12–18.
6. Lannoye, E.; Flynn, D.; O'Malley, M. Transmission, variable generation, and power system flexibility. *IEEE Trans. Power Syst.* **2015**, *30*, 57–66. [CrossRef]
7. Wheelan, S.A. *Harnessing Variable Renewables*; International Energy Agency: Paris, France, 2011.

8. Lannoye, E.; Flynn, D.; O'Malley, M. Evaluation flexibility. *IEEE Trans. Power Syst.* **2012**, *27*, 922–931. [CrossRef]
9. Nosair, H.; Bouffard, F. Flexibility envelopes for power system operational planning. *IEEE Trans. Sustain. Energ.* **2015**, *6*, 800–809. [CrossRef]
10. Ma, J.; Silva, V.; Belhomme, R.; Kirschen, D.S.; Ochoa, L.F. Evaluating and planning flexibility in sustainable power systems. *IEEE Trans. Sustain. Energ.* **2013**, *4*, 200–209. [CrossRef]
11. Zhuo, Z.; Du, E.; Zhang, N.; Kang, C.; Xia, Q.; Wang, Z. Incorporating Massive Scenarios in Transmission Expansion Planning with High Renewable Energy Penetration. *IEEE Trans. Power Syst.* **2020**, *35*, 1061–1074. [CrossRef]
12. Li, X.; Fan, J.; Cao, L.; Wang, L.; Hu, T.; Wu, T.; Zhou, Z. Analysis on the adaptability of large-scale grid-connected PV station. *Power Sys. Prot. Control* **2018**, *46*, 164–169.
13. Liu, W.; Cai, W.; Yangqing, D.; Liang, C.; Wang, J.; Wang, N.; Ma, Y. The evolution of grid's self-organized criticality state based on entropy theory under the circumstance of large-scale wind power centralized grid. *Power Syst. Technol.* **2013**, *37*, 3392–3398.
14. Zhang, C.; Cheng, H.; Liu, L.; Wang, Z.; Lu, J.; Zhang, X. Adaptability index and evaluation method for power transmission network structure with integration of high penetration of renewable energy. *Autom. Electr. Power Syst.* **2017**, *41*, 55–61.
15. Dai, W.; Yang, Z.; Yu, J.; Zhao, K.; Wen, S.; Lin, W.; Li, W. Security region of renewable energy integration: Characterization and flexibility. *Energy* **2019**, *187*, 115975. [CrossRef]
16. Papavasiliou, A.; Oren, S.S. Large-Scale integration of deferrable demand and renewable energy sources. *IEEE Trans. Power Syst.* **2014**, *29*, 489–499. [CrossRef]
17. Khan, S.U.; Mehmood, K.K.; Haider, Z.M.; Rafique, M.K.; Kim, C. A bi-level EV aggregator coordination scheme for load variance minimization with renewable energy penetration adaptability. *Energies* **2018**, *11*, 2809. [CrossRef]
18. Sun, W.; Wang, C.; Zeng, P.; Han, J. Review on evaluation method and index of power system homogeneity. *Power Syst. Technol.* **2015**, *39*, 1205–1212.
19. Liu, W.; Cai, W.; Zhang, N.; Dan, Y.; Liang, C.; Wang, W.; Liang, C. Evolution of self-organizing of grid critical state based on united weighted entropy theory. *Proc. CSEE* **2015**, *35*, 1363–1370.
20. Sheng, S.; Gu, Q. A day-ahead and day-in decision model considering the uncertainty of multiple kinds of demand response. *Energies* **2019**, *12*, 1711. [CrossRef]
21. Ji, W.; Wu, J.; Zhang, M.; Liu, Z.; Shi, G.; Xie, X. Blind image quality assessment with joint entropy degradation. *IEEE Access* **2019**, *7*, 30925–30936. [CrossRef]
22. Bai, H.; Miao, S. Hybrid flow betweenness approach for identification of vulnerable line in power system. *IET Gener. Transm. Distrib.* **2015**, *9*, 1324–1331. [CrossRef]
23. Teimourzadeh, S.; Aminifar, F. MILP Formulation for Transmission Expansion Planning with Short-Circuit Level Constraints. *IEEE Trans. Power Syst.* **2016**, *31*, 3109–3118. [CrossRef]
24. Cheng, Y.; Zhang, N.; Wang, J.; Li, H.; Wang, Z.; Xie, L.; Kang, C. Comprehensive Evaluation of Transmission Network Planning for Integration of High-penetration Renewable Energy. *Autom. Electr. Power Syst.* **2019**, *43*, 33–42; 57.
25. Huang, Y.; Li, Y.; Gao, C. Multi-objective transmission network planning based on non-dominated sorting differential evolution. *Dianwang Jishu/Power Syst. Technol.* **2011**, *35*, 85–89.

Article

Optimising a Microgrid System by Deep Reinforcement Learning Techniques

David Domínguez-Barbero *, Javier García-González, Miguel A. Sanz-Bobi and Eugenio F. Sánchez-Úbeda

Institute for Research in Technology (IIT), ICAI School of Engineering, Comillas Pontifical University, 28015 Madrid, Spain; javiergg@comillas.edu (J.G.-G.); masanz@comillas.edu (M.A.S.-B.); eugenio.sanchez@comillas.edu (E.F.S.-Ú.)

* Correspondence: david.dominguez@iit.comillas.edu

Received: 16 April 2020; Accepted: 28 May 2020; Published: 2 June 2020

Abstract: The deployment of microgrids could be fostered by control systems that do not require very complex modelling, calibration, prediction and/or optimisation processes. This paper explores the application of Reinforcement Learning (RL) techniques for the operation of a microgrid. The implemented Deep Q-Network (DQN) can learn an optimal policy for the operation of the elements of an isolated microgrid, based on the interaction agent-environment when particular operation actions are taken in the microgrid components. In order to facilitate the scaling-up of this solution, the algorithm relies exclusively on historical data from past events, and therefore it does not require forecasts of the demand or the renewable generation. The objective is to minimise the cost of operating the microgrid, including the penalty of non-served power. This paper analyses the effect of considering different definitions for the state of the system by expanding the set of variables that define it. The obtained results are very satisfactory as it can be concluded by their comparison with the perfect-information optimal operation computed with a traditional optimisation model, and with a Naive model.

Keywords: machine learning; microgrids; optimisation methods; power systems; reinforcement learning

---

1. Introduction

The transformation of the electric power industry to reduce carbon emissions is changing the roles of both generators and consumers. On the one hand, the generation system is hosting larger amounts of renewable energy sources (RES), with an increasing share of distributed generation, i.e., small generators placed at the distribution level. On the other hand, end consumers are no longer passive agents of the system, but rather active ones that can modulate their consumption during the day, and reduce their net energy balance by means of more efficient and smarter operation of the loads. In this increasingly decentralised power system, involved agents must make decisions continuously in a limited information environment. This decentralisation can be materialised by the creation of micro-networks (microgrids) that are groupings of distributed generation resources, consumption and storage systems that can be partially controlled and that can work both connected to the conventional network or in an isolated manner. The development of smart grids suggests that despite the existing barriers, the penetration of these microgrids will increase considerably soon [1].

Finding the optimal operation of the different elements of the microgrid is not a simple task, and there are several possible approaches in the literature such as robust optimisation [2], stochastic optimisation [3], Mixed Integer-Linear Programming (MILP) combined with a Particle Swarm Optimisation [4] or Reinforcement Learning (RL) [5]. It is important to highlight that robust or stochastic optimisation models require an explicit representation of the microgrid, a mathematical formulation that includes the objective function to be optimised and the set of all the constraints that define the feasible domain. In addition, the input parameters required to forecast the random variables of interest, such as loads and RES generation, are also a practical need for those optimisation approaches. Therefore, the development of an optimisation-based controller requires the precise selection and estimation of all the parameters and models used to represent each component. Obtaining such information could be extremely difficult in the case of a massive deployment of microgrids. For instance, in a hypothetical future scenario of the power system hosting millions of microgrids (from residential prosumers to larger microgrids in industrial areas), the diversity of consumption patterns and the intermittent nature of renewable generation might represent a serious barrier, as each specific microgrid may require different parameters and even different models [6] leading to unacceptable implementation costs.

In this context, the main advantage of RL is that it allows making decisions in a limited information framework based on the actions taken in the past, and on the observed effects of such actions by automatically adapting the control strategy over time. Unlike standard model-based control approaches, RL techniques do not require the identification of the system in the form of a model. In this way, RL is able to learn from the experience similarly to human behaviour. For example, in [7], it is shown through numerical examples that the proposed learning algorithms can work effectively despite not having prior information on the dynamics of the system. Another example can be seen in [8] where an RL algorithm is proposed to coordinate the loading and unloading of several storage systems with different characteristics. The authors in [9] present a comprehensive review of the main research works found in the literature where RL is applied to a variety of typical problems of power systems, and it presents an RL framework to operate real systems meanwhile a simulation model is used to learn.

Over the last years, a new version of RL methods has emerged. It is called Deep Reinforcement Learning (DRL) because it combines RL with Deep Learning (DL). These methods have an important impact on the academic field, but the industrial applications are currently at the early stages of development. As this is a very active research field, there is an increasing trend of papers that study the application of DRL to power systems. For instance, in [10] several research works that apply RL to power systems are discussed, and some of them are focused on DRL methods. Among them, Deep Q-Network is the most popular technique given its simplicity in contrast to other more complex DRL methods. It should be noted that despite being a recent publication, none of the papers reviewed in [10] address the problem of finding the optimal steady state operation of a microgrid by using DRL.

A more updated and very recent survey paper is [11], where the authors present a broad review of RL applications to power systems, although the categorisation differs from [10]. In particular, one of the categories is the energy supply side, which is very related to the topic studied in this paper, i.e., the microgrid optimal scheduling problem. Under this research area, reference [12] should be highlighted as it addresses the optimisation of the operation of the microgrid using a Deep Q-Network (DQN). The Artificial Neural Network (ANN) used in [12] to implement the DQN is a Multi-Layer Perceptron (MLP). Nevertheless, there are more advanced ANN architectures such as Convolutional Neural Networks (CNN), with one-dimension convolution layers, and Recurrent Neural Networks (RNN), with memory cell layers. These architectures are very suitable for improving the train phase when the input data has a time series structure. For instance, the authors in [5] developed a CNN to find the optimal operation of a microgrid that could be operated both isolated or connected to grid.

The microgrid can be modelled as a Markov Decision Process (MDP). Therefore, beyond the selection of the ANN architecture, the application of DRL (and also RL) requires a definition of a state space of the environment so that the agent can select the optimal actions based on which state the system is in. However, deciding what defines the state of the system may not be trivial. For example, in chess, the position of all the pieces on the chessboard at any given time defines the state of the system, and it is perfectly observable by the player. However, in a microgrid, the decision of what defines the state of the system is not so direct, since the number of variables to be measured can be as large as desired and the evolution over time is subject to uncertainty. The work in [13] presents the theoretical consequences of varying the configuration of the state and discusses the bias-overfitting tradeoff of DRL. This paper delves also on the effect of the state configuration, and following the suggested research guidelines of [5], the impact of different state definitions of the microgrid are analysed. Therefore this paper takes [5] as a starting point using a DQN based on CNN. In addition, as the time window used to define the state can have different lengths, the main contribution of this paper is the comparative analysis of the impact of the duration of such time window on the performance of the model. In addition, the considered microgrid is more complex than the one studied in [5] as it includes a diesel generator that can act as backup power. Finally, another contribution of this paper is that the economic assessment of the operation of the batteries does not require fixing any price, as the opportunity cost is properly modelled in the sense that the charge/discharge decisions have a direct impact on the usage of the diesel generator. Therefore, this approach is more realistic and can make its application easier to any particular system.

The paper is organised as follows. Section 2 introduces the microgrid elements, and also the modelling fundamentals. The DRL algorithm is described in Section 3, and Section 4 presents its application to a study case where the microgrid structure is described in detail. Section 5 analyses the RL behaviour, comparing the results among the different state configurations. Finally, the main conclusions are summarised in Section 6.

2. The Microgrid Framework

2.1. General Overview of a Microgrid

Figure 1 shows a schematic representation of a microgrid that can be connected to the network through the Point of Common Coupling (PCC). The arrows indicate the possible directions of power flows. The control system is in charge of determining the operation of all the manageable elements of the microgrid. The distributed generation can be fossil fuel generators (for example diesel generators), or renewable sources such as wind or solar generation. Regarding the loads, the microgrid includes elements that consume electricity, such as lighting devices, heating and cooling systems of buildings or charging of electric vehicles. Electricity consumption can change over time and, if the loads are controllable, be susceptible to demand management programs. Finally, the storage systems such as electric batteries, flywheels, heat storage and more, can allow the microgrid to perform multiple functions. Depending on the technology used, the storage capacity can vary and can have a fast or slow dynamic response. In addition, thanks to the use of power electronics and control devices, the storage systems can provide ancillary services such as frequency regulation, and voltage control, although this is out of the scope of this paper. Under a steady-state perspective where the objective is to find the hourly scheduling of all the elements, the primary function of the storage systems is to absorb the inherent variability of electricity consumption and RES generation and to provide backup generation when needed. Finally, in this paper, it is assumed that the microgrid is not connected to the network, and the elements with the dashed line in Figure 1 are not included in the studied configuration.

2.2. Description of the Studied Microgrid

The main elements to be defined are the energy demand, the solar panel, the diesel generator and the energy storage system that consists of a Li-ion battery and a hydrogen tank. Both the physical characteristics of these elements in the studied microgrid, as well as the data used for fitting and testing the models, are discussed in later Section 4.1.

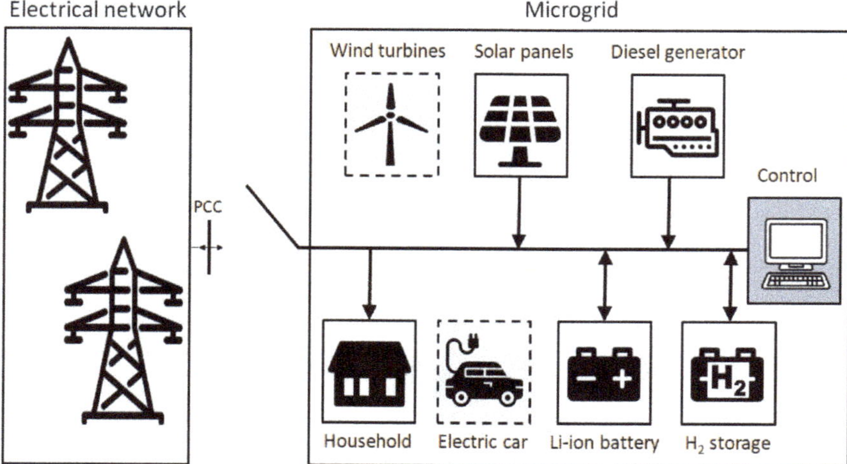

Figure 1. Schematic example of a microgrid.

2.2.1. Demand

The load is modelled as a time series with hourly intervals that represent the average consumption of each hour, and it is assumed that historical data is available. Furthermore, this demand is not dependent on the behaviour of all the other elements of the microgrid. In case there is not enough generation to supply the microgrid demand, the non-served power will take the required positive value to ensure the demand balance equation. This non-served power will be severely penalised. However, in case there is an excess of RES generation, it will be assumed that such extra production can be curtailed without any cost (RES spillages).

2.2.2. Generation

The microgrid generation system consists of a solar panel and a diesel generator. Regarding the solar panel, the maximum PV generation profile can be represented by an hourly time series, and its variable operation cost can be neglected. On the other hand, the diesel generator can provide backup energy in case of a lack of RES production. This diesel generator can be dispatched by the control system that establishes the power to be produced in a deterministic manner, i.e., possible failures of the equipment is out of the scope of this paper. Two different operation modes will be considered. In the first one, the output power is allowed to take any continuous value between the minimum and the maximum power. In the second one, only discrete values are possible. For simplicity, the input-output cost curve is modelled as a quadratic function where the independent term, i.e., the no-load cost, is only incurred when the generator is committed. Start-up and shutdown costs are neglected, and further details can be seen in Section 4.

2.2.3. Storage

With respect to the storage, a Li-ion battery is used as short-term storage supporting the microgrid in periods when RES cannot supply the demand. It is characterised by the maximum capacity R_t^B [kWh], the maximum power P_t^B [kW] for charge and discharge and the charge and discharge efficiencies μ_t^B, ζ_t^B. It is important to highlight, under the proposed RL approach, that it would be easy to take into account a more detailed model of the battery to capture the non-linear relationship among the state of charge (SoC), the maximum power, the efficiencies and other characteristics. However, as the classical optimisation model used to compare the results cannot include all those non-linear relationships, a simple model has been preferred. The other storage device, i.e., a hydrogen system, is used as long-term storage, supporting large periods of high demand and low RES. The characterisation of the hydrogen storage is analogous to the Li-ion battery: maximum capacity R_t^H [kWh], maximum power P_t^H [kW] for charge and discharge and the charging and discharging efficiencies μ_t^H and ζ_t^B.

3. Reinforcement Learning

RL belongs to the area of Machine Learning benefiting from ideas of optimal control of incompletely-known Markov decision process, and it has been applied to many fields [14]. An RL problem is modelled as an agent and an environment that interact with each other as in Figure 2. The agent makes decisions following a policy, exploring the environment dynamics and obtaining rewards. The agent updates its policy π guided by these rewards r_t over a discrete space of time divided in time steps t, improving the cumulative reward.

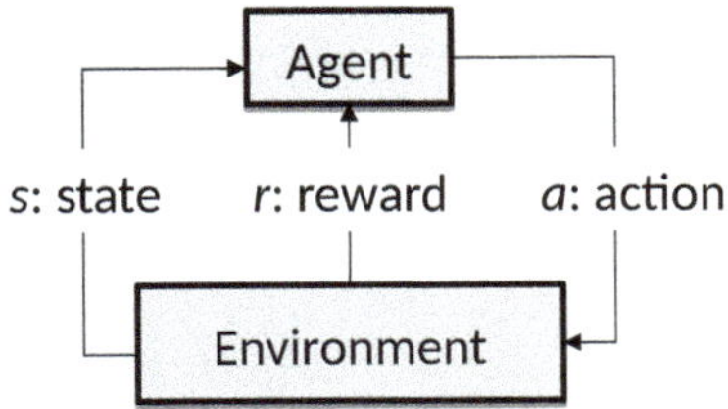

Figure 2. General interactions in Reinforcement Learning (RL).

In the RL context, a state space S and an action space A has to be defined. Given an initial state s_0, a policy $\pi \in \Pi$ performs a trajectory $\tau = (s_0, a_0, s_1, a_1, ..., s_n)$ where s_n is a terminal state. A reward function $R(s_t, a_t, s_{t+1})$ and a return function $G(\tau)$ have to be defined also according to the RL problem. The optimal policy π^* is the one that maximises the expected return. An intrinsic feature of an RL algorithm is the policy-based and the value-based characteristic, which affects the method of computing the next action and the learning behaviour. Some methods have both characteristics, but the proposed method in this paper has only the value-based characteristic.

3.1. Deep Reinforcement Learning (DRL)

Due to the curse of dimensionality, approximation functions are commonly used in more complex problems to approximate the policy, and given the significant impact of Deep Learning, a collection of RL algorithms using Artificial Neural Networks (ANN) has emerged, [15–17].

In this paper, a Deep Learning version of Q-Learning [18] called Deep Q-Network (DQN) is used [19], with a configuration and architecture similar to [5]. This method uses an ANN to approximate its value function, called Q-function. This method is model-free, i.e., the model does not use experiences to create an internal model to generate synthetic experiences and learn with, unlike model-based such as Dyna-Q [20] or Monte Carlo Tree Search (MCTS) [21]. Another characteristic of some of these algorithms is the capacity of work with both continuous and discrete action spaces, but DQN works only with discrete action spaces, hence, environments with continuous action spaces have to be discretised to be able to apply this algorithm in this kind of environments.

Deep Q-Network (DQN)

One of the most known algorithms in RL is Q-Learning. Every method in RL needs a policy function to know what action the agent has to take to perform the best trajectory and earn the best return from a state. In Q-Learning, the policy function is computed using the $Q(\cdot)$ function

$$Q(s,a) = r + \gamma \max_{a'} Q(s',a'), \tag{1}$$

and with this function, the algorithm could perform the best believed action:

$$best\,a = \underset{a}{\operatorname{argmax}}\, Q(s,a). \tag{2}$$

The Q-function is updated as follows:

$$Q(s,a) = Q(s,a) + \alpha[r + \gamma \max_{a'} Q(s',a') - Q(s,a)], \tag{3}$$

with convergence to the optimal $Q^*(s,a)$. s' and a' are the next state and the action to be performed in s' respectively.

The curse of dimensionality phenomena makes the Q-Learning algorithm unsuitable to solve high dimensional problems, and approximations of the $Q(\cdot)$ function have to be introduced to deal with this drawback [14]. Since ANNs are universal approximators, DQN can be implemented to model the $Q_\theta(\cdot)$ function, being θ the parameters of the ANN. The updating process of the Q-values has to be computed to update each parameter θ of the ANN. Additionally, DQN uses a replay memory to help to efficiently update the parameters and get a good trade-off between exploration and exploitation [22].

3.2. Application of RL in the Microgrid

3.2.1. Action Definition

At every time period t, the RL algorithm must decide how much power should be charged or discharged at the Li-ion battery and hydrogen storage, and the power that should be produced by the diesel generator. Therefore, the agent action at time t is defined as:

$$a_t = (P_t^{h2}, P_t^d). \tag{4}$$

3.2.2. State Definition

The microgrid state can be defined as a tuple of k temporal slices. The idea is not only to rely on the instantaneous information of the microgrid at time step t, but also to consider the information recorded from the most recent stages. Therefore, a window-based state s_t is modelled for each time step as in [13]:

$$s_t = (slice_{t-k}, slice_{t-k+1}, ..., slice_{t-1}, slice_t). \tag{5}$$

Regarding the variables to be monitored, in this paper it is proposed that each temporal slice is defined by the next variables:

$$slice_t = (P_{t-1}^{pv}, D_{t-1}, S_t^b, S_t^{h2}). \tag{6}$$

Notice that the state of charge (SoC) corresponds to the value of energy stored at the end of each hour, and that the generated or consumed power represents its average value during each hourly period.

In order to compute the SoC of both storage devices, the following energy balance equations must be taken into account:

$$S_t^b = \begin{cases} S_{t-1}^b - \mid P_t^b \mid / \zeta^b & \Leftrightarrow P_t^b \geq 0 \\ S_{t-1}^b + \mid P_t^b \mid \eta^b & \Leftrightarrow P_t^b < 0 \end{cases}, \tag{7}$$

$$S_t^{h2} = \begin{cases} S_{t-1}^{h2} - \mid P_t^{h2} \mid / \zeta^{h2} & \Leftrightarrow P_t^{h2} \geq 0 \\ S_{t-1}^{h2} + \mid P_t^{h2} \mid \eta^{h2} & \Leftrightarrow P_t^{h2} < 0 \end{cases}. \tag{8}$$

Notice that both storage devices must be operated within their feasible limits, and this is ensured by a continuous monitoring at each decision step.

3.2.3. Reward Definition

The reward function, expressed in €, is directly related to the generation cost of the diesel generator, plus the possible penalty of the non-served power:

$$r_t = R(s_t, a_t, s_{t+1}) = -C^d(P_t^d) - c_{\text{pns}} P_t^{\text{pns}}, \tag{9}$$

where

$$C^d(x) = \delta_2 x^2 + \delta_1 x + \delta_0 \tag{10}$$

is the diesel cost function [€] with quadratic coefficients δ_2, δ_1 and δ_0. P^b, P_t^{pns} and P_t^{curt} are calculated using the demand balance-constraint equation given by

$$(D_t - P_t^{\text{pns}}) = (P_t^{pv} - P_t^{curt}) + P_t^b + P_t^{h2} + P_t^d, \tag{11}$$

prioritising the usage of the battery to reduce the not-supplied power and the curtailment.

4. Case Study

This section describes the case study used in this paper, as well as the different configurations of the proposed Deep Reinforcement Learning (DRL) that have been used to perform the comparative analysis.

4.1. Description of the Data

The data set consists of three consecutive years with hourly values. Therefore, $|T| = 3 \times 8760 = 26,280$. The input data are the solar panel maximum generation profile, the hourly load, and all the technical characteristics of the other components. All these parameters are explained in detail hereafter and summarised in Table 1.

Table 1. Microgrid parameters.

Component	Parameter	Value
Solar Panel	P_{max}^{pv} [kW]	6
Load	D_{max} [kW]	2.1
Diesel	P_{max}^d [kW]	1.0
	δ_2 [€/kW^2]	0.31
	δ_1 [€/kW]	0.108
	δ_0 [€]	0.0157
Li-ion battery	P_{max}^b [kW]	2.9
	S_0^b [kWh]	0
	S_{max}^b [kWh]	2.9
	ζ^b	0.95
	η^b	0.95

Table 1. *Cont.*

Component	Parameter	Value
H_2 storage	P^{h2}_{max} [kW]	1.0
	S^{h2}_0 [kWh]	100
	S^{h2}_{max} [kWh]	200
	ζ^{h2}	0.65
	η^{h2}	0.65
cost of not supplied power	c_{pns} [€/kWh]	1

4.1.1. Photovoltaic Panels

The solar generation profile is given by the irradiation data used from [5] that can be obtained from the the Deep Reinforcement learning framework (DeeR) python library [23]. This data was collected from a Belgium location and the factor of the aggregated irradiation by month is 1:5 between the lowest and the highest monthly available solar generation. The optimal sizing of the microgrid is out of the scope of this paper. Therefore, the number of panels has been chosen just by dividing the annual demand by the total irradiation taking into account a standard efficiency for the PV panels. As a result, the PV system dimension for the case study is 30 m^2, with a maximum installed power of 6 kW, which is half of the size proposed by [24]. Furthermore, some websites as [25] discuss that a 6 kW PV system is a commonly used figure in the U.S.

4.1.2. Load

The hourly data for the load profile has also been obtained from [23]. This load profile represents the typical residential consumption of a consumer with an average daily demand of 18.33 kWh. In case the microgrid is unable to supply all the load, it is assumed a cost 1€ per not supplied kWh.

4.1.3. Energy Storage Elements

The Li-ion battery charge and discharge efficiency rates are 0.95 for both processes. For the hydrogen (H_2) electrolyser (charge) and fuel cell (discharge) the efficiency rates are 0.65 for both cases, [24]. The battery capacity and hydrogen storage size are 2.9 kWh and 200 kWh. The battery model has the characteristics of an LG Chem RESU3.3 used for households, with a maximum power of 3.0 kW but clipped to 2.9 kW for the microgrid (due to the hourly step). The hydrogen fuel cell maximum power is 1.0 kW based on a Horizon 1000W PEM fuel cell, and for symmetry 1.0 kW for the electrolyser.

4.1.4. Non-Renewable Generation

The proposed microgrid has a diesel generator with a nominal rate of 1.0 kW. The cost curve of this diesel generator has been adjusted to a quadratic-curve (see Equation (10)), adapting the parameters of the IEEE 30 bus system generators used in [26]. Table 1 shows the obtained coefficients of the obtained cost polynomial.

4.2. *Optimisation-Based Model (MIQP) Used as a Reference Model and Benchmark*

In the hypothetical case where the microgrid's hourly demand and solar power for the entire time horizon were known, it would be possible to formulate a deterministic optimisation problem in order to obtain the most favourable scheduling of all the elements of the microgrid. The value of the objective function in this setting would represent the lower bound (if the objective is to minimise the operation cost), and it could serve as a benchmark value to compare the results obtained by alternative methods. The decision variables of this optimisation problem are the charge and discharge power of the batteries, the possible generation of the diesel group, the not supplied power and the possible spillages. This benchmark model was implemented in GAMS [27], where the detailed equations were omitted here for the sake of simplicity, but added in Appendix A. The structure of the resulting optimisation problem is the following one:

$$\begin{aligned} \min_{x} \quad & c(x) \\ \text{subject to} \quad & g(x) \leq 0 \end{aligned} , \tag{12}$$

where the objective function $c(x)$ is the sum of the diesel generation cost along the entire time horizon plus the penalty term related to the possible non-supplied energy. The cost of the diesel generator is assumed to be a quadratic function, where the independent term is only incurred when the generator is on. This fact requires the use of unit-commitment binary variables.

The constraints $g(x)$ are the energy balance at the microgrid, and at the storage devices taking into account their charging and discharging efficiencies, their maximum storage levels and the maximum rated power.

As the objective is a quadratic function and some binary variables are needed, the resulting model is a Mixed Integer Quadratic Programming model (MIQP).

4.3. *Naive Strategy*

A Naive algorithm has also been implemented to mimic the results that a simple strategy could achieve. The insight of this method is to charge the batteries when there is a surplus of energy, and discharge them otherwise. The pseudocode of this strategy is shown in Algorithm 1, where PVGEN is the PV power [kW] and the LOAD is the household consumed power [kW].

Algorithm 1 Naive algorithm

```
if PVGEN > LOAD then
    extra = PVGEN − LOAD
    if extra > batt capacity then
        batt charge = min(batt capacity, P^b)
    else
        batt charge = min(extra, P^b)
    end if
    extra = extra − batt charge
    if extra > hyd capacity then
        hyd charge = min(hyd capacity, P^h2)
    else
        hyd charge = min(extra, P^h2)
    end if
    extra = extra − hyd charge
else
    lack = LOAD − PVGEN
    if lack > batt capacity then
        batt discharge = min(batt capacity, P^b)
    else
        batt discharge = min(lack, P^b)
    end if
    lack = lack − batt discharge
    if lack > hyd capacity then
        hyd discharge = min(hyd capacity, P^h2)
    else
        hyd discharge = min(lack, P^h2)
    end if
    lack = lack − hyd discharge
    if lack > 0 then
        diesel = min(max diesel, lack)
    end if
end if
```

It is assumed that the Naive algorithm could be implemented in a microcontroller making decisions on a continuous manner based on the instantaneous status of the microgrid. In this sense, this differs from the DRL approach of this paper that would require at least the information of the last hour in case the size of the time window was 1 h.

4.4. DQN Configuration

This paper uses a DQN due to its implementation simplicity and powerful performance [19]. It consists of three modules:

(1) A Convolutional Neural Network (CNN) that it is shown in Figure 3.
(2) A replay memory to store the experiences and train the CNN. The implemented memory is called Experience Replay Memory [28], being the easiest to implement.
(3) A module to process the agent observations and combine them with internal agent data (the last state), to make the next state and store it in the memory.

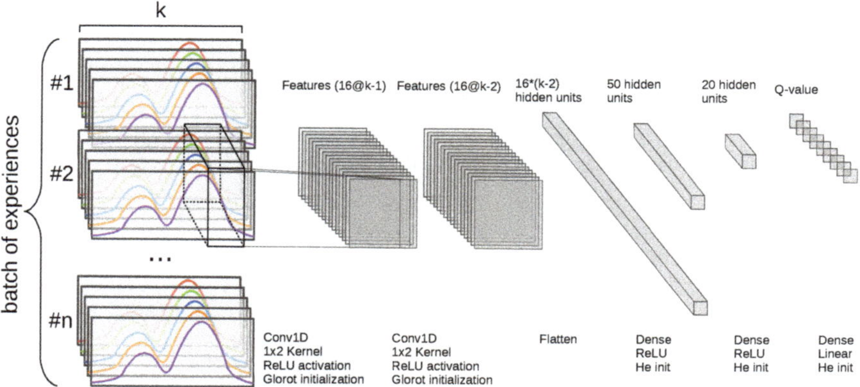

Figure 3. Convolutional Neural Network architecture used in the proposed Deep Q-Network (DQN).

5. Results

The CNN architecture proposed is similar but not equal to the one proposed in [5]. It takes a whole tensor as the input of the first layer instead of scatter time-series input in 1D-convolution layers and being merged with the remain inputs after the convolution computation. This configuration brings more simplicity and scalability to the model without losing performance in the results despite the information redundancy. This CNN has two 1-dimension convolution layer. The initialisation procedures of these layers are the ones from Glorot [29] for Convolutional layers and He [30] for Dense layers. All the parameter values used are shown in Table 2.

Table 2. DQN parameters.

Parameter	Value
Batch size	20
Memory size	10,000
Optimiser	Nadam [31]
Error measure	MSE
Exploration function	$f(s) = 0.1 + 0.9e^{-s*10^{-6}}$
Early stop scope	200
Discount factor (γ)	0.99

During the training phase, to deal with the lack of data, a regularisation technique called early stop is applied. In this technique, the training set is split into a smaller training set and the development set, and after a fixed number of training steps, the model is evaluated with the development set saving the best snapshot of the model during the training.

This paper carries out the analysis of the window size of the state throughout different configurations of the parameter k in Equation (5). The window size ranges from $k = 3$ to $k = 24$ in steps of 3 h. Following the state definition of Equation (5), some parameters in $t < 0$ are required (state s_0 contains some temporal slices in $t < 0$), and these are set to zero.

This section presents the results obtained with the DQN method applied to the microgrid described previously. The DQN has been implemented in Python 3.7.7 using Tensorflow 2.1.0. The MIQP model has been coded in GAMS using the solver CPLEX 12.9.0.0. The computer system was an Intel Core i7-8550U (1.80 GHz–4.00 GHz) with 16 GB RAM running under Ubuntu 18.04 LTS x64.

The results gather several options for the definition of the state of the microgrid varying the window size parameter k in Equation (5). As stated before, the motivation of this analysis is to find how much recent information should be included in the state definition in order to obtain the best performance of the microgrid. To put the results obtained with the DQN in perspective, other models have been included in this analysis: the perfect-information MIQP model, the Naive strategy and a Random Policy. The first one represents a lower bound of the operation cost that can be achieved with such microgrid. The third one can be seen as an upper bound, where the average value of ten samples (runs of the model) are included in the assessment. Based on the historical three-year data, the return $G(\tau)$ of the trajectory τ followed by the policy of the proposed DQN is used to measure its goodness. The return match the cost of the microgrid in € and the trajectory τ is the operation of the microgrid. Note that the first two years of available data are used to train the DQN model, separated in the training set of the first year and the development set of the second year, whereas the last year is used for testing. Concerning the reference provided by the MIQP, the model operates the microgrid without any discretisation of the decision variables. The same applies for the Naive and the Random strategies.

The MIQP model generates the lower bound given by a three-year accumulated cost of 2677.43 € with a relative GAP of 6.06 as GAMS defines $(BestFeasible - BestPossible)/BestFeasible$, taking 24 h to find this solution. The accumulated cost for every year is shown in Table 3. The Naive strategy achieves an expenses of 11,138.60 €, and the Random strategy 14,066.59 €, 316.02% and 425.38% more than MIQP, respectively. Notice that the Naive strategy is not much better than the Random strategy, and this fact calls for smarter policies to operate the microgrid, as the one proposed in this paper.

Table 3 shows the accumulated cost obtained after three years of operation of the microgrid using the policy learned, for the different configurations of the proposed DQN. These results are significantly better than the other models, achieving results between 3653.59 € for the best case with a window size of 9, and 462,722 € for the worst-case with a window size of 21. Even the worst-case is much better than a Naive or Random strategies. Relative to the results of the MIQP model, the worst-case performance is two times worse than the best case. These results show that the choice of the window size parameter has significant consequences in the performance of the DQN. Larger window size involves learning difficulties in the policy for the DQN.

Another observation is the comparative between the best case DQN and the MIQP results. The DQN achieves a good performance, being a method easy to implement and with near-optimal results.

Analysing the agent behaviour, Figure 4 shows an example of the hourly microgrid operation carried out by the proposed RL method. It consists of six days of the first year (three consecutive days of winter and three of summer).

Table 3. Accumulated cost of each algorithm [€].

Cost [€]	Year 1	Year 2	Year 3	Total	Relative $(X - BEST)/BEST \cdot 100$
MIQP (GAP 6%)	967.34	864.05	846.04	2677.43	0.00%
RL k = 3	1305.94	1126.49	1239.35	3671.77	37.14%
RL k = 6	1355.80	1140.92	1248.61	3745.33	39.89%
RL k = 9	1299.56	1123.53	1230.50	3653.59	36.46%
RL k = 12	1389.60	1198.63	1308.11	3896.33	45.52%
RL k = 15	1348.61	1174.02	1268.02	3790.64	41.58%
RL k = 18	1503.69	1307.67	1443.58	4254.94	58.92%
RL k = 21	1634.33	1441.21	1551.68	4627.22	72.82%
RL k = 24	1515.46	1320.44	1439.12	4275.02	59.67%
Naive	3778.74	3681.04	3678.82	11,138.60	316.02%
Random	4816.64	4554.78	4695.17	14,066.59	425.38%

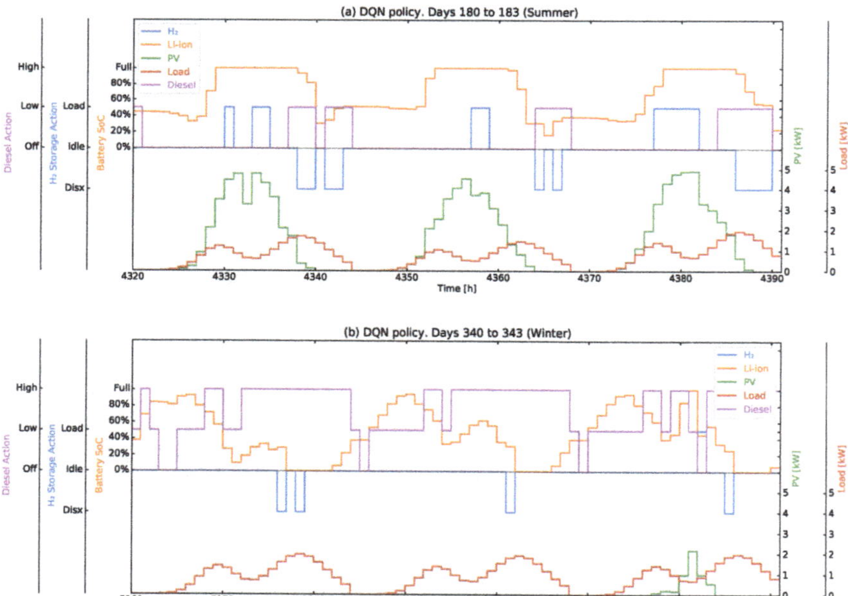

Figure 4. DQN operation of the microgrid, for summer (**a**) and winter (**b**). Hourly steps in a 3-day window.

Since the DQN only works with discrete action spaces, the continuous action space defined in the MDP model of the microgrid in Equation (4) has to be discretised. Each hour t, the action a_t consists of two variables (P_t^d and P_t^{h2}) corresponding to the power generated by the diesel generator and the power generated (consumed if negative) by the hydrogen fuel cell respectively. The required discretisation of the diesel output power P_t^d leads to the following cases:

- $P_t^d = P_{max}^d$, i.e., the agent configures the diesel generator to dispatch the maximum power (1 kW in the example, and "High" label in Figure 4).
- $P_t^d = P_{max}^d/2$, i.e., the agent configures the diesel generator to dispatch half of the maximum power it can (0.5 kW in the example, and "Low" label in Figure 4).
- $P_t^d = 0$, i.e., the agent configures the diesel generator in off. (0 kW in the example, and "Off" label in Figure 4).

The agent considers also three different configurations for the values that P_t^{h2} can take:

- $P_t^{h2} = -P_{max}^{h2}$, i.e., the agent configures the hydrogen electrolyser to charge at maximum rate from the system (load at 1 kW in the example, and "Load" label in Figure 4).
- $P_t^{h2} = 0$, i.e., the agent configures the hydrogen components in an idle state (load/discharge at 0 kW, and "Idle" label in Figure 4).
- $P_t^{h2} = P_{max}^{h2}$, i.e., the agent configures the hydrogen fuel cell to inject the maximum power into the system (discharge at 1 kW in the example, and "Disx" label in Figure 4).

To summarise, in each hour the agent can take nine different configurations. The battery SoC takes values between 0 kWh (0% in the figure) and 2.9 kWh (Full in the figure that correspond to 100% of the maximum SoC).

According to Figure 4, there are no significant differences between the summer and winter load profiles. In contrast, the available solar irradiation makes the difference between these opposite seasons, being the main responsible for the change in the operation policy decided automatically by the RL method.

There is a clear daily pattern operation of the Li-ion battery, but it differs with the season. During summer the daily energy surplus provided by the solar panels is used to fill this battery during daylight hours, being systematically discharged during the rest of the hours. However, the lack of solar irradiation in winter is compensated by the continuous generation of the diesel to cover the demand in the main hours, using the diesel surplus of the off-peak hours to charge the battery for it usage during non-sunlight hours. On the contrary, during summer the use of the diesel is limited to some hours where there is not enough solar irradiation. Concerning the hydrogen fuel cell, it is used in a similar diesel pattern, to cover the demand when there is not enough solar irradiation.

In order to illustrate the behaviour of the model, a numerical example is presented hereafter. Assuming that the state s_t considers a tensor of shape $(k, 4)$ with $k = 9$, from the output results shown in Figure 4a, the detailed values that correspond to hour $t = 4380$ are the next ones:

$$
\begin{aligned}
s_t &= \left(slice_{t-k},\ slice_{t-k+1},\ \ldots,\ slice_{t-1},\ slice_t\right) \\
&= \begin{pmatrix} P^{pv}_{t-k-1}, & P^{pv}_{t-k}, & \ldots, & P^{pv}_{t-2}, & P^{pv}_{t-1} \\ D_{t-k-1}, & D_{t-k}, & \ldots, & D_{t-2}, & D_{t-1} \\ S^{b}_{t-k}, & S^{b}_{t-k+1}, & \ldots, & S^{b}_{t-1}, & S^{b}_{t} \\ S^{h2}_{t-k}, & S^{h2}_{t-k+1}, & \ldots, & S^{h2}_{t-1}, & S^{h2}_{t} \end{pmatrix} \\
&= \begin{pmatrix} 0.002, & 0.151, & 0.461, & 1.122, & 1.973, & 3.301, & 4.295, & 4.767, & 4.891 \\ 0.061, & 0.186, & 0.447, & 0.837, & 1.222, & 1.398, & 1.270, & 0.963, & 0.711 \\ 1.028, & 0.989, & 1.001, & 1.257, & 1.932, & 2.260, & 2.900, & 2.900, & 2.900 \\ 36.000, & 36.000, & 36.000, & 36.000, & 36.000, & 36.650, & 37.300, & 37.950, & 38.600 \end{pmatrix}.
\end{aligned}
$$

Then, the action taken by the DQN using Equation (2) is:

$$
\begin{aligned}
a_t = \left(P_t^d,\ P_t^{h2}\right) = \\
\left(0.000,\ -1.000\right)
\end{aligned}
$$

As a consequence of this action, the new state in hour $t+1 = 4381$ is reached, with its corresponding new slice:

$$\begin{aligned} slice_{t+1} &= \left(P_t^{pv},\ D_t,\ S_{t+1}^b,\ S_{t+1}^{h2}\right) \\ &= \left(4.899,\ 0.672,\ 2.900,\ 39.250\right), \end{aligned}$$

in which S_{t+1}^b, that is calculated using the Equation (A3), takes the value 0.0 kW because it is full and the curtailment P_t^{curt}, that is calculated using the balance Equation (11), takes the value of 3.217 kW.

Another interesting result that can be highlighted is the management of the hydrogen storage along the year. One could expect that since there is less solar generation in winter, it would be better to store energy during summer in order to have it available when necessary. Figure 5 shows the hydrogen storage SoC for the whole 3-year period since this profile is barely appreciable in the 3-day window of Figure 4. It can be seen that the proposed DQN is capable of finding an optimal yearly pattern, charging the hydrogen tank in summer and discharging it in winter, and this optimal behaviour is obtained just with the most recent information at each step. A summary of the usage of the hydrogen storage is shown in Table 4.

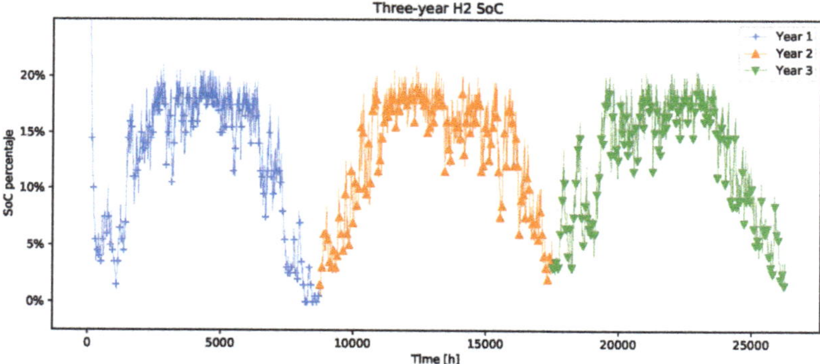

Figure 5. Hydrogen state of charge (SoC) percentage over three years.

Table 4. Hydrogen storage SoC fluctuation.

Hydrogen Storage	Year 1	Year 2	Year 3
Charged	546.00 kWh	587.50 kWh	578.50 kWh
Discharged	594.50 kWh	586.00 kWh	578.00 kWh

6. Conclusions

This paper presents the application of DRL techniques to manage the elements of an isolated microgrid. The advantage of the proposed approach is the ability of the algorithm to learn from its own experience, and this can be an interesting feature to foster the up-scaling of these type of solutions that do not require a tailor-made specific optimisation model adapted to the particularities of each microgrid. However, as the application of RL and DRL to power systems is an emerging research field, the selection of the most appropriate ANN architecture, or the definition of what variables should be included in the the so-called system state, are still open questions. With respect to the first one, the proposed CNN architecture provides simplicity and a good performance. Regarding the second question, this paper provides a sensible set of variables to configure the system state. In addition, the effect of increasing the time window of such state has been analysed, and the numerical results show that 9 h is the optimal size of such time window for the studied microgrid. In order to measure the quality of the DRL results, this paper presents a traditional Mixed Integer Quadratic Programming optimisation model to compute the theoretical best operation possible, i.e., the hypothetical case of having complete information for the whole time period. Using such model as benchmark, whereas a Naive algorithm similar to the ones used in practical implementation results in an operation 316.02% worse, the developed DRL model reaches a value of 36.46%. Future work will be oriented to the implementation of Transfer Learning and Robust Learning to address multi-agent microgrid problems.

Author Contributions: Conceptualisation, J.G.-G. and M.A.S.-B.; methodology, D.D.-B., J.G.-G. and M.A.S.-B.; software, D.D.-B.; validation, D.D.-B., J.G.-G., M.A.S.-B. and E.F.S.-Ú.; formal analysis, D.D.-B., J.G.-G. and M.A.S.-B.; data curation, D.D.-B. and E.F.S.-Ú.; writing—original draft preparation, D.D.-B.; writing—review and editing, D.D.-B., J.G.-G., M.A.S.-B. and E.F.S.-Ú.; visualisation, D.D.-B.; supervision, J.G.-G. and M.A.S.-B. All authors have read and agreed to the published version of the manuscript.

Funding: This research has been funded by the Strategic Research Grants program of Comillas University, and by "Programa microrredes inteligentes, Comunidad de Madrid (P2018/EMT4366)".

Acknowledgments: The authors want to express their gratitude to Francisco Martín Martínez and Miguel M. Martín-Lopo for their valuable help during the development of this research.

Conflicts of Interest: The authors declare no conflict of interest.

Notation

To facilitate the mathematical representation of the elements of the microgrid, the following nomenclature is used throughout the paper.

Sets and indexes:

$t \in T$ time periods from 1 to $|T|$

Parameters:

P_t^{pv} Maximum available PV power generation at time period t, [kW]
P_{max}^{pv} Nominal rate of the solar panel, [kW]
D_t Load of the microgrid at time period t, [kW]
D_{max} Maximum power the microgrid can consume, [kW]
ζ^b, Discharge efficiency of the Li-ion battery
η^b, Charge efficiency of the Li-ion battery
ζ^{h2} Discharge efficiency constants of the hydrogen storage
η^{h2} Charge efficiency hydrogen storage
δ_2 Quadratic term of the diesel generator cost function, [€/kW^2]
δ_1 Linear term of the diesel generator cost function, [€/kW]
δ_0 No-load term of the diesel generator cost function when committed, [€]
c_{pns} Cost of not supplied energy, [€/kW]
P_{max}^d Maximum output power of the diesel generator, [kW]
P_{max}^b Maximum power the Li-ion battery can dispatch, [kW]
S_{max}^b Maximum SoC of the battery, [kWh]
S_0^b Initial SoC of the battery, [kWh]
P_{max}^{h2} Maximum power the hydrogen fuel cell can dispatch, [kW]
S_{max}^{h2} Maximum SoC of the hydrogen storage, [kWh]
S_0^{h2} initial SoC of the hydrogen storage, [kWh]

Variables:

P_t^d	Output power of the diesel generator at time period t, [kW]
P_t^b	Generated (+) and consumed (-) power from the Li-ion battery at time period t, [kW]
P_t^{h2}	Generated (+) and consumed (-) power from the hydrogen storage at time period t, [kW]
S_t^b	State of Charge (SoC) of the Li-ion battery at time period t, [kWh]
S_t^{h2}	State of Charge (SoC) of the hydrogen storage at time period t, [kWh]
P_t^{pns}	Not-supplied power at time period t, [kW]
P_t^{curt}	Non-negative variable that represents the PV curtailment at time period t, [kW]

Variables (only for MIQP):

U_t^d	Unit commitment of diesel at t
P_t^{b+}	Consumed power from the Li-ion battery at time period t, [kW]
P_t^{b-}	Generated power from the Li-ion battery at time period t, [kW]
P_t^{h2+}	Hydrogen electrolyser consumption at t, [kW]
P_t^{h2-}	hydrogen fuel cell generation at t, [kW]

Abbreviations

The following abbreviations are used in this manuscript:

ANN	Artificial Neural Network
CNN	Convolutional Neural Network
DQN	Deep Q-Network
DRL	Deep Reinforcement Learning
GAMS	General Algebraic Model System
LTS	Long-Term Support
MCTS	Monte Carlo Tree Search
MIQP	Mixed Integer Quadratic Programming
MSE	Mean Squared Error
PCC	Point of Common Coupling
PV	Photovoltaic Panel
RES	Renewable energy sources
RL	Reinforcement Learning
SoC	State of Charge
pns	power not supplied

Appendix A. Mixed Integer Quadratic Programming Formulation

$$\min \quad \sum_t [\delta_0 U_t^d + \delta_1 P_t^d + \delta_2 (P_t^d)^2 + c_{pns} P_t^{pns}] \tag{A1}$$

$$\text{subject to} \quad D_t - P_t^{pns} = (P_t^{pv} - P_t^{curt}) + (P_t^{b-} - P_t^{b+}) + (P_t^{h2-} - P_t^{h2+}) + P_t^d \quad \forall t \in T \tag{A2}$$

$$S_t^b = -\frac{P_t^{b-}}{\eta^b} + \zeta^b P_t^{b+} + S_{t-1}^b \quad \forall t \in T \tag{A3}$$

$$S_t^{h2} = -\frac{P_t^{h2-}}{\eta^{h2}} + \zeta^{h2} P_t^{h2+} + S_{t-1}^{h2} \quad \forall t \in T \tag{A4}$$

$$S_0^{h2} \leq S_t^{h2} \quad t = |T| \tag{A5}$$

$$P_t^d \leq U_t^d * \overline{P}^d \quad \forall t \in T \tag{A6}$$

$$U_t^d \in \{0, 1\} \quad \forall t \in T \tag{A7}$$

$$P_t^{pns}, P_t^d, P_t^{curt}, P_t^{b-}, P_t^{b+}, P_t^{h2-}, P_t^{h2+} \geq 0 \quad \forall t \in T \tag{A8}$$

$$P_t^d \leq P_{\max}^d \quad \forall t \in T \tag{A9}$$

$$P_t^{curt} \leq P_t^{pv} \quad \forall t \in T \tag{A10}$$

$$P_t^{b-}, P_t^{b+} \leq P_{\max}^b \quad \forall t \in T \tag{A11}$$

$$P_t^{h2-}, P_t^{h2+} \leq P_{\max}^{h2} \quad \forall t \in T, \tag{A12}$$

where Equation (A1) is the objective function, Equation (A2) is the demand balance constraint, Equations (A3) and (A4) model the dynamics of the battery and the hydrogen storage, respectively, Equation (A5) establishes a minimum hydrogen storage level in the last hour, and Equation (A6) indicates that unit commitment of the diesel generator is a binary variable. The remaining equations impose the upper and lower bounds to the decision variables.

References

1. Ali, A.; Li, W.; Hussain, R.; He, X.; Williams, B.; Memon, A. Overview of Current Microgrid Policies, Incentives and Barriers in the European Union, United States and China. *Sustainability* **2017**, *9*, 1146. [CrossRef]
2. Gupta, R.A.; Gupta, N.K. A robust optimization based approach for microgrid operation in deregulated environment. *Energy Convers. Manag.* **2015**, *93*, 121–131. [CrossRef]
3. Reddy, S.; Vuddanti, S.; Jung, C.M. Review of stochastic optimization methods for smart grid. *Front. Energy* **2017**, *11*, 197–209. [CrossRef]
4. Kim, R.K.; Glick, M.B.; Olson, K.R.; Kim, Y.S. MILP-PSO Combined Optimization Algorithm for an Islanded Microgrid Scheduling with Detailed Battery ESS Efficiency Model and Policy Considerations. *Energies* **2020**, *13*, 1898. [CrossRef]
5. François-Lavet, V.; Taralla, D.; Ernst, D.; Fonteneau, R. Deep Reinforcement Learning Solutions for Energy Microgrids Management. In Proceedings of the European Workshop on Reinforcement Learning (EWRL 2016), Barcelona, Spain, 3–4 December 2016.
6. Mbuwir, B.V.; Ruelens, F.; Spiessens, F.; Deconinck, G. Battery Energy Management in a Microgrid Using Batch Reinforcement Learning. *Energies* **2017**, *10*, 1846. [CrossRef]
7. Kim, B.; Zhang, Y.; van der Schaar, M.; Lee, J. Dynamic Pricing and Energy Consumption Scheduling with Reinforcement Learning. *IEEE Trans. Smart Grid* **2016**, *7*, 2187–2198. [CrossRef]
8. Qiu, X.; Nguyen, T.A.; Crow, M.L. Heterogeneous Energy Storage Optimization for Microgrids. *IEEE Trans. Smart Grid* **2016**, *7*, 1453–1461. [CrossRef]
9. Glavic, M.; Fonteneau, R.; Ernst, D. Reinforcement Learning for Electric Power System Decision and Control: Past Considerations and Perspectives. *IFAC-PapersOnLine* **2017**, *50*, 6918–6927. [CrossRef]
10. Glavic, M. (Deep) Reinforcement learning for electric power system control and related problems: A short review and perspectives. *Annu. Rev. Control* **2019**, *48*, 22–35. [CrossRef]

11. Yang, T.; Zhao, L.; Li, W.; Zomaya, A.Y. Reinforcement learning in sustainable energy and electric systems: A survey. *Annu. Rev. Control* **2020**. [CrossRef]
12. Ji, Y.; Wang, J.; Xu, J.; Fang, X.; Zhang, H. Real-Time Energy Management of a Microgrid Using Deep Reinforcement Learning. *Energies* **2019**, *12*, 2291. [CrossRef]
13. Francois-Lavet, V.; Rabusseau, G.; Pineau, J.; Ernst, D.; Fonteneau, R. On Overfitting and Asymptotic Bias in Batch Reinforcement Learning with Partial Observability. *J. Artif. Intell. Res.* **2019**, *65*, 1–30. [CrossRef]
14. Sutton, R.S.; Barto, A.G. *Reinforcement Learning: An Introduction*, 2nd ed.; The MIT Press: Cambridge, UK, 2018.
15. Arulkumaran, K.; Deisenroth, M.P.; Brundage, M.; Bharath, A.A. A Brief Survey of Deep Reinforcement Learning. *IEEE Signal Process. Mag.* **2017**, *34*, 26–38. [CrossRef]
16. Li, Y. Deep Reinforcement Learning: An Overview. *arXiv* **2017**, arXiv:1701.07274.
17. Francois-Lavet, V.; Henderson, P.; Islam, R.; Bellemare, M.G.; Pineau, J. An Introduction to Deep Reinforcement Learning. *Found. Trends® Mach. Learn.* **2018**, *11*, 219–354. [CrossRef]
18. Watkins, C.J.C.H.; Dayan, P. Q-learning. *Mach. Learn.* **1992**, *8*, 279–292. [CrossRef]
19. Mnih, V.; Kavukcuoglu, K.; Silver, D.; Graves, A.; Antonoglou, I.; Wierstra, D.; Riedmiller, M. Playing Atari with Deep Reinforcement Learning. *arXiv* **2013**, arXiv:1312.5602.
20. Sutton, R.S. Integrated Architectures for Learning, Planning, and Reacting Based on Approximating Dynamic Programming. In Proceedings of the Seventh International Conference on Machine Learning, Morgan Kaufmann, Austin, TX, USA, 21–23 June 1990; pp. 216–224.
21. Chaslot, G.; Bakkes, S.; Szita, I.; Spronck, P. Monte-Carlo Tree Search: A New Framework for Game AI. In Proceedings of the Fourth Artificial Intelligence and Interactive Digital Entertainment Conference, Palo Alto, CA, USA, 22–24 October 2008; Mateas, M., Darken, C., Eds.; AAAI Press: Menlo Park, CA, USA, 2008; pp. 216–217.
22. Mnih, V.; Kavukcuoglu, K.; Silver, D.; Rusu, A.A.; Veness, J.; Bellemare, M.G.; Graves, A.; Riedmiller, M.; Fidjeland, A.K.; Ostrovski, G.; et al. Human-level control through deep reinforcement learning. *Nature* **2015**, *518*, 529–533. [CrossRef] [PubMed]
23. François-Lavet, V.; Taralla, D. DeeR. 2016. Available online: http://deer.readthedocs.io (accessed on 16 April 2020).
24. François-Lavet, V.; Gemine, Q.; Ernst, D.; Fonteneau, R. *Towards the Minimization of the Levelized Energy Costs of Microgrids using Both Long-Term and Short-Term Storage Devices*; CRC Press: Boca Raton, FL, USA, 2016.
25. EnergySage. EnergySage: 6 kW Solar System. 2019. Available online: http://news.energysage.com/6kw-solar-system-compare-prices-installers (accessed on 16 April 2020).
26. Jasmin, E.A.; Imthias Ahamed, T.P.; Jagathy Raj, V.P. Reinforcement Learning approaches to Economic Dispatch problem. *Int. J. Electr. Power Energy Syst.* **2011**, *33*, 836–845. [CrossRef]
27. Corporation, G.D. General Algebraic Modeling System (GAMS) Release 24.2.1. 2013. Available online: http://www.gams.com/ (accessed on 16 April 2020).
28. Lin, L.J. *Reinforcement Learning for Robots Using Neural Networks*; Technical Report CMU-CS-93-103; School of Computer Science, Carnegie Mellon University: Pittsburgh, PA, USA, 1993.
29. Glorot, X.; Bengio, Y. Understanding the difficulty of training deep feedforward neural networks. In Proceedings of the Thirteenth International Conference on Artificial Intelligence and Statistics, Chia Laguna Resort, Sardinia, Italy, 13–15 May 2010; Teh, Y.W., Titterington, M., Eds.; Proceedings of the Machine Learning Research (PMLR); 2010; Volume 9, pp. 249–256. Available online: http://proceedings.mlr.press/v9/glorot10a.html (accesed on 16 April 2020).
30. He, K.; Zhang, X.; Ren, S.; Sun, J. Delving Deep into Rectifiers: Surpassing Human-Level Performance on ImageNet Classification. In Proceedings of the IEEE International Conference on Computer Vision (ICCV), Santiago, Chile, 7–13 December 2015.
31. Dozat, T. Incorporating Nesterov Momentum into Adam. In Proceedings of the ICLR 2016 Workshop, San Juan, Puerto Rico, 2–4 May 2016.

Review

Review of Positive and Negative Impacts of Electric Vehicles Charging on Electric Power Systems

Morsy Nour [1,2,*], José Pablo Chaves-Ávila [1], Gaber Magdy [2] and Álvaro Sánchez-Miralles [1]

1 Institute for Research in Technology (IIT), ICAI School of Engineering, Comillas Pontifical University, 28015 Madrid, Spain; jose.chaves@iit.comillas.edu (J.P.C.-Á.); alvaro.sanchez@iit.comillas.edu (Á.S.-M.)
2 Department of Electrical Engineering, Faculty of Energy Engineering, Aswan University, Aswan 81528, Egypt; gabermagdy@aswu.edu.eg
* Correspondence: morsy.mohammed@iit.comillas.edu

Received: 23 July 2020; Accepted: 4 September 2020; Published: 8 September 2020

Abstract: There is a continuous and fast increase in electric vehicles (EVs) adoption in many countries due to the reduction of EVs prices, governments' incentives and subsidies on EVs, the need for energy independence, and environmental issues. It is expected that EVs will dominate the private cars market in the coming years. These EVs charge their batteries from the power grid and may cause severe effects if not managed properly. On the other hand, they can provide many benefits to the power grid and get revenues for EV owners if managed properly. The main contribution of the article is to provide a review of potential negative impacts of EVs charging on electric power systems mainly due to uncontrolled charging and how through controlled charging and discharging those impacts can be reduced and become even positive impacts. The impacts of uncontrolled EVs charging on the increase of peak demand, voltage deviation from the acceptable limits, phase unbalance due to the single-phase chargers, harmonics distortion, overloading of the power system equipment, and increase of power losses are presented. Furthermore, a review of the positive impacts of controlled EVs charging and discharging, and the electrical services that it can provide like frequency regulation, voltage regulation and reactive power compensation, congestion management, and improving power quality are presented. Moreover, a few promising research topics that need more investigation in future research are briefly discussed. Furthermore, the concepts and general background of EVs, EVs market, EV charging technology, the charging methods are presented.

Keywords: electric vehicles; uncontrolled charging; delayed charging; controlled charging; V2G; V2B; V2H; peak shaving; valley filling; congestion management; renewable energy sources

---

1. Introduction

Greenhouse gases (GHGs) emissions, global warming, and climate change are getting significant attention worldwide [1]. Countries aim to diminish the use of fossil fuel which is the main reason behind these issues. Most of fossil fuel consumption is in electricity generation and transportation sectors [2]. In 2014, 35% of the total energy consumption was by the transportation sector [3]. Electricity generation can cut fossil fuel usage by moving to renewable energy sources (RESs) instead of traditional nonrenewable generation and transportation sector can cut fossil fuel usage by moving to electrified transportation. With this transformation in the transportation sector, carbon dioxide (CO2) emissions can be reduced [4,5]. The advancement in EV technology can lower the reliance on fossil fuels and leads to emission reduction [5]. Charging the EVs from a power grid with high share of RESs generation contributes more to decreasing pollutant emissions [6].

Electrified transportation like metros, electric trains, trolleybuses, trams, etc., are widely used in many countries public transportation and is considered a mature technology, but private electrified transportation like private EVs have not had large adoption in the past due to limitations in battery technology in terms of heavy weight, high price, short life, and long charging duration. Due to the breakthroughs and recent developments in battery technology, EVs are proposed as a competitor to traditional vehicles driven by gasoline due to the fast decline of EV prices [7]. Moreover, EVs produce no pollution (locally) and have a very high energy conversion efficiency compared to traditional vehicles because the internal combustion engine (ICE) is less efficient than the electric motor (EM). Furthermore, several countries already initiated governmental programs to increase EVs market share by financial subsidies and tax exemptions to motivate people to buy EVs. Additionally, many car manufacturers produce various types of EVs that have different specifications and prices. This gives the buyer a lot of options to choose from depending on his budget and car usage (i.e., long distance or short distance).

These EVs can be charged from large charging stations, street chargers, workplace chargers, and private home chargers [8], which will be supplied from the distribution network. It is expected that this new load which draws a large amount of electric energy from the power system in a short time will have harmful effects on the distribution network if not managed optimally and will foster the need for large infrastructure upgrades, which are an economic burden for utility companies. Therefore, comprehensive studies should be executed to understand and assess the characteristics of EVs load and probable effects they may have on the electric power system, especially the distribution networks [9], and consider this new load in the design, operation, and planning processes. Additionally, techniques to minimize and alleviate these negative effects and for optimal integration of EVs into the power grid should be developed [10].

A large number of research studies have investigated and assessed the impacts of uncontrolled EV charging on distribution networks due to uncertainties related to these studies. Many uncertainties should be considered like charging start time, the charger power rating, the charging location, EVs battery capacity, EVs battery state of charge (SoC) [11] when started charging, penetration level, and distribution network status. The studies assessed the impacts of uncontrolled EVs charging on distribution networks total power demand [12], transformer loading and life [13], cable loading [14], voltage profile [15], power losses [16], voltage and current unbalance [17], and harmonics distortion [18]. Moreover, various studies proposed solutions for optimal integration of EVs by delayed charging, smart charging, vehicle to grid (V2G) technology, vehicle to building (V2B) technology, and vehicle to home (V2H) technology, and how it can provide electrical services like frequency regulation, voltage regulation and reactive power compensation, peak shaving and valley filling, integration of RESs, spinning reserves, and improving power quality [19,20].

This paper is organized as follows: Section 2 presents a general background of EVs, EVs market, EVs types, the benefits and challenges of EVs, and EV charging technology. Section 3 presents the charging and discharging methods; for instance, uncontrolled charging, delayed charging, controlled charging, V2G technology, V2B technology, and V2H technology. Section 4 presents the negative impacts of uncontrolled EVs charging on electric power systems. Section 5 presents the positive impacts of controlled charging and discharging of EVs on electric power systems. Section 6 presents a few promising research topics that need more investigation in future research. Finally, conclusions are presented in Section 7.

2. Electric Vehicles Technology

An EV is a vehicle that is powered or driven, at least in part by electricity such as trams, metros, electric cars, electric trains, trolleybuses, etc. Most of the mentioned technologies are mature and widely used, except the private EVs (i.e., electric cars), but they are getting huge attention in the last years. EV technology is not new as it appeared in the 19th century powered by lead–acid batteries and it was not used a lot because the fuel-powered cars, i.e., internal combustion engine vehicles (ICEVs), showed better performance at that time because the energy density of fuel was better than lead–acid batteries. Usually, EV terminology is widely used to refer to private electric cars and not to other electrified transportation methods like trams, metros, etc.

2.1. Electric Vehicles Market

Based on the 2019 global EV outlook report of the International Energy Agency (IEA) [7], there is a very rapid growth of EVs market. The report shows that the EVs stock crossed 3 million in 2017 and exceeded 5 million in 2018. China represents the largest EVs market, then Europe and the United States after that. Norway achieved the highest deployment of EVs with a market share of 46% followed by Iceland 17% and Sweden 8%. It is expected that the number of EVs will increase significantly in the coming years. The forecasts for EVs number at 2030 between 130 and 250 million. There are many EVs in the market now from several manufacturers (BMW, Chevrolet, Ford, Hyundai, Kia, Mercedes-Benz, Mitsubishi, Nissan, Tesla, Renault, Volkswagen, etc.) with different specifications (battery technology, battery capacity, electric motor power, electric range, and onboard charger power rating,) which gives a wide range of options for people interested in buying an EV. Although the fast increase in the EV market, it still represents a small percentage of the global passenger light-duty vehicles market currently. This means that EVs have a long way until they become capable of making a significance in GHGs emissions and oil demands.

It should be mentioned that large scale adoption of EVs will have economic impacts and they should be viewed from two points of view, the EV owners' point of view and the power grid point of view [21]. From the power grid point of view, a high-power load represented in EVs will be introduced, which means additional expenses in fuel and generation capacity [22]. The grid power losses will rise, although they can be reduced by the use of controlled charging approaches [23]. Several studies showed that the electric power system will have a significant power loss due to EVs charging with different penetration levels. The lifespan of transformers and cables in the distribution grid may decrease due to excessive uncontrolled charging. A study in [24] showed that controlled charging of EVs results in reducing the peak demand and 60% saving in the system cost. Therefore, the improvement in EVs charging infrastructure and charging strategies need attention to enhance the power grid economic aspects. From EV owners' point of view, various benefits can be achieved like reducing the operating costs due to lower electricity prices compared with fuel and higher efficiency of EMs used in EVs compared with ICEs used in ICEVs [25]. On the other hand, the high initial cost of EVs due to the expensive batteries used compared with ICEVs represents a considerable challenge. Mass production of EVs, advancement in battery technology, new charging infrastructure and smart charging strategies, and providing rewards and incentives to EVs owners may control the high initial cost of EVs [26–28].

2.2. Types of Electric Vehicles

There are many types of EVs categorized according to the energy converter (i.e., ICE or EM) used to propel the vehicle, the power source (i.e., battery, fuel cell, or gasoline), and if it charges from an external source (i.e., charging station or home charger) [4]. A brief explanation of those types is presented in the following subsections. The basic structure of different EVs types is shown in Figure 1.

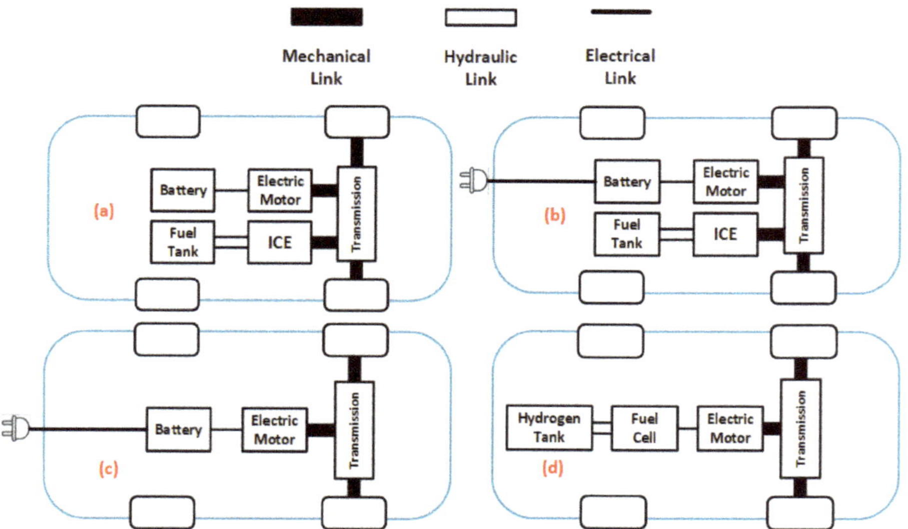

Figure 1. Basic structure of different electric vehicles (EVs) types. (**a**) Hybrid Electric Vehicle (HEV); (**b**) Plug-in Hybrid Electric Vehicle (PHEV); (**c**) Battery Electric Vehicle (BEV); and (**d**) Fuel Cell Electric Vehicle (FCEV).

2.2.1. Hybrid Electric Vehicle (HEV)

HEV is similar to ICEV but with oversized EM and battery. The battery can be charged by regenerative braking and by the ICE at light loads. Usually, battery and EM drive the vehicle at lower speeds and the ICE drives the vehicle at higher speeds. Furthermore, EM can assist the ICE at high load and enhance vehicle performance and efficiency. HEV has lower GHGs emissions and fuel consumption than ICEVs. In this type, no charging from distribution network by EV charger is used [29,30]. Therefore, it does not result any negative impacts on the power system due to battery charging and cannot provide any electrical services. There are different structures of this type [29]: series, parallel, series/parallel, mild, and complex HEVs, in addition to series/parallel plug-in hybrid electric vehicle (PHEV) which is explained in the next subsection. The basic configuration for parallel HEV is shown in Figure 1a.

2.2.2. Plug-in Hybrid Electric Vehicle (PHEV)

This is a HEV, but its battery can be charged by regenerative braking, by ICE, and EV charger supplied from distribution network as well. It is characterized by larger EM power, smaller ICE, and larger battery capacity compared with HEV in order to extend the electric range. It can operate in all-electric mode and use EM only which results in zero GHGs emissions. The capacity of the battery specifies the range of electric operation [29,30]. This type usually has a small battery capacity. Therefore, it is expected that it has a limited negative impact on electric power system. Moreover, its ability to provide electrical services is limited. PHEV can be in any hybrid configuration. The basic configuration for parallel PHEV is shown in Figure 1b.

2.2.3. Battery Electric Vehicle (BEV)

BEV is expected to dominate the EVs market with future advancements in battery technology and price reduction. BEV contains only EM which is powered by a battery and does not contain ICE. The electric driving range depends on battery capacity. The main advantage of this type is that it produces no emissions locally, which is very crucial in big cities. There are many configurations of this type [29] depending on different companies' designs but the basic configuration and principal components are shown in Figure 1c. The battery is charged from the distribution network by different kinds of EV chargers. Moreover, during deceleration and braking of EV the motor operates as a generator and produces electric energy that charges the battery. This is called regenerative braking which is used in other public electrified transportation like metros and trams too. To enable regenerative braking bidirectional DC/AC converter is used. It allows electric energy to flow from battery to AC motor in the driving mode and from the AC motor to the battery in braking mode [29,30]. The main challenge for BEV is the long charging period and limitation of public charging infrastructure. This issue will be tackled in the near future with the advancement in battery technology which will extend the electric range that the EV can drive before the need to be recharged. Furthermore, the large-scale deployment of public fast charging stations will reduce EV owners range anxiety. Moreover, in the future, BEV manufacturers may enable battery swapping. This means replacing the empty battery with a fully charged one in battery swapping station which needs a very short time. More details about battery swapping are given in Section 2.4.3. This type of EVs has the largest battery capacity and its charging can result in a severe negative impact on the power system if not managed properly. Additionally, its ability to provide electrical services is higher than other EV types.

2.2.4. Fuel Cell Electric Vehicle (FCEV)

FCEV is powered by EM like BEV, but it uses a fuel cell instead of a battery. The vehicle is refueled with hydrogen and the fuel cell converts hydrogen gas chemical energy to electric energy which powers EV motor. Hydrogen can be produced from fossil fuels like natural gas or water electrolysis. FCEV has short refueling time like ICEVs [29,30]. The basic configuration of FCEV is shown in Figure 1d. It can also work with a combination of battery and supercapacitors. If no battery is used, this vehicle type will have no impact on the power system because it does not need electric charging from distribution system.

2.3. Benefits and Challenges of Electric Vehicles

It is essential to introduce the advantages and disadvantages of conventional vehicles which are dominant in the market now. Table 1 shows the advantages and disadvantages of ICEVs. The disadvantages of ICEVs might be a big problem in the future of the transportation sector. This was the reason to look for some alternatives. EVs appeared as a promising solution to some of these disadvantages.

Table 1. Advantages and disadvantages of internal combustion engine vehicles (ICEVs).

Advantages	Disadvantages
Long range	Low efficiency (15 to 35%)
Fast refueling	No regenerative braking
Low cost per unit of power	Pollution due to exhaust gases
	Noise

EVs show many benefits compared with ICEVs but few challenges must be faced to increase EVs adoption. Table 2 shows both the benefits and challenges of EVs [29,31]. Because both traditional vehicles and EVs have their strengths and weaknesses, hybrid vehicles (i.e., HEVs and PHEVs) with several configurations appeared, which try to combine the strengths of both types and avoid weaknesses.

Table 2. Benefits and challenges of EVs.

Benefits	Challenges
Emissions are localized at power stations and not in cities which improves citizens' health	Limited electric range
High efficiency because of high electric motor efficiency	Limited number of charging stations and long charging (refueling) time
Low (for PHEV) or no emissions (for BEV)	Negative impacts on the power grid if uncontrolled charging is used
Fuel or energy independence from countries which produce fuel	Consumer acceptance
Simple transmission, fewer moving parts and low maintenance	High prices
Less noise pollution	Battery low energy density
Provide electrical services to the power system and helps the RESs integration.	Expensive battery and needs replacement after few years

2.4. Electric Vehicles Charging Technologies

EVs charging can be classified into three main technologies: conductive charging, wireless (i.e., contactless) charging (WC), and battery swapping. Conductive charging is the simplest and the currently used charging method. For conductive charging, there is physical contact (i.e., cable) between the power supply and battery but for WC there is no physical contact. WC and battery swapping are not widely used like conductive charging and still under study and development. The following subsections will provide more details about these technologies. Figure 2 presents classification of different charging technologies.

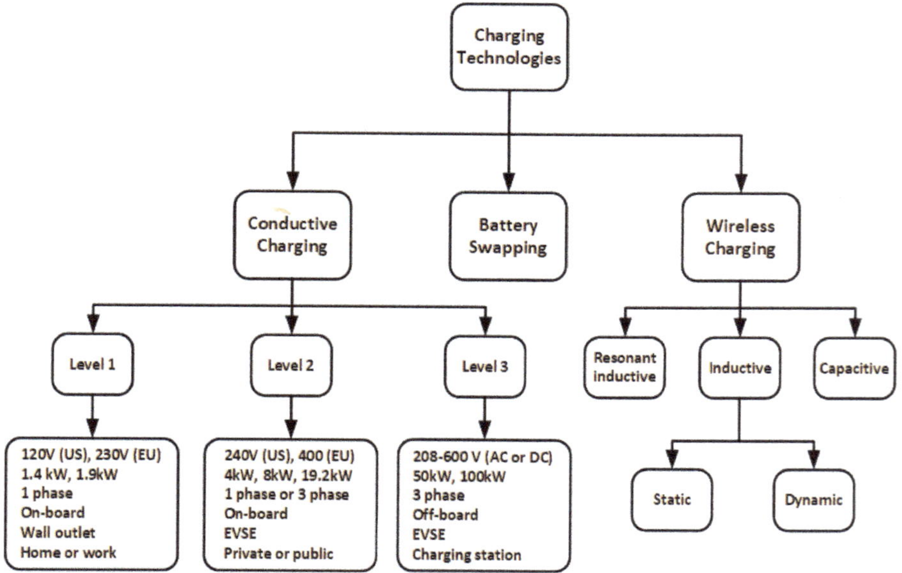

Figure 2. Classification of EVs charging technologies.

2.4.1. Conductive Charging

EV battery chargers have a significant responsibility in the advancement of EVs because the EVs adoption and social acceptance depends on the effortless access to charging stations or street chargers. Several topologies were presented for single phase and three phase EV chargers [32,33]. It consists of AC/DC converter, power factor correction elements, and DC/DC converter as shown in Figure 3. Charger systems are classified to on-board (i.e., inside vehicle for slow charging) and off-board (i.e., outside vehicle for fast charging). Moreover, they can be classified to unidirectional or bidirectional chargers [34]. Unidirectional charging has simple charging hardware and allows power flow from grid to EV only. Bidirectional charging allows power flow from the grid to EV and can inject power from the EV battery to the grid, building, or home.

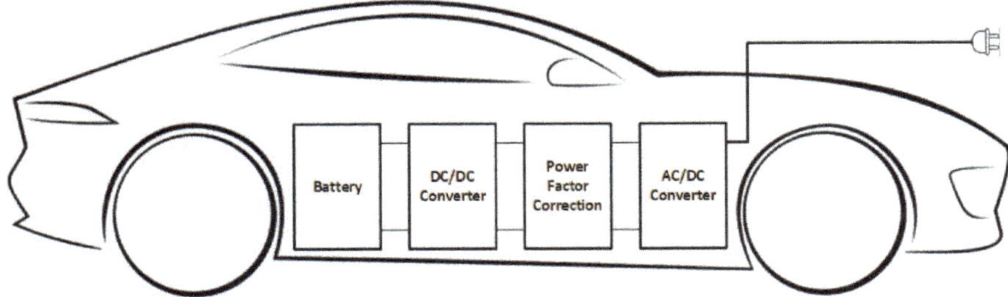

Figure 3. Onboard EV charger components.

Availability and advancement of EV charging infrastructure can decrease the required onboard energy storage and decrease the range anxiety of EV owners. According to the Society of Automotive Engineers (SAE) Standard J1772 [35], there are three charging levels. Most of the EV owners are expected to charge at home overnight according to the Electric Power Research Institute (EPRI) so Level 1 and Level 2 chargers will be the primary option [36], while level 3 will be used for commercial charging stations.

- **Level 1 Charging**

This is the slowest charging way and the simplest because no additional infrastructure is needed, and any wall outlet can be used. In the US, a standard 120 V/15 A wall outlet is used for Level 1. It is available only as an on-board charger. Although its cost is less than other charging levels, the EV needs a long time to be fully charged. Due to its low power rating, this charging level has the lowest impacts on distribution systems.

- **Level 2 Charging**

Level 2 charging uses 208 V or 240 V at currents up to 80 A, and 19.2 kW charging power. EV owners prefer Level 2 compared with Level 1 because of its shorter charging time. It may need dedicated electric vehicle supply equipment (EVSE) installation for public or home charging. Some EVs like Nissan Leaf have an on-board charger of this charging level.

- **Level 3 Charging**

Level 3 is for fast charging and it operates as a commercial refueling station (i.e., less than 1-h charging time) similar to the conventional gas station which can be installed at city main roads and highways. It is supplied from a three phase circuit with 480V or higher voltages. It is available only as an off-board charger because the charging power is high and may exceed 100 kW. It is clear that level 3 charging is not suitable for home charging. It has a high installation cost, and this represents a potential issue. It is expected that public chargers will use Level 2 or Level 3 for fast charging in shopping centers, parking lots, restaurants, hotels, theaters, etc. High charging power represents an advantage from charging time point of view, but it may generate a peak demand and overload the distribution network equipment in addition high installation cost.

2.4.2. Wireless Charging (WC)

WC enables EVs charging without physical contact or cable between the power supply and battery. Advancement of WC will reduce the required onboard battery capacity which will decrease EVs price and mass, which will result in reduction of EVs energy consumption. WC may become a future alternative for traditional conductive charging. WC has a potential to be used for charging electric bus batteries [37]. It can operate at different voltages (level 1, 2, and 3). The highest efficiency recorded for WC is 90% [38]. There are three main technologies of WCS: inductive, resonant inductive, and capacitive WC [39,40].

Inductive wireless charging (IWC) will be explained to provide the basic idea of this technology. The IWC contains AC/DC converter that converts AC power supplied from electricity grid to DC. Then, it is converted again to AC power with high frequency fed to transmitting (i.e., primary) coil. All these components are in the street underground. The EV contains receiving (i.e., secondary) coil that receives power from transmitting coil by electromagnetic induction through the air gap. Then the AC power is converted to DC by AC/DC converter and charge the battery [41]. A simplified diagram of IWC that explains the main concept is shown in Figure 4. IWC can be classified into static inductive charging and dynamic inductive charging. For static inductive charging, EV must be stationary during charging. However, dynamic inductive charging allows WC while EV is moving [42].

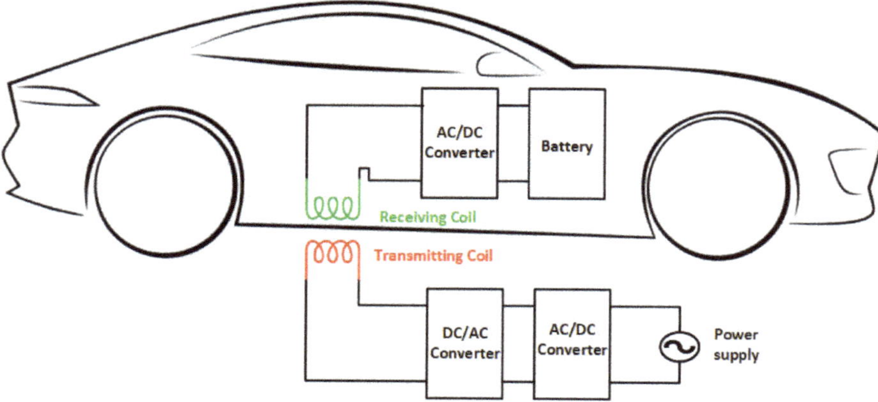

Figure 4. A simplified diagram of inductive charging.

Current WC is designed for unidirectional power flow from grid to vehicle, but future development of this technology is to enable EVs to discharge power to the grid wirelessly to provide electrical services. The advantages of this technology are electrical safety, no cables needed, and user convenience. The challenges of this technology are the high infrastructure cost compared to conductive charging, and low power transfer efficiency between coils [40].

2.4.3. Battery Swapping

Battery swapping station (BSS) is a charging station at which the empty EV battery will be replaced by a fully charged battery in few minutes [43]. Battery swapping may be used with electric buses that have a high capacity battery which will take a long time to be charged by traditional conductive charging. This technology requires a large stock of batteries owned by the BSS or a third party and rented to EV owner. BSS contains a distribution transformer, AC/DC converters to charge the batteries, batteries, and battery swapping equipment [44,45]. Some studies considered that BSS can use bidirectional chargers to provide electrical services by V2G mode [46,47]. The challenges of this technology are battery standardization, high infrastructure cost, and large space for BSS. In 2013, Tesla company revealed a battery swapping system that can swap the battery in 90 s [48].

3. Charging and Discharging Methods

Charging methods can be classified into two main categories, unidirectional and bidirectional charging. In unidirectional charging the energy flow is only from the grid to EV (i.e., uncontrolled charging, delayed charging, and controlled charging). In bidirectional charging, energy can flow from the grid to EV and from EV to the grid or loads as buildings, or homes (i.e., V2G, V2B, and V2H). Figure 5 presents the classification of different charging and discharging methods. Figure 6 shows how the total power demand profile of distribution system will change with different EV charging or discharging methods [10].

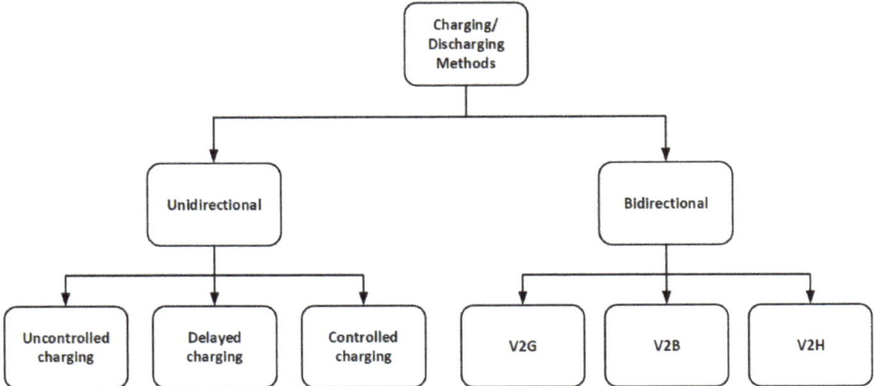

Figure 5. Classification of different charging and discharging methods.

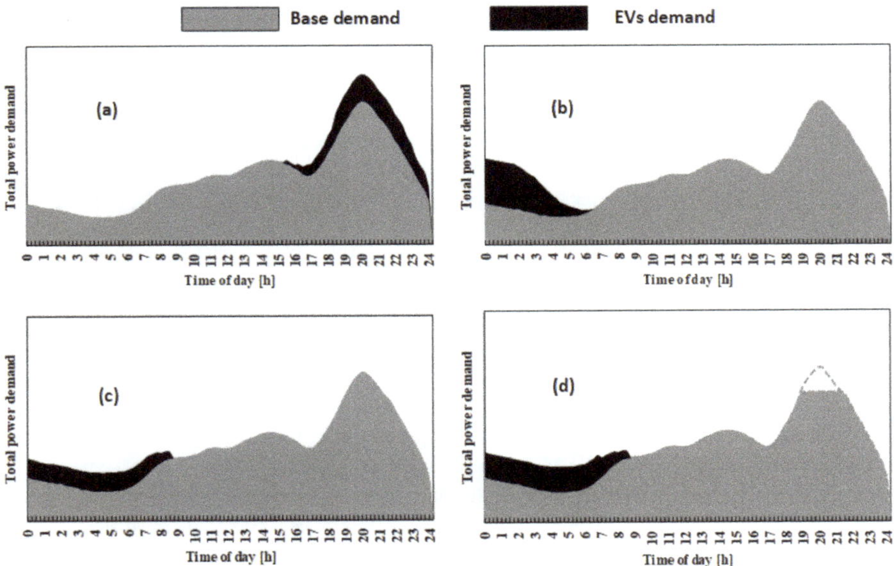

Figure 6. Total power demand profile with different charging and discharging methods. (**a**) Uncontrolled charging; (**b**) Delayed Charging; (**c**) Controlled Charging; and (**d**) V2G, V2B, and V2H.

3.1. Uncontrolled Charging

This is the simplest method to charge EVs and the current used way. The EV is plugged in for charging at the maximum power rating of EV charger until the EV battery is fully charged (i.e., state of charge (SoC) is 100%), similar to any electric device with a battery (i.e., laptop, smart phone, etc.). Several studies concluded that uncontrolled charging of EVs may result in severe negative impacts on distribution networks such as increase in peak load demand, overloading of transformers and cables and shorten their life, increase voltage drop, increase system unbalance due to single phase chargers, increase power losses, and increase harmonic distortion [49,50]. Moreover, this type of charging limits the EVs acceptable penetration level, because EV owners charge their vehicles when arriving home from work, which usually coincides with peak hours as shown in Figure 6a. The figure shows the total power demand of residential consumers during the day [51] before connecting EVs (in grey) and how the total power demand will change when EVs are connected to the distribution network for charging. Furthermore, it will accelerate the need for infrastructure upgrade. This charging method is also known as dumb charging, uncoordinated charging, and unregulated charging.

3.2. Delayed Charging

The severe impacts of uncontrolled charging can be alleviated with the use of delayed charging. In this case, the utility has different electricity prices during the day with low price at off peak time (i.e., time utility wants EV owners to charge at) and high price at peak time (i.e., time utility does not want EV owners to charge at). This method controls the charging time and not the charging power (i.e., EV charge at charger maximum power rating). With the optimal design of electricity tariff prices during the day, it can work as an incentive for EV owners to charge their vehicles on low price times which maximize both utility (i.e., distribution system operator (DSO) or distribution company) and EV owners benefit and result in valley filling as shown in Figure 6b. Although this method is a very simple technique to flatten the load profile and shift the load to off peak time, non-optimal design of electricity price, can incentivize a large number of EV owners to charge at off peak times, which may result in a second peak especially at the beginning of off peak time. Moreover, this method ignores EV owners' preferences and needs to charge their EVs at different times during the day. This charging method is also known as off peak charging, and indirect controlled charging with time of use pricing [52,53].

3.3. Controlled Charging

Although delayed charging has less impacts on distribution networks than uncontrolled charging, it has limitations. These limitations can be surpassed using controlled charging. This method controls the charging time and charging power of EV depending on some distribution network parameters like total power demand, transformer loading, voltage stability, power losses, etc., or to minimize the charging cost. In this technique, EV acts as a controllable load. Various studies proposed controlled charging algorithms for maximizing EV owner benefit by charging cost reduction and maximize utility benefit by distribution network stress and losses reduction, enhancement of power quality, and shifting the EV load to off peak hours which result in valley filling [54,55] as shown in Figure 6c. This charging method is also known as coordinated charging and smart charging.

Controlled charging can be classified as centralized, decentralized, and autonomous control architecture [56]. In centralized control, a central controller collects data such as electricity prices, system loading condition, EVs status, and owner preferences; based on this data the controller determines the set points of controlled EVs. Although centralized charging control usually results in optimal use of the system and enables EVs to participate in many electrical services, it has few disadvantages. There are privacy issues because all users data can be accessed by the central controller, the need for expensive two way communication infrastructure, large data amount must be processed, which is a high computational burden, any problem in the central controller or loss of communication can lead to severe consequences, and a large number of messages should be communicated in a short period which can result in communication issues. Moreover, most of the utility companies do not use real time pricing, which makes this method inapplicable currently [56].

In decentralized charging control, the control entity (i.e., DSO or aggregator) sends signals to EVs like electricity prices to incentivize them to take a specific action such as reducing their charging power or stop charging at peak hours. In this case no need to send EVs private information to the control entity. Although decentralized control methods need a cheaper processing and communication infrastructure compared with centralized control methods, they have few disadvantages. The optimal utilization of the system is not ensured, their ability to participate in ancillary service markets is limited and they are vulnerable to variations in customers behavior [56].

In autonomous charging control, the charging power is regulated based on local inputs such as voltage and battery SoC without any communication between EVs and control entity [57,58]. Usually, autonomous charging control is classified as decentralized control with no communication. Although this control architecture is the single option for distribution systems with no communication facilities and it can enable the integration of a larger number of EVs to the power grid in its current status, it has few disadvantages. Their ability to participate in ancillary service markets is limited, they are vulnerable to variations in customers behavior and the system cannot be operated optimally [56].

3.4. Vehicle to Grid (V2G)

This refers to the capability of EVs to supply power to the distribution network and operate as distributed energy storage devices. Bidirectional EV chargers are used to enable the electric power to flow in both directions from grid to vehicle (i.e., charging) and from vehicle to grid (i.e., discharging). When there is surplus electric energy (off-peak times) EVs will charge, which is called grid to vehicle (G2V) mode, and when there is a deficit in electric energy and consumption is higher than generation, EVs will supply power to the distribution network and this enhances the system reliability and efficiency [59]. So EVs can be seen by utilities as load and source. Because EV power and energy are limited and approximately have no effect on the power system, usually an aggregator is the responsible for aggregating EVs power or energy, which are located in geographical area to participate in electrical services.

The interaction in this case is between EV and power system management organization like DSO, which has to operate the power system in a reliable way. V2G can provide many grid services such as frequency regulation, spinning reserve, enabling the integration of more RESs, and peak load shaving and valley filling [60] as shown in Figure 6d. Other benefits of V2G is generating revenues for EV owners, and decrease emission and operating cost if grid services were provided by traditional nonrenewable generation [60]. Although this is a very promising technology regarding providing electrical services to the power system, it has many challenges and barriers. This technology is very complex and needs many infrastructure changes because it needs bidirectional chargers and continuous two-way communication between EVs and system operator or aggregator. Moreover, it is expected that continuous charge and discharge of EV battery may lead to battery degradation and shortening its life. In contrast, new studies concluded that EV can participate in V2G without battery degradation [61].

3.5. Vehicle to Building (V2B)

V2B is similar to V2G but in V2B there is no communication between the vehicle and the grid, and it only communicates with the building so the energy stored in the EV battery can be used to supply the building loads only. The use of V2B mode can be very effective during peak load times and outage conditions. EVs are used as an energy storage device and operates in two modes: G2V and V2B. It operates in a G2V mode to charge the battery at low cost when the grid is lightly loaded and there is surplus electric power generation. It operates in a V2B mode to supply the building loads when the electricity price is high at peak hours [62]. By doing this V2B is providing peak shaving and valley filling which is utility benefit and reduce building expenses by discharge at periods with high electricity prices and reduce building demand from the grid. This method is simpler than V2G as it operates behind the building meter, but it provides less electrical services to the power system. V2B involves one building (i.e., many homes) and many EVs.

3.6. Vehicle to Home (V2H)

This is very similar to V2B but in this case, only one home is involved and not a whole building as in V2B and only one EV. Normally EV absorbs power from home and can supply power to home when needed. It needs a simple architecture compared to V2G and V2B and can provide less electrical services to the power system compared to them. It can flatten the house daily load profile and reduce house consumption during peak hours with high electricity prices. Local energy production from photovoltaic (PV) or small wind generation can be effectively used by storing the excess generation in EV battery and use it when needed [63,64].

4. Negative Impacts of Electric Vehicles on Electric Power Systems

Uncontrolled charging of EVs with high penetration levels is expected to introduce unwanted negative impacts on the power system. The possible negative impacts are the impact due to increase of peak demand, voltage deviation from the acceptable limits, phase unbalance due to the single-phase chargers, harmonics injection, overloading of the power system equipment, and increase of power system power losses [65]. A large number of studies have investigated these impacts because the estimation of EVs charging impacts is based on several conditions [65] and have many uncertainties that must be considered and modeled properly in the study to accurately estimate the impact of EVs charging on the power systems, the main key variables are [66,67]:

- **EV battery charger**

Depending on the charging infrastructure and the EV model single phase or three phase chargers may be used for charging. Moreover, there are many power ratings for EVs chargers operating at different voltage levels. So, the fast chargers are expected to increase the peak demand in a larger value compared with slow chargers.

- **Time**

The time of connecting the EVs for charging is not certain so it is unknown how many EVs may be charging at the same time and the probability that the EV charging time interferes with the peak demand time of the distribution networks. Some studies assume the charging start time of EVs, and other studies make surveys in a geographical area or a city to know the home arriving time of vehicles and model it as probability distribution.

- **Location**

The place at which an EV charge is random. EV owner may charge at home, friend's home, street charger, or charging station. It may be different from distribution network to another where are the charging locations and what are the EV charger type (i.e., private or commercial chargers).

- **Battery capacity**

There are a wide range of EV batteries capacity. PHEV usually contains a small battery capacity while BEV contains higher battery capacity. EVs with high battery capacity will draw a larger amount of energy from the power grid.

- **Battery SoC**

The battery SoC at the plug-in instant is stochastic. Many studies assume the SoC and others consider it as a probability distribution.

- **Penetration level**

EVs still have a very small market share which will increase quickly for few countries like Norway and slowly for many other countries. Therefore, many studies considered different EVs penetration levels ranging from 10% to 100%.

- **Distribution system status**

There are many differences between distribution systems in terms of structure (i.e., radial or ring), equipment loading condition before connecting EVs, voltage level (i.e., medium voltage or low voltage), voltage profile, daily load profile of loads, etc.

Therefore, numerous studies were executed to assess the impacts of EVs charging on the power system. Most of the studies focus on the distribution network at which EVs charging occurs, because the most severe impacts are expected to occur at the distribution level. Classification of positive and negative impacts of EVs charging/discharging on electric power system can be seen in Figure 7 and they are explained in the following sections.

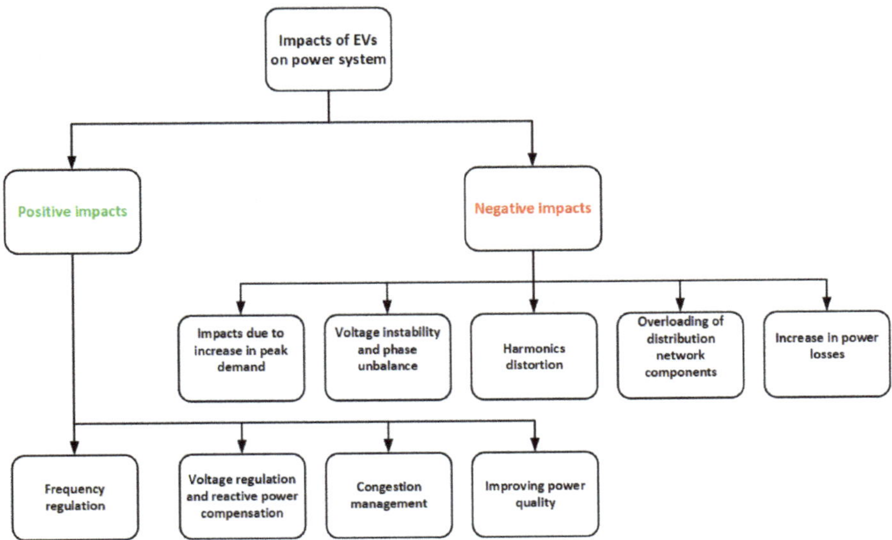

Figure 7. Classification of EVs charging impacts on electric power systems.

4.1. *Impacts Due to Increase in Peak Demand*

Various studies have been performed for assessment of EVs charging impact on the peak demand and load profile on distribution systems. Many distribution networks in different countries and with different topologies and characteristics were chosen to conduct these studies. In [68], the peak demand increased by 53% when uncontrolled EVs charging was used with 30% penetration level. In another study [69], with only 10% penetration level of uncontrolled EVs charging in residential network, the peak demand increased significantly. In [70], the effect of charging light duty vehicles up to 100% penetration level on system peak demand was investigated. For 100% penetration level, uncontrolled charging resulted in a high increase in peak demand, which must be minimized by delayed charging. In [71], it was found that EV rapid charging at peak hours will result in a significant rise in peak demand and equipment limits were exceeded even with very small penetration levels. However, by the use of slow charging at off peak hours, distribution network could integrate up to 50% penetration level without violating equipment limits.

In [72], the impact of uncontrolled charging of EVs on total power demand for three distribution networks (i.e., urban, suburban, and rural) with different capacities were evaluated. The study considered penetration levels up to 25%. For the highest penetration level, the peak demand increased by 9% for both urban and rural networks and 11% for suburban network. This increase made the rural network exceed the maximum capacity limit while urban and suburban did not exceed the capacity limit. In [73], it was found that uncontrolled charging increased the peak demand. The study proposed a demand response method to control EVs demand and other flexible home devices. This method enabled the integration of EVs without increase in peak demand from the base case (i.e., without EVs).

In a recent study [74], the impacts of uncontrolled EVs charging on Great Britain power system was investigated. The study also considered how controlled charging can mitigate these impacts. For a 100% penetration level, it was found that the peak demand will increase by 8 GW at generation and transmission level. Moreover, the results showed that controlled charging can significantly reduce the increase in peak demand and eliminate the need for new generation infrastructure. At the distribution level 100% penetration level will require 28% upgrades at distribution network. Controlled charging can reduce the distribution network upgrade requirements to only 9%. Another interesting conclusion in this study is that load profile at distribution level and load profile at transmission level cannot be flattened at the same time using controlled charging. For instance, flattening the load at transmission level (i.e., national demand) will require 19% upgrades at the distribution level. Moreover, flattening the load at distribution level will result in 6 GW increase in national demand.

In [75,76], it was concluded that the increase in peak demand can be crucially mitigated by the use of optimized charging and time of use (ToU) charging. A study on the Estonian grid [77] with 30% penetration level of the passenger' cars available was conducted and the results showed an increase in the peak load with 5% for uncontrolled charging and 4% increase for controlled charging. Section 5.3 gives more details on how controlled charging can flatten the load profile and how controlled charging/discharging can cut the peak demand (i.e., peak shaving).

From the previous studies, it is clear that large integration of EVs charging in uncontrolled manner may lead to a significant increase in peak demand. This increase in peak demand will result in higher operation costs because expensive generation must be operated at peak hours for short duration to supply peak load. In addition, infrastructure upgrade is required if the peak demand is higher than components capacity at generation, transmission, or distribution levels. Moreover, the use of delayed charging and controlled charging is effective in minimizing peak demand increase due to EVs demand and the accompanied negative impacts; this can enable higher EVs penetration.

4.2. Voltage Instability and Phase Unbalance

Voltage instability represents a challenging issue and can result in system disruptions [78]. The reason for that is the operation at high load demand and near the stability limit. The grid voltage stability is crucially affected by the characteristics of the load. EVs load characteristics are different from the conventional loads (i.e., residential, industrial, and commercial) characteristics and the earlier estimation of its power and energy demands are difficult. In addition, the EVs consume more power in a short time to fully charge the battery. Furthermore, single phase EVs chargers may increase phase unbalance at distribution network. Phase unbalance results in unwanted negative effects at distribution network operation and connected loads and should remain in the acceptable limits.

Numerous studies have been executed to evaluate the impacts of EVs charging on voltage instability, voltage deviation, and phase unbalance. In [79], the voltage stability was investigated in a study implemented in the Institute of Electrical and Electronics Engineers (IEEE) 43 bus distribution system. The results showed that voltage stability of distribution network is highly reduced by EVs fast charging. The impact of uncontrolled charging on voltage deviation at different daily durations was assessed in [80]. Obtained results showed a large increase in voltage deviation which was close to exceeding the acceptable limits especially if EVs were charging at peak period. The study proposed stochastic programming strategies to minimize voltage deviation. In [72], the impact of uncontrolled charging of EVs on voltage drop for three distribution networks (i.e., urban, suburban, and rural) was estimated. The study considered penetration levels up to 25%. For rural network which has long feeders, higher voltage drop was recorded due to EVs charging which fosters the need for voltage regulation devices. The impact of uncontrolled charging of EVs on voltage deviation in primary and secondary distribution systems was examined in [81]. The study considered different penetration levels (i.e., 30% and 50%), different EVs types (i.e., BEV and PHEV), and different charging levels (i.e., level 1 and level 2). Results showed no voltage limits violation at any node of primary distribution system. However, voltage limits violations were recorded in the secondary system. Moreover, it showed that BEV caused more voltage drop than PHEV due to its higher battery capacity and level 2 charging caused more voltage drop than level 1. In another work [82], the effect of large-scale integration of EVs with high power charging in IEEE 39 bus distribution system and how the distribution network reliability was improved by optimal charging of EVs was investigated. Another method for reducing the voltage instability was proposed in [83]; the method is based on voltage control by the tap-changing transformer.

In [84], the impact on voltage unbalance due to uneven EVs distribution on the three phases was investigated. The study considered two scenarios for EVs distribution on phases. For scenario A, 50% of EVs are connected to phase a, 30% to phase b, and 20% to phase c. For scenario B, 80% of EVs were connected to phase a, 20% to phase b, and 0% to phase c. For scenario A, the voltage unbalance factor (VUF) did not violate the limits until 50% penetration level. However, for scenario B, the VUF reached the limits at 25% penetration level. Another study [85] investigated the impact of single phase EVs charging (i.e., G2V) and discharging (i.e., V2G) on voltage unbalance of low voltage distribution network. The study considered many cases and results showed that in some cases VUF exceeded the limit for both charging and discharging modes. In [86] a significant phase unbalance occurs due to the EVs charging with single phase chargers at level 1 at the residential network which results from the unequal distribution of EVs chargers in the three phases. However, a small impact on voltage and current unbalance was observed due to EVs charging in [87]. It should be mentioned that by using controlled charging and discharging of EVs with considering VUF minimization, the voltage unbalance at distribution network can be minimized. More details are provided in Section 5.4.

Previous studies showed that usually low EVs penetration levels can have a small impact on voltage values even if uncontrolled charging was used. However, high EVs penetration levels can cause high voltage drop, and voltage value may exceed the acceptable limits especially at the end of long feeders (i.e., rural network), which will require the installation of voltage regulation devices. Studies showed that secondary distribution networks have a higher probability of exceeding voltage limits than primary distribution networks. Moreover, phase unbalance represents a challenging issue since most EVs are expected to be charged by single phase private chargers. Very small EVs penetration levels can cause small phase unbalance. However, high EVs penetration levels and high uneven distribution of chargers on the three phases may result in VUF higher than acceptable limits.

4.3. Harmonics Distortion

Power quality problems may arise due to EVs charging. Because power electronic devices are used in EV chargers, so high EVs integration can affect the power network's power quality. Harmonics injected by EV chargers into the power grid will lead to negative effects on electric power system components which are designed to be supplied by pure sinusoidal waveform and increase system losses. Few studies found that EV chargers had a non-significant effect on harmonic distortion. For instance, in [88], a comprehensive harmonic study was implemented using the Monte Carlo method and the results showed a minor impact of harmonics in the distribution network. Another study [89], found that commercial EV chargers resulted in a small increase in the total harmonic distortion of voltage (THDv), and it was less than 0.8%.

In contrast, many studies concluded that EVs charging had a large impact on the distribution system. In [90], the THDv increased to 11.4% due to uncontrolled rapid charging and this value exceeded the limit of 8% based on EN 50,160 standard and a solution to the harmonic problem was proposed by using the control of PV inverter as an active filter. In [91], the impact of EVs charging with slow and fast charger on total harmonic distortion (THD) was assessed for different EVs models. A high total harmonic distortion of current (THDi) was recorded for fast charging between 12% to 24%. In [92], the impact of traditional EV charger on the quality of system voltage and drawn current was evaluated. Traditional charger caused very high THD. The authors proposed a smart charger that draws sinusoidal current and has a unity power factor. The smart charger significantly reduced THD compared with traditional charger. The negative effects of EV home chargers on distribution network power quality and transformer life were studied in [93]. The study found a quadratic relation between THDi and life consumption of the transformer. Moreover, the study concluded that THDi should not be more than 25 to 30% to have acceptable increase in life consumption of the transformer. In [94], the impact of EV chargers' current harmonics on distribution system capacity was studied. Results showed that the 10 kV cable was overloaded at 27.25% penetration level when current harmonics due to EV chargers was considered. However, the cable was overloaded at 30.74% penetration level when current harmonics due to EV chargers was neglected. For harmonics reduction from EVs integration, filters must be added to the EV chargers.

Only a few studies found that EV chargers harmonics result in a minor impact in the distribution network. Most of the studies showed that traditional EV chargers can cause unacceptable harmonics values. These high harmonics will result in decreasing the life cycle of distribution network components (i.e., transformers and cables). However, by proper design of EV charger circuits, control strategy, and filters integrated into the charger circuit, the charger harmonics can be alleviated significantly. More details can be found in Section 5.4.

4.4. Overloading of Distribution Network Components

The high EVs energy demand requires a large amount of electric energy to be transmitted from the generation stations to the distribution networks. The distribution networks' equipment such as transformers and cables may get overloaded due to the new EVs load and this will lead to stress these components and reduce their lifespan and foster the need for infrastructure upgrade. Several researches were executed to analyze the impacts of EVs charging on distribution system components. In [95], the distribution transformer aging due to uncontrolled charging of EVs with level 1 and 2 charging power was investigated. The results showed that level 2 charging has a higher aging impact on the transformer in comparison with level 1 charging. The impacts of EVs charging on transformer and underground cable were assessed in [96] for low and high penetration levels (i.e., 12.5% and 70%). The transformer and cable were overloaded in both penetration levels. The impact of uncontrolled charging of EVs on distribution transformers with 25 kVA and 50 kVA power rating was examined in [81]. The study considered different penetration levels (i.e., 30% and 50%), different EVs types (i.e., BEV and PHEV), and different charging levels (i.e., level 1 and level 2). Results showed that 50% penetration level resulted in overloading 50% of the 25 kVA transformers and 35% of the 50 kVA transformers. Furthermore, BEV with level 2 charging caused 10% increase in the number of overloaded transformers compared with PHEV with level 1 charging.

In [72], the impact of uncontrolled charging of EVs on the loading of secondary transformers installed in three distribution networks (i.e., urban, suburban, and rural) with different capacities was executed. The study considered penetration levels up to 25%. The study counted the number of transformers overloaded above 20%. EVs charging resulted in increasing the number of transformers overloaded above 20% for suburban area, while urban and rural areas did not have high overloading percentages. The impact of uncontrolled charging on distribution networks transformers and cables was investigated in [97]. Results showed a large increase in the number of overloaded transformers and cables. The number of overloaded transformers and cables decreased by 25% and 8%, respectively, when controlled charging was used. In [98], it was found that the presence of EVs will lead to reducing life duration of distribution transformer. In [99], it was concluded that the uncontrolled charging of EVs will lead to aging of 25 kVA distribution transformer. Moreover, it was found that transformer aging can be reduced using controlled charging strategies.

Another study [100], concluded that uncontrolled level 1 EVs charging has insignificant impact on transformer life but the massive penetration of EVs may have a severe effect on transformer lifespan. The transformer lifespan can be enhanced by using off-peak EV charging and load management [95,100,101]. In [102] the cable loading was examined for EVs peak charging hours. The results demonstrate that the cable can handle up to 25% penetration level for slow charging and up to 15% penetration level for fast charging and cannot handle massive EVs penetration easily.

The high energy demand of EVs will increase the loading at different parts of the power system (i.e., generation, transmission, and distribution). Distribution level is highly affected by EVs charging compared to transmission and generation levels, and most of the studies focused on the impacts of EVs charging on distribution network components. The studies showed that the acceptable EVs penetration level before overloading of network components varies depending on the network components' capacity and their loading condition before connecting EVs. Studies concluded that the uncontrolled charging of EVs result in overloading of many transformers and cables at the distribution network and can result in reducing their lifespan and requires components upgrade which represents economical challenge to electric utilities. Results showed that many distribution networks can only allow 10% penetration level before overloading if level 2 charging was used. The network can handle higher penetration levels if level 1 slow charging was used. Using proper charging and discharging methods (i.e., delayed charging, controlled charging, V2G, V2B, and V2H), distribution networks can integrate higher EVs penetration levels before they reach their capacity limits. More details can be found in Section 5.3.

4.5. Increase in Power Losses

Extra power demand represented in EVs charging will lead to higher currents flowing and extra power losses in different system components, such as generators, transformers, and cables, which is the main concern for utilities. Various studies were performed to examine the EVs charging impact on system losses. In [103], the impact of uncontrolled charging of EVs on two large scale distribution system was executed under three penetration levels (i.e., 35%, 51%, and 62%). Uncontrolled charging resulted in a large increase in energy losses and required investment cost. Losses and investment costs reduced when delayed charging or controlled charging were used. The impact of uncontrolled charging on power losses at different daily durations was assessed in [80]. Obtained results showed a large increase in power losses especially if EVs were charging at peak period. The study proposed stochastic programming strategies to minimize power losses. In [104], a study examined the EV charging impact on a Danish distribution network. The obtained results showed that for uncontrolled charging with 50% penetration level the grid losses increased by 40% and increased only 10% for controlled charging.

In [105], the impact of EVs charging on distribution transformer power losses was investigated. It was found that for penetration levels ranging from 2% to 40%, the transformer losses increased to more than 300% mainly due to windings copper losses increase. The increase in power losses at the IEEE 33 bus distribution system due to EVs fast charging station was evaluated in [106]. The study investigated many cases by changing the charging station bus and the charging station power consumption. It was found that installing charging stations at weak buses (i.e., far from the main transformer) increased system power losses. Moreover, the power losses could be reduced by distributing charging station load at two buses instead of one bus. In [107], the increase in energy losses of a distribution network due to EVs charging was examined. The distribution network supplies residential and commercial loads and located on a Korean island. The results showed that daily energy losses increased by 66% for 40% penetration level.

Previous studies focused on assessing the impact of uncontrolled EVs charging on the power losses at distribution networks. The studies concluded that uncontrolled EVs charging resulted in a high increase in distribution network power losses. Using proper charging and discharging methods (i.e., delayed charging, controlled charging, V2G, V2B, and V2H), power losses due to EVs charging can be reduced significantly. Additionally, the installation of distributed generation (DG) near the charging location can decrease the energy supplied from the grid, and hence reduce the power losses.

5. Positive Impacts of Electric Vehicles on Electric Power Systems

EVs are parked for most of the daytime [108], and they are connected to the charger for a longer duration than required recharging duration. Therefore, EV battery can be used to provide grid services and gain revenues for EV owners by injecting power to the grid to keep demand–supply balance or by controlling the charging time and power to reduce the charging cost and electricity bill. Many studies showed that controlled EVs charging can improve power system efficiency, reduce operation cost, and minimize RESs curtailment. Moreover, EVs controlled discharging can provide additional benefits and electrical services [109]. EVs can provide short time scale electrical services due to the fast response of battery chargers such as primary frequency control (PFC), medium time scale electrical services, such as secondary frequency control and long time scale electrical services, such as congestion management and minimization of power losses due to the high battery capacity [110]. The focus in reviewing the literature will be in operational aspects and services provided by EVs and not in the optimization algorithms and control methods used due to the broad spectrum of approaches used in these studies [111].

5.1. Frequency Regulation

Power system frequency should be maintained at nominal value (i.e., 50 or 60 Hz) for normal operation. It is considered an important indicator of active power supply–demand balance. In normal operation, the power imbalance occurs due to continuous load variation or fluctuation of RESs generation that depend on weather conditions. In emergency conditions, the power imbalance occurs due to sudden outages of loads, transmission lines, or generating units. Failing to maintain frequency in the specified limits will result in load shedding in case of under frequency or disconnection of generating units in case of over frequency. In traditional power system, frequency regulation is achieved by synchronous generators in large power plants (e.g., hydro and thermal power plants) [112]. In future power systems, controllable loads like heat pumps and EVs will have a significant role in frequency control [113]. EV batteries have a faster response compared to traditional generation units due to the fast response of EV power electronic interface (i.e., EV charger). Therefore, controlled charging and discharging of EVs can be an effective option for frequency regulation. Moreover, frequency regulation is becoming more challenging due to the reduction of system inertia and increase of fluctuation due to the increasing share of RESs with power electronic interface [114–116].

The ability of available commercial EVs (i.e., Nissan Leaf) to provide PFC by only changing the charging power and with no V2G capability was tested experimentally in [117]. A small isolated power system with renewable generation was used as a test system. The results proved the technical feasibility of EVs to provide PFC with fast response time. In [118], the authors studied how EVs can participate in PFC in two ways. The first is to switch off EVs charging and the second way is to inject power to the grid in V2G mode. Another study [119] proposed a control method to provide PFC in three area power system by coordinating EVs charging and discharging while minimizing battery degradation cost. The effectiveness of EVs to provide primary frequency regulation was tested in [120] for a small isolated power system containing wind turbines, diesel generators, and hydro generators. The study verified that EVs were very effective in reducing frequency oscillation with a small change of EVs consumed energy and negligible variation of the required charging time. It also showed that EVs can enable more wind energy share while keeping normal operation.

In [121], the authors studied two modes of EVs charging control for frequency regulation while considering EV owner diving behavior. The first mode only controls the charging power and the second mode controls charging and discharging power. Results showed that EVs were effective in reducing frequency fluctuation. A coordinated control strategy between EV operating in V2G mode and traditional generation for load frequency control (LFC) was proposed in [122]. The proposed control strategy was tested on the Great Britain power system. The results showed the effectiveness of the proposed strategy in improving frequency regulation and reduction of power mismatch. Moreover, due to EV participation, the traditional power generation output variations were reduced. The EVs ability to participate in LFC in a microgrid operating in isolated mode was investigated in [123]. The microgrid contains both renewable and nonrenewable generation. The studies proved that EVs can enhance frequency stability in addition to reducing emissions from nonrenewable generation and increase microgrid operator profit. In [124], EVs operating at V2G mode were used to provide LFC in a multi-area power system containing traditional nonrenewable generation (i.e., thermal, hydro, and gas turbines).

Previous studies results proved the feasibility of using EVs to provide frequency regulation due to its fast response. Results showed that the use of EVs can achieve rapid control action in balancing the generated power and the power demand during load and source variations and the system frequency perturbation controlled by EVs is much lower than the other generation units. EVs can provide frequency control either by regulating charging power with no V2G capability or by regulating charging and discharging power (i.e., V2G capability). In the first approach, frequency regulation can be achieved by only change the EVs charging power. This approach is simple, requires simple infrastructure, and have a limited effect on battery degradation. In the second approach, frequency regulation is provided by controlling both charging and discharging power. This approach is more effective than the first approach. However, this approach is more complex, requires infrastructure upgrade (i.e., ICT and bidirectional chargers), and affect battery life cycle due to continuous charging and discharging. The studies showed that EVs are effective in providing frequency regulation at traditional power systems containing only nonrenewable generation. It resulted in reducing frequency fluctuation and reducing the variation of generation units output power. Furthermore, it is effective when there is a renewable generation installed in traditional power systems. It can reduce fluctuations due to intermittent renewable generation, enable the integration of more renewable generation, and reduce curtailment of renewable generation production. Moreover, EVs proved their effectiveness in providing frequency regulation in microgrids with a high share of renewable generation. It can reduce frequency fluctuation, increase microgrid profit, reduce curtailment from renewable generation, and enable integrating high share of intermittent renewable generation while operating in acceptable limits.

5.2. *Voltage Regulation and Reactive Power Compensation*

The voltage at any point of the power system must be maintained within acceptable limits. Voltage is an indicator of the loading status of distribution system. Voltage is high if the network is lightly loaded and low if the network is highly loaded. Although the common problem in distribution system is exceeding the lower limit, excessive DG may cause voltage rise and exceeding the voltage upper limit. Keeping the voltage within normal values at distribution level, which usually has a radial structure, is a challenge especially for long feeders and may require voltage regulation devices. Violation of these limits will result in improper operation or damage to the connected loads and may lead to voltage instability. Voltage control can be achieved by active or reactive power control. Active power can be controlled by DGs, energy storage technologies, and controllable devices like EVs and heat pumps. Likewise, reactive power can be controlled by transformers on-load tap changer (OLTC), capacitors, and static var compensators.

Many researchers studied voltage regulation by EVs active power control. An online controlled charging method was tested in [125]. The objective was to maximize EV owners satisfaction while considering distribution network limits. The algorithm minimized voltage deviation, transformer loading, and power losses. In [126], a decentralized/autonomous controlled charging method was proposed. It regulates the charging power based on the local voltage and battery SoC. It charges at a high charging power if the voltage is normal and decrease the charging power or stop charging if the voltage is low. EVs with low SoC have a charging priority. The proposed method reduced the voltage drop and improved the voltage profile compared to uncontrolled charging. In [127], an optimization algorithm was developed to provide peak shaving and valley filling in addition to improving voltage profile by controlling the charging and discharging of EVs. In [128], a controlled EVs charging technique was used to solve the voltage rise issue resulted from excessive PV generation at distribution system.

Another option for EVs based voltage regulation is by using capacitors in the DC link of the EV bidirectional chargers to supply reactive power. EV charger can do this even if the EV is not connected for charging. The ability of level 1 EV bidirectional charger to provide reactive power support was tested in [129]. The study showed that the DC link capacitor can provide reactive power support with no effect on battery degradation. In [130] a direct voltage control method was used to enable EV charger at DC fast charging station to inject reactive power to the grid to regulate bus voltage and reduce system power losses. In [131], a strategy for voltage regulation at distribution network was proposed. It is based on the coordination of DG, OLTC, and EVs providing reactive power support by operating at V2G mode. The proposed algorithm was effective in voltage regulation, minimizing OLTC operation times, and reduce the active power curtailment of DG. A limited number of studies investigated the use of EV chargers for reactive power compensation, which makes it a promising research area to be investigated in future research studies.

From studies results, it can be concluded that uncontrolled charging of EVs can result in a higher voltage drop at distribution networks and voltage values exceeding the acceptable limits, especially in long feeders, and will require infrastructure upgrade by installing voltage regulation devices. However, using controlled charging and discharging methods the voltage at all parts of the distribution network can be kept within acceptable limits and daily voltage profile can be improved without voltage regulation devices. Moreover, studies showed that the EV charger DC link capacitor is effective in providing reactive power support and voltage regulation.

5.3. Congestion Management

Load demand varies during the day and usually have peak hours in the evening. During peak hours expensive generators should be turned on for few hours for supply–demand balance. Uncontrolled charging of EVs will result in enlarging the peak power demand, which results in operating expensive power generation in addition to transmission and distribution networks stress. If the installed generation capacity is less than the required demand at peak period, new power plants must be constructed. This issue can be eliminated or alleviated by using delayed charging, controlled charging, V2G, V2B, and V2H. Using delayed charging and controlled charging can shift EV charging to off peak hours, which results in valley filling as shown in Figure 6b,c. Using V2G, V2B, and V2H technology enables injecting power to the grid, building, or home at peak hours which results in peak shaving and charge at off peak hours which results in valley filling as shown in Figure 6d. The process of peak shaving and valley filling is called load flattening or load leveling, which means reducing the difference between the maximum demand and the minimum demand during the day. By doing this congestion at the power system can be handled (i.e., congestion management).

Delayed charging was proposed in [132], and compared to uncontrolled charging. The results indicated that delayed charging is effective in reducing the stress of distribution system components, reducing voltage drop and power losses compared to uncontrolled charging. A decentralized controlled charging method for valley filling was proposed in [133]. This method needs simple unidirectional communication between the system operator and EV to broadcast day-ahead electricity prices. Based on the price, EV can autonomously control the charging behavior and charge at off peak period to reduce charging costs. This resulted in valley filling and reduction in generation cost by

28% compared to uncontrolled charging. The impact of uncontrolled EVs charging on total power demand was investigated in [134]. Results showed that uncontrolled charging will increase peak demand. Delayed charging and controlled charging were proposed for alleviating the impact of uncontrolled charging. The results showed that the proposed methods were very efficient in shifting the EVs charging from peak period and postponing any network upgrade.

Another study showed that the German transmission system cannot handle large scale uncontrolled EV charging even with future expansion plans [135]. It also showed that V2G technology is effective in reducing transmission system congestion and enhance grid stability. The effectiveness of V2G in providing peak shaving and valley filling was proved in [136]. By allowing some of the EVs to inject power back to the grid at peak period resulted in reducing the original peak demand (i.e., peak shaving). By allowing EVs to charge at off peak period only resulted in valley filling. In [137], the integration of EVs in a weak isolated grid in a Spanish island was studied. The study proposed a strategy for coordinated charging and discharging of EVs. Results proved the effectiveness of EVs in the efficient management of the grid, in addition to filling the valley and shaving the peak demand.

In [138], V2B was used in a building-integrated microgrid which contains EVs, battery storage, and PV. The objective was to reduce the peak load, which will decrease subscribed power rating and reducing the building electricity bill. The study showed that V2B can enable peak load shaving. By encouraging more buildings to have a similar management system it can reduce the peak load demand in the distribution system or even the whole grid. The effectiveness of V2H in reducing home peak demand was tested in [64]. Results showed that V2H was very effective in minimizing home peak demand and home electricity bills.

Previous studies' results showed that uncontrolled charging of EVs will result in an increase of peak demand at different levels of the power system (i.e., generation, transmission, and distribution) and can overload many network components and will require components upgrade which represents high investment costs for system operators. A more cost-effective solution to these issues is using proper charging methods. The studies proved that simple charging method like delayed charging, which only depends on variable electricity prices during the day, can shift most of the EVs load to off-peak hours by motivating EV owners to decrease the charging costs. Delayed charging can lead to reducing the increase in peak demand, reduce the number of overloaded components, postpone infrastructure upgrades, and valley filling. A more advanced charging method is controlled charging, where EVs can change their charging time and power depending on many variables such as electricity prices, load demand, components loading, grid constraints, etc., depending on the proposed control strategy. Controlled charging can lead to reduced charging costs, reduced increase in peak demand, reduction in the number of overloaded components, delay in network components upgrade, and valley filling. Moreover, the use of V2G, V2B, and V2B can provide more benefits to the grid and EV owners. By injecting stored energy in EV batteries to the grid, the peak demand can be shaved (i.e., peak shaving) and reduce power system operation costs by shutting down generation units with high operating costs. Furthermore, they can defer the need for infrastructure upgrades, gain revenue for EV owners by charging when the electricity price is low and discharge when the electricity price is high (i.e., energy arbitrage), and reduce home or building electricity bills.

5.4. *Improving Power Quality*

Although single phase uncontrolled charging of EVs may cause a severe effect on system unbalance, controlled charging methods can decrease system unbalance. In [139], the reduction of voltage unbalance by controlling EVs charging current was tested experimentally. The proposed control method is autonomous and depends only on local voltage measurement without the need for infrastructure upgrade and expensive communication infrastructure. The results proved that the proposed EVs controlled charging method can reduce voltage unbalance and minimize voltage drop. In another study, a method based on PV and EVs interaction for voltage unbalance minimization was tested in [140]. In [141], an optimization algorithm was used to control the charging and discharging of EVs to minimize VUF. An unbalanced distribution network with 1.93% VUF was used to test the proposed strategy. Uncontrolled charging of EVs increased the VUF significantly to 7.7%. The controlled charging of EVs declined the VUF to 0.71%. Controlled charging and discharging of EVs dropped the VUF to 0.5%. To control the voltage unbalance, the number of EVs connected at each phase was optimized using a genetic algorithm in [142]. The authors of [143] proposed a hybrid method combining centralized controlled charging of EVs and decentralized controlled discharging of reactive power for voltage unbalance minimization. The proposed strategy was effective in reducing VUF compared to uncontrolled charging case.

It was presented in Section 4.3 that EV chargers can result in significant harmonic distortion. However, the proper design of charger power electronics circuits and control methods can minimize or eliminate this issue. Few studies proposed EV chargers that cause no harmonic distortion or can provide harmonic filtering. A novel control strategy applied to onboard bidirectional three phase EV charger was proposed in [144]. The proposed method enabled the charger to draw or inject sinusoidal current with no harmonics regardless of the power quality of the grid. Another study [144] proposed EV charger that can operate as active filter and eliminate the harmonics caused by other loads at the charger connection point in addition to providing reactive power compensation. In [145], the EV charging station was used to eliminate the harmonics in distribution network by acting as active filter.

The high penetration of PV generation at distribution network can cause power quality issues due to its intermittent nature and dependence on weather conditions. The variation of weather conditions (i.e., clouds) results in fluctuation of PV power output which can cause voltage fluctuation and light flicker. Therefore, DSOs need to develop mechanisms to mitigate power quality issues caused by the fluctuation of PV output. Several studies proposed controlled charging of EVs as a possible solution to alleviate the fluctuations caused by PV generation [146–148]. For instance, reference [148] assessed the impacts of rapid variation of PV output on voltage fluctuation at low voltage distribution network for three scenarios. The study proposed a controlled charging of EVs to mitigate voltage fluctuations. The results proved the effectiveness of the proposed strategy in reducing voltage fluctuations and accompanied light flicker.

The previous sections demonstrated that EVs can provide many benefits and services to electric power systems. A very important secondary benefit that EVs can provide is helping the integration of RESs. It is well known that to face environmental challenges and depletion of fossil fuel, large capacities of RESs are being installed worldwide. Most of RESs installations are PV and wind due to the technology advancement and fast drop in their prices. Electricity generation from RESs like PV and wind are variable and intermittent depending on environmental conditions (i.e., sun and wind) and hard to predict or forecast. This increases fluctuation and uncertainty in power system besides load fluctuation [149]. Moreover, they are non-dispatchable generation and they cannot change the generated power following demand variations like traditional power plants, which increase the need for more flexibility in the power system which cannot be provided by central power plants only. More flexibility in the power system can be achieved from distribution systems by demand response (i.e., controllable or dispatchable loads) and energy storage.

Till now there is no economical utility-scale energy storage technology. EVs can act as controllable loads by controlled charging and as energy storage by V2G, V2B, and V2H. Several studies investigated how EVs can enable integrating more RESs [150]. EVs can absorb excess RESs generation and deliver it to the grid when needed. In [151], it was concluded that EVs and heat pumps can reduce the cost of integrating RESs, such as required balancing cost and required back-up generation cost. In [152], based on the German 2030 scenario of RESs penetration, it was found that by controlling EVs charging more RESs can be integrated into the power system. In [153], two case studies were considered for Germany and California 2030 scenario with a high share of EVs and RESs. It concluded that the smart charging of EVs can mitigate RESs fluctuation.

Reference [154], investigated the impact of V2G on two power systems and with different penetration of wind generation, from 0% to 100%. It was found that V2G can enable a higher share of wind power generation in the power system. An optimization strategy was tested in [155] to schedule the charging of EVs which were connected to microgrid with PV and wind generation. The proposed strategy achieved load leveling in addition to reducing microgrid operating costs and EVs charging costs. The impact of EVs operating at controlled charging mode or V2G mode on the operation of a power system with high share of RESs was examined in [156]. Results showed that controlled EVs charging and discharging reduced deployed reserves usage, better usage of renewable generation by reducing wind spillage, and reduced operation cost. In [157], it was found that controlled charging and discharging of EVs can reduce operation costs and reduce emissions, besides maximizing RESs utilization. In [158], an optimization algorithm was used to manage EVs charging and discharging to increase the penetration of RESs. The proposed strategy was able to increase PV penetration up to 50%.

6. Discussions and Future Research Directions

The main contribution of this article is to provide a review of potential negative impacts of EVs charging on electric power systems mainly due to uncontrolled charging, and how through proper charging and discharging methods, those impacts can be reduced and become even positive impacts as discussed in Sections 4 and 5. For each potential negative impact or positive impact of EVs, the findings and conclusions of a few research papers were discussed as examples of the research done at each point to clarify it. Figure 8 provides a summary of the negative impacts of uncontrolled EVs charging on the electric power system and how these impacts can be mitigated or even become positive impacts using suitable charging and discharging methods.

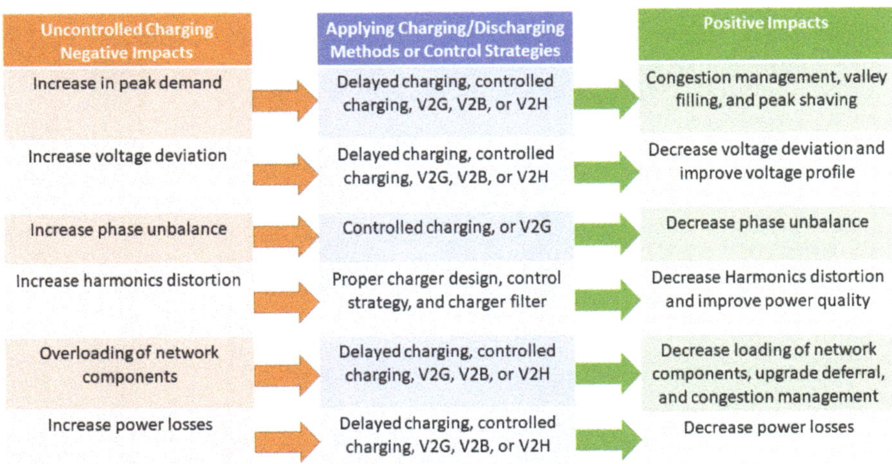

Figure 8. Uncontrolled charging negative impacts on the power system and how it can be mitigated and become positive impact using the proper charging/discharging method.

As described in previous sections, there is a lot of ongoing research on the integration of EVs to electric power system, focusing on accurate evaluation of the negative impacts of uncontrolled EV charging or the services and benefits EVs may provide if other smart charging and discharging methods are used. Other research areas that are being investigated by researchers are:

- **Cost–benefit analysis of different charging and discharging methods**

Most of the studies that investigate the benefits that smart charging and discharging methods can provide focus on the technical feasibility and charging cost. However, there is a need to assess the cost of providing these services on EV batteries degradation, because EV battery is a very expensive component in EVs. After that, the economic feasibility of different charging and discharging methods can be accurately estimated. Reference [159] executed a cost–benefit analysis of controlled charging and V2G implementation considering EV batteries degradation cost. The results showed that controlled charging is economical. Moreover, the study concluded that V2G implementation is not economical without wind generation presence, while it is economical in the presence of wind generation. This area requires more studies to evaluate the economic feasibility of different EV smart charging and discharging methods on different case studies and different scenarios. The different scenarios can consider the economic feasibility of using EVs to provide various services (i.e., frequency regulation, voltage regulation, congestion management, etc.; or consider different power systems with different characteristics (i.e., presence of different types of RESs, different penetration levels of RESs, types of power plants, etc.).

- **Coordination between transmission system operator (TSO) and distribution system operator (DSO) for providing EV services**

As explained in previous sections, EVs can provide many local and system-wide power and energy services. Figure 9 summarizes the services provided to different power system parties (i.e., transmission system operator (TSO), DSO, and loads such as buildings or homes) [160]. It is worth mentioning that the provision of system-wide services by EVs may result in issues at distribution system at which EVs are connected which may result in conflict of interests between TSO and DSO. For instance, the use of EVs to provide a TSO service such as frequency regulation which requires continuous change at charging and discharging power may cause negative impacts at the distribution system managed by DSO such as overloading of distribution network components, phase unbalance, etc. Therefore, there is a need for coordination between TSO and DSO to guarantee reliable and cost-efficient EVs based services [161,162]. This topic is rarely investigated in the literature and there many open questions that require study. Therefore, it requires more attention in future research.

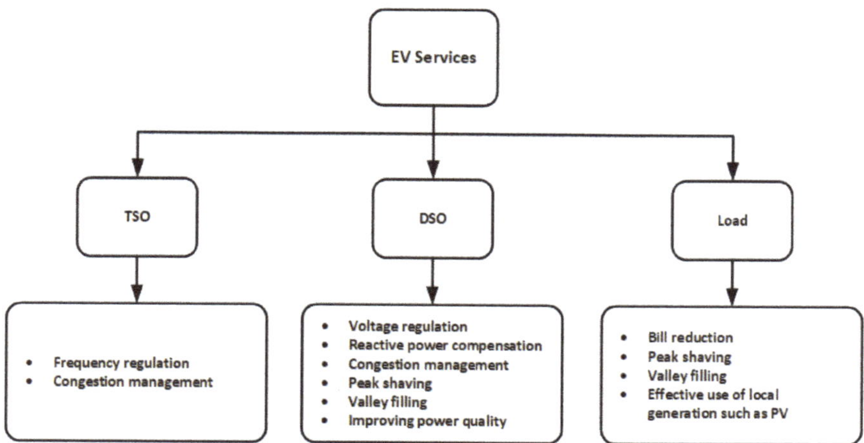

Figure 9. EV services provided to different power system parties.

- **Planning of public charging Infrastructure**

With the large-scale adoption of EVs in the near future, many public EV charging stations will be deployed in streets, highways, workplaces, shopping centers, etc. One of the main challenges for EVs acceptance is its limited range compared to ICEVs and driver range anxiety. Therefore, the EV chargers deployment must be planned accurately and efficiently to achieve both transportation and power system objectives and needs. Many studies investigated the planning of future EV charging infrastructure. The studies considered the optimal location and capacity of EV chargers [163]. Some studies considered the transportation network only without considering power system conditions. Other studies considered the power system only without considering transportation networks. These studies consider power system economic and operation constraints while reducing investments required for power system infrastructure upgrade. It is worth mentioning that EV charging stations couple both transportation network and power system; therefore, both must be taken into account at EV charging infrastructure planning. Few studies considered both transportation network and power system at the planning of EV chargers [164], because EV infrastructure planning studies require real data for both the transportation sector and power system, which varies between countries. There is a room for more research in this area considering different case studies.

7. Conclusions

The paper presented EVs technology and the current need for it as well as its benefits compared to traditional vehicles in addition to challenges it must tackle to achieve high adoption and social acceptance. Furthermore, it presented the current EVs market and future predictions. Different charging technologies were presented such as conductive charging, which is the current charging method and other charging methods such as wireless charging and battery swapping which may have future potential. A review of the negative impacts that EV may cause on electric power systems if uncontrolled EV charging is used was presented. Conclusions of many studies that assessed these impacts were discussed. All the studies showed that uncontrolled EV charging will result in unwanted negative impacts on the power system especially the distribution networks and it will foster the need for infrastructure upgrade. The severity of these impacts varies between different studies due to many uncertainties in EV charging impacts studies (e.g., distribution system status, EV battery capacity, EV battery SoC, time and location of charging, EV charger power rating, and EV penetration level).

The paper also showed that these impacts can be mitigated using delayed charging and controlled charging methods, which can benefit both the power system and EV owner. More benefits and electrical services can be obtained from EVs with advanced charging control methods like V2G, V2B, and V2H. Different electrical services that can be provided with controlled charging/discharging of EVs were presented and discussed. This review article briefly discussed the main research areas that are being investigated for the integration of EVs to sustainable future power system, which can be helpful for engineers and researchers. It further discussed a few interesting research topics that need more study in future research.

Author Contributions: M.N. wrote the manuscript first draft. J.P.C.-Á., G.M., and Á.S.-M. reviewed, edited, and suggested modifications to the manuscript. All authors have read and agreed to the published version of the manuscript.

Funding: This research received no external funding.

Conflicts of Interest: The authors declare no conflict of interest.

Abbreviations

List of abbreviations used in this paper.

AC	Alternating Current
BEV	Battery Electric Vehicles
BSS	Battery Swapping Stations
CO_2	Carbon Dioxide
DC	Direct Current
DG	Distributed Generation
DSO	Distribution System Operator

EM	Electric Motor
EV	Electric Vehicle
EVSE	Electric Vehicle Supply Equipment
FCEV	Fuel Cell Electric Vehicle
G2V	Grid to Vehicle
GHGs	Greenhouse gases
h	Hour
HEV	Hybrid Electric Vehicles
ICE	Internal Combustion Engine
ICEV	Internal Combustion Engine Vehicle
IEA	International Energy Agency
IEEE	Institute of Electrical and Electronics Engineers
IWC	Inductive Wireless Charging
km	Kilometer
kVA	Kilo volt ampere
kW	Kilowatt
kWh	Kilowatt hour
LFC	Load Frequency Control
Li-ion	Lithium-ion
OLTC	On-Load Tap Changer
PFC	Primary Frequency Control
PHEV	Plug-in Hybrid Electric Vehicles
PLDV	passenger light-duty vehicles
PV	Photovoltaic
RES	Renewable Energy Source
SoC	State of Charge
THD	Total Harmonic Distortion
THDi	Total Harmonic Distortion of current
THDv	Total Harmonic Distortion of Voltage
ToU	Time of Use
TSO	Transmission System Operator
V2B	Vehicle to Building
V2G	Vehicle to Grid
V2H	Vehicle to Home
VUF	Voltage Unbalance Factor
WC	Wireless Charging

References

1. UNFCCC; Conference of the Parties (COP). Adoption of the Paris Agreement Proposal by the President. In Proceedings of the Paris Climate Change Conference, Paris, France, 30 November–12 December 2015.
2. Labatt, S.; White, R.R. *Carbon Finance: The Financial Implications of Climate Change*; John Wiley & Sons: Hoboken, NJ, USA, 2007.
3. OECD. *World Energy Statistics 2017*; OECD: Paris, France, 2017.
4. Shaukat, N.; Khan, B.; Ali, S.; Mehmood, C.; Khan, J.; Farid, U.; Majid, M.; Anwar, S.M.; Jawad, M.; Ullah, Z. A survey on electric vehicle transportation within smart grid system. *Renew. Sustain. Energy Rev.* **2018**, *81*, 1329–1349. [CrossRef]
5. Hedegaard, K.; Ravn, H.; Juul, N.; Meibom, P. Effects of electric vehicles on power systems in Northern Europe. *Energy* **2012**, *48*, 356–368. [CrossRef]
6. Nanaki, E.A.; Koroneos, C.J. Comparative economic and environmental analysis of conventional, hybrid and electric vehicles—The case study of Greece. *J. Clean. Prod.* **2013**, *53*, 261–266. [CrossRef]
7. Agency, I.E. Global EV Outlook 2019 to electric mobility. *IEA Publ.* **2019**.
8. Knez, M.; Zevnik, G.K.; Obrecht, M. A review of available chargers for electric vehicles: United States of America, European Union, and Asia. *Renew. Sustain. Energy Rev.* **2019**, *109*, 284–293. [CrossRef]

9. Dubey, A.; Santoso, S. Electric Vehicle Charging on Residential Distribution Systems: Impacts and Mitigations. *IEEE Access* **2015**, *3*, 1871–1893. [CrossRef]
10. García-Villalobos, J.; Zamora, I.; Martin, J.S.; Asensio, F.J.; Aperribay, V. Plug-in electric vehicles in electric distribution networks: A review of smart charging approaches. *Renew. Sustain. Energy Rev.* **2014**, *38*, 717–731. [CrossRef]
11. Haus, B.; Mercorelli, P. Polynomial Augmented Extended Kalman Filter to Estimate the State of Charge of Lithium-Ion Batteries. *IEEE Trans. Veh. Technol.* **2020**, *69*, 1452–1463. [CrossRef]
12. Nour, M.; Said, S.M.; Ali, A.; Farkas, C. Smart Charging of Electric Vehicles According to Electricity Price. In Proceedings of the 2019 International Conference on Innovative Trends in Computer Engineering ITCE, Aswan, Egypt, 2–4 February 2019; pp. 432–437. [CrossRef]
13. Nafisi, H. Investigation on distribution transformer loss-of-life due to plug-in hybrid electric vehicles charging. *Int. J. Ambient. Energy* **2019**, 1–7. [CrossRef]
14. Nour, M.; Ramadan, H.; Ali, A.; Farkas, C. Impacts of plug-in electric vehicles charging on low voltage distribution network. In Proceedings of the 2018 International Conference on Innovative Trends in Computer Engineering (ITCE), Aswan, Egypt, 19–21 February 2018; pp. 357–362.
15. Tong, X.; Ma, Q.; Tang, K.; Liu, H.; Li, C. Influence of electric vehicle access mode on the static voltage stability margin and accommodated capacity of the distribution network. *J. Eng.* **2019**, *2019*, 2658–2662. [CrossRef]
16. Nour, M.; Ali, A.; Farkas, C. Mitigation of Electric Vehicles Charging Impacts on Distribution Network with Photovoltaic Generation. In Proceedings of the 2019 International Conference on Innovative Trends in Computer Engineering (ITCE), Aswan, Egypt, 19–21 February 2018; pp. 384–388.
17. Ul-Haq, A.; Azhar, M.; Mahmoud, Y.; Perwaiz, A.; Al-Ammar, E.A. Probabilistic Modeling of Electric Vehicle Charging Pattern Associated with Residential Load for Voltage Unbalance Assessment. *Energies* **2017**, *10*, 1351. [CrossRef]
18. Leou, R.-C.; Teng, J.-H.; Lu, H.-J.; Lan, B.-R.; Chen, H.-T.; Hsieh, T.-Y.; Su, C.-L. Stochastic analysis of electric transportation charging impacts on power quality of distribution systems. *IET Gener. Transm. Distrib.* **2018**, *12*, 2725–2734. [CrossRef]
19. Yong, J.Y.; Ramachandaramurthy, V.K.; Tan, K.M.; Mithulananthan, N. A review on the state-of-the-art technologies of electric vehicle, its impacts and prospects. *Renew. Sustain. Energy Rev.* **2015**, *49*, 365–385. [CrossRef]
20. Hota, A.R.; Juvvanapudi, M.; Bajpai, P. Issues and solution approaches in PHEV integration to the smart grid. *Renew. Sustain. Energy Rev.* **2014**, *30*, 217–229. [CrossRef]
21. Richardson, D.B. Electric vehicles and the electric grid: A review of modeling approaches, Impacts, and renewable energy integration. *Renew. Sustain. Energy Rev.* **2013**, *19*, 247–254. [CrossRef]
22. Talebizadeh, E.; Rashidinejad, M.; Abdollahi, A. Evaluation of plug-in electric vehicles impact on cost-based unit commitment. *J. Power Sources* **2014**, *248*, 545–552. [CrossRef]
23. Lyon, T.P.; Michelin, M.; Jongejan, A.; Leahy, T. Is "smart charging" policy for electric vehicles worthwhile? *Energy Policy* **2012**, *41*, 259–268. [CrossRef]
24. Weis, A.; Jaramillo, P.; Michalek, J.J. Estimating the potential of controlled plug-in hybrid electric vehicle charging to reduce operational and capacity expansion costs for electric power systems with high wind penetration. *Appl. Energy* **2014**, *115*, 190–204. [CrossRef]
25. Windecker, A.; Ruder, A. Fuel economy, cost, and greenhouse gas results for alternative fuel vehicles in 2011. *Transp. Res. Part. D Transp. Environ.* **2013**, *23*, 34–40. [CrossRef]
26. Karabasoglu, O.; Michalek, J.J. Influence of driving patterns on life cycle cost and emissions of hybrid and plug-in electric vehicle powertrains. *Energy Policy* **2013**, *60*, 445–461. [CrossRef]
27. Lunz, B.; Yan, Z.; Gerschler, J.B.; Sauer, D.U. Influence of plug-in hybrid electric vehicle charging strategies on charging and battery degradation costs. *Energy Policy* **2012**, *46*, 511–519. [CrossRef]
28. Gass, V.; Schmidt, J.; Schmid, E. Analysis of alternative policy instruments to promote electric vehicles in Austria. *Renew. Energy* **2014**, *61*, 96–101. [CrossRef]
29. Tie, S.F.; Tan, C.W. A review of energy sources and energy management system in electric vehicles. *Renew. Sustain. Energy Rev.* **2013**, *20*, 82–102. [CrossRef]
30. Un-Noor, F.; Padmanaban, S.; Mihet-Popa, L.; Mollah, M.N.; Hossain, E. A Comprehensive Study of Key Electric Vehicle (EV) Components, Technologies, Challenges, Impacts, and Future Direction of Development. *Energies* **2017**, *10*, 1217. [CrossRef]

31. Andwari, A.M.; Pesiridis, A.; Rajoo, S.; Martinez-Botas, R.; Esfahanian, V. A review of Battery Electric Vehicle technology and readiness levels. *Renew. Sustain. Energy Rev.* **2017**, *78*, 414–430. [CrossRef]
32. Singh, B.; Chandra, A.; Al-Haddad, K.; Pandey, A.; Kothari, D.P. A review of single-phase improved power quality ac~dc converters. *IEEE Trans. Ind. Electron.* **2003**, *50*, 962–981. [CrossRef]
33. Singh, B.; Singh, B.; Chandra, A.; Al-Haddad, K.; Pandey, A.; Kothari, D. A Review of Three-Phase Improved Power Quality AC–DC Converters. *IEEE Trans. Ind. Electron.* **2004**, *51*, 641–660. [CrossRef]
34. Yılmaz, M.; Krein, P.T.; Yilmaz, M. Review of charging power levels and infrastructure for plug-in electric and hybrid vehicles. In Proceedings of the 2012 IEEE International Electric Vehicle Conference (IEVC), Greenville, SC, USA, 4–8 March 2012; pp. 1–8. [CrossRef]
35. Young, K.; Wang, C.; Wang, L.Y.; Strunz, K. *Electric Vehicle Integration into Modern Power Networks*; Springer: Berlin/Heidelberg, Germany, 2013.
36. Botsford, C.; Szczepanek, A. Fast Charging vs. Slow Charging: Pros and cons for the New Age of Electric Vehicles. *Int. Battery Hybrid Fuel Cell Electr. Veh. Symp.* **2009**.
37. Yang, Y.; El Baghdadi, M.; Lan, Y.; Benomar, Y.; Van Mierlo, J.; Hegazy, O. Design Methodology, Modeling, and Comparative Study of Wireless Power Transfer Systems for Electric Vehicles. *Energies* **2018**, *11*, 1716. [CrossRef]
38. Cao, Y.; Ahmad, N.; Kaiwartya, O.; Puturs, G.; Khalid, M. Intelligent Transportation Systems Enabled ICT Framework for Electric Vehicle Charging in Smart City. In *Handbook of Smart Cities*; Springer Science and Business Media LLC: Berlin, Germany, 2018; pp. 311–330.
39. Wang, Z.; Wei, X.; Dai, H. Design and Control of a 3 kW Wireless Power Transfer System for Electric Vehicles. *Energies* **2015**, *9*, 10. [CrossRef]
40. Musavi, F.; Eberle, W. Overview of wireless power transfer technologies for electric vehicle battery charging. *IET Power Electron.* **2014**, *7*, 60–66. [CrossRef]
41. Li, S.; Mi, C.C. Wireless power transfer for electric vehicle applications. *IEEE J. Emerg. Sel. Top. Power Electron.* **2015**, *3*, 4–17.
42. Ahmad, A.; Khan, Z.A.; Alam, M.S.; Khateeb, S. A Review of the Electric Vehicle Charging Techniques, Standards, Progression and Evolution of EV Technologies in Germany. *Smart Sci.* **2017**, *6*, 36–53. [CrossRef]
43. Sarker, M.; Pandžić, H.; Ortega-Vazquez, M. Optimal operation and services scheduling for an electric vehicle battery swapping station. In Proceedings of the 2015 IEEE Power & Energy Society General Meeting, Denver, CO, USA, 26–30 July 2015.
44. Zheng, Y.; Dong, Z.; Xu, Y.; Meng, K.; Zhao, J.H.; Qiu, J. Electric Vehicle Battery Charging/Swap Stations in Distribution Systems: Comparison Study and Optimal Planning. *IEEE Trans. Power Syst.* **2013**, *29*, 221–229. [CrossRef]
45. Sarker, M.R.; Pandžić, H.; Ortega-Vazquez, M. Electric vehicle battery swapping station: Business case and optimization model. In Proceedings of the 2013 International Conference on Connected Vehicles and Expo (ICCVE), Las Vegas, NV, USA, 2–6 December 2013; pp. 289–294. [CrossRef]
46. Rao, R.; Zhang, X.; Xie, J.; Ju, L. Optimizing electric vehicle users' charging behavior in battery swapping mode. *Appl. Energy* **2015**, *155*, 547–559. [CrossRef]
47. Yang, S.; Yao, J.; Kang, T.; Zhu, X. Dynamic operation model of the battery swapping station for EV (electric vehicle) in electricity market. *Energy* **2014**, *65*, 544–549. [CrossRef]
48. Electric Car Maker Tesla Unveils 90-s Battery Pack Swap—Reuters. Available online: https://www.reuters.com/article/us-tesla-swap/electric-car-maker-tesla-unveils-90-s-battery-pack-swap-idUSBRE95K07H20130621 (accessed on 31 December 2019).
49. Shareef, H.; Islam, M.; Mohamed, A. A review of the stage-of-the-art charging technologies, placement methodologies, and impacts of electric vehicles. *Renew. Sustain. Energy Rev.* **2016**, *64*, 403–420. [CrossRef]
50. Alshahrani, S.; Khalid, M.; AlMuhaini, M.M. Electric Vehicles Beyond Energy Storage and Modern Power Networks: Challenges and Applications. *IEEE Access* **2019**, *7*, 99031–99064. [CrossRef]
51. Papathanassiou, S.; Nikos, H.; Kai, S. A benchmark low voltage microgrid network. In Proceedings of the CIGRE Symposium: Power Systems with Dispersed Generation, Athens, Greece, 13–16 April 2005; pp. 1–8.
52. Shao, S.; Zhang, T.; Pipattanasompom, M.; Rahman, S. Impact of TOU rates on distribution load shapes in a smart grid with PHEV penetration. *IEEE PES T&D 2010* **2010**, 1–6. [CrossRef]

53. Gao, Y.; Wang, C.; Wang, Z.; Liang, H. Research on time-of-use price applying to electric vehicles charging. In Proceedings of the IEEE PES Innovative Smart Grid Technologies, Berlin, Germany, 14–17 October 2012; pp. 1–6.
54. Tan, K.M.; Ramachandaramurthy, V.K.; Yong, J.Y. Integration of electric vehicles in smart grid: A review on vehicle to grid technologies and optimization techniques. *Renew. Sustain. Energy Rev.* **2016**, *53*, 720–732. [CrossRef]
55. Rahman, I.; Vasant, P.M.; Singh, M.S.B.; Abdullah-Al-Wadud, M.; Adnan, N. Review of recent trends in optimization techniques for plug-in hybrid, and electric vehicle charging infrastructures. *Renew. Sustain. Energy Rev.* **2016**, *58*, 1039–1047. [CrossRef]
56. Faddel, S.; Al-Awami, A.T.; Mohammed, O.A. Charge Control and Operation of Electric Vehicles in Power Grids: A Review. *Energies* **2018**, *11*, 701. [CrossRef]
57. Faddel, S.; Mohamed, A.A.S.; Mohammed, O.A. Fuzzy logic-based autonomous controller for electric vehicles charging under different conditions in residential distribution systems. *Electr. Power Syst. Res.* **2017**, *148*, 48–58. [CrossRef]
58. Al-Awami, A.T.; Sortomme, E.; Akhtar, G.M.A.; Faddel, S. A Voltage-Based Controller for an Electric-Vehicle Charger. *IEEE Trans. Veh. Technol.* **2015**, *65*, 4185–4196. [CrossRef]
59. Saldaña, G.; Martin, J.S.; Zamora, I.; Asensio, F.J.; Oñederra, O. Electric Vehicle into the Grid: Charging Methodologies Aimed at Providing Ancillary Services Considering Battery Degradation. *Energies* **2019**, *12*, 2443. [CrossRef]
60. Habib, S.; Kamran, M.; Rashid, U. Impact analysis of vehicle-to-grid technology and charging strategies of electric vehicles on distribution networks—A review. *J. Power Sources* **2015**, *277*, 205–214. [CrossRef]
61. Thompson, A. Economic implications of lithium ion battery degradation for Vehicle-to-Grid (V2X) services. *J. Power Sources* **2018**, *396*, 691–709. [CrossRef]
62. Pang, C.; Dutta, P.; Kezunovic, M. BEVs/PHEVs as Dispersed Energy Storage for V2B Uses in the Smart Grid. *IEEE Trans. Smart Grid* **2011**, *3*, 473–482. [CrossRef]
63. Shin, H.; Baldick, R. Plug-In Electric Vehicle to Home (V2H) Operation Under a Grid Outage. *IEEE Trans. Smart Grid* **2017**, *8*, 2032–2041. [CrossRef]
64. Alahyari, A.; Fotuhi-Firuzabad, M.; Rastegar, M. Incorporating Customer Reliability Cost in PEV Charge Scheduling Schemes Considering Vehicle to Home Capability. *IEEE Trans. Veh. Technol.* **2014**, *64*, 1. [CrossRef]
65. Habib, S.; Khan, M.M.; Abbas, F.; Sang, L.; Shahid, M.U.; Tang, H. A Comprehensive Study of Implemented International Standards, Technical Challenges, Impacts and Prospects for Electric Vehicles. *IEEE Access* **2018**, *6*, 13866–13890. [CrossRef]
66. Bohn, S.; Feustel, R.; Agsten, M. MC-based Risk Analysis on the Capacity of Distribution Grids to Charge PEVs on 3-ph 0.4-kV Distribution Grids Considering Time and Location Uncertainties. *SAE Int. J. Passeng. Cars Electron. Electr. Syst.* **2015**, *8*, 394–400. [CrossRef]
67. Ahmadian, A.; Mohammadi-Ivatloo, B.; Elkamel, A. A Review on Plug-in Electric Vehicles: Introduction, Current Status, and Load Modeling Techniques. *J. Mod. Power Syst. Clean Energy* **2020**, *8*, 412–425. [CrossRef]
68. Wang, Z.; Paranjape, R. An Evaluation of Electric Vehicle Penetration under Demand Response in a Multi-Agent Based Simulation. In Proceedings of the 2014 IEEE Electrical Power and Energy Conference, Calgary, AB, Canada, 12–14 November 2014; pp. 220–225.
69. Putrus, G.A.; Suwanapingkarl, P.; Johnston, D.; Bentley, E.C.; Narayana, M. Impact of electric vehicles on power distribution networks. In Proceedings of the 2009 IEEE Vehicle Power and Propulsion Conference, Dearborn, MI, USA, 7–10 September 2009; pp. 827–831.
70. McCarthy, D.; Wolfs, P. The HV system impacts of large scale electric vehicle deployments in a metropolitan area. In Proceedings of the AUPEC 2010—20th Australasian Universities Power Engineering Conference 2010, Christchurch, New Zealand, 5 December 2010.
71. Schneider, K.P.; Gerkensmeyer, C.; Kintner-Meyer, M.; Fletcher, R. Impact assessment of plug-in hybrid vehicles on pacific northwest distribution systems. In Proceedings of the 2008 IEEE Power and Energy Society General Meeting—Conversion and Delivery of Electrical Energy in the 21st Century, Pittsburgh, PA, USA, 20–24 July 2008; pp. 1–6. [CrossRef]
72. Kelly, L.; Rowe, A.; Wild, P. Analyzing the impacts of plug-in electric vehicles on distribution networks in British Columbia. In Proceedings of the 2011 IEEE Electrical Power and Energy Conference, Montreal, QC, Canada, 22–23 October 2009; Institute of Electrical and Electronics Engineers (IEEE); pp. 1–6.

73. Shao, S.; Pipattanasomporn, M.; Rahman, S. Grid Integration of Electric Vehicles and Demand Response with Customer Choice. *IEEE Trans. Smart Grid* **2012**, *3*, 543–550. [CrossRef]
74. Crozier, C.; Morstyn, T.; McCulloch, M. The opportunity for smart charging to mitigate the impact of electric vehicles on transmission and distribution systems. *Appl. Energy* **2020**, *268*, 114973. [CrossRef]
75. Kristoffersen, T.K.; Capion, K.; Meibom, P. Optimal charging of electric drive vehicles in a market environment. *Appl. Energy* **2011**, *88*, 1940–1948. [CrossRef]
76. Denholm, P.; Short, W. *Evaluation of Utility System Impacts and Benefits of Optimally Dispatched Plug-In Hybrid Electric Vehicles*; NREL: Golden, CO, USA, 2006.
77. Drovtar, I.; Rosin, A.; Landsberg, M.; Kilter, J. Large scale electric vehicle integration and its impact on the Estonian power system. In Proceedings of the 2013 IEEE Grenoble Conference, Grenoble, France, 16–20 June 2013; Institute of Electrical and Electronics Engineers (IEEE); pp. 1–6.
78. Van Cutsem, T. Voltage instability: Phenomena, countermeasures, and analysis methods. *Proc. IEEE* **2000**, *88*, 208–227. [CrossRef]
79. Dharmakeerthi, C.; Mithulananthan, N.; Saha, T.K. Impact of electric vehicle fast charging on power system voltage stability. *Int. J. Electr. Power Energy Syst.* **2014**, *57*, 241–249. [CrossRef]
80. Clement-Nyns, K.; Haesen, E.; Driesen, J. The Impact of Charging Plug-In Hybrid Electric Vehicles on a Residential Distribution Grid. *IEEE Trans. Power Syst.* **2009**, *25*, 371–380. [CrossRef]
81. Gray, M.K.; Morsi, W.G. Power Quality Assessment in Distribution Systems Embedded With Plug-In Hybrid and Battery Electric Vehicles. *IEEE Trans. Power Syst.* **2014**, *30*, 663–671. [CrossRef]
82. Xiong, J.; Zhang, K.; Liu, X.; Su, W. Investigating the impact of plug-in electric vehicle charging on power distribution systems with the integrated modeling and simulation of transportation network. In Proceedings of the ITEC Asia-Pacific 2014—IEEE Transportation Electrification Conference & Expo 2014, Beijing, China, 31 August–3 September 2014; pp. 1–5. [CrossRef]
83. Soroudi, A.; Keane, A. *Plug in Electric Vehicles in Smart Grids*; Springer: Berlin/Heidelberg, Germany, 2015; p. 89.
84. Ul-Haq, A.; Cecati, C.; Strunz, K.; Abbasi, E. Impact of Electric Vehicle Charging on Voltage Unbalance in an Urban Distribution Network. *Intell. Ind. Syst.* **2015**, *1*, 51–60. [CrossRef]
85. Shahnia, F.; Ghosh, A.; Ledwich, G.; Zare, F. Predicting Voltage Unbalance Impacts of Plug-in Electric Vehicles Penetration in Residential Low-voltage Distribution Networks. *Electr. Power Compon. Syst.* **2013**, *41*, 1594–1616. [CrossRef]
86. Richardson, P.; Flynn, D.; Keane, A. Impact assessment of varying penetrations of electric vehicles on low voltage distribution systems. In Proceedings of the 2008 IEEE Power and Energy Society General Meeting—Conversion and Delivery of Electrical Energy in the 21st Century, Providence, RI, USA, 25–29 July 2010; pp. 1–6. [CrossRef]
87. Liu, R.; Dow, L.; Liu, E. A survey of PEV impacts on electric utilities. In Proceedings of the Innovative Smart Grid Technologies (ISGT), 2011 IEEE PES, Perth, Australia, 13–16 November 2011.
88. Jiang, C.; Torquato, R.; Salles, D.; Xu, W. Method to assess the power quality impact of plug-in electric vehicles. In Proceedings of the 2014 16th International Conference on Harmonics and Quality of Power, Bucharest, Romania, 25–28 May 2014; pp. 177–180. [CrossRef]
89. Bass, R.; Harley, R.; Lambert, F.; Rajasekaran, V.; Pierce, J. Residential harmonic loads and EV charging. In Proceedings of the 2001 IEEE Power Engineering Society Winter Meeting Conference Proceedings (Cat No 01CH37194), Columbus, OH, USA, 28 January–1 February 2001; pp. 803–808. [CrossRef]
90. Nguyen, V.-L.; Tran-Quoc, T.; Bacha, S. Harmonic distortion mitigation for electric vehicle fast charging systems. In Proceedings of the 2013 IEEE Grenoble Conference, Grenoble, France, 16–20 June 2013; pp. 1–6.
91. Melo, N.; Mira, F.; De Almeida, A.; Delgado, J. Integration of PEV in Portuguese distribution grid: Analysis of harmonic current emissions in charging points. In Proceedings of the 11th International Conference on Electrical Power Quality and Utilisation, Lisbon, Portugal, 17–19 October 2011; pp. 1–6. [CrossRef]
92. Monteiro, V.; Gonçalves, H.; Afonso, J.L. Impact of Electric Vehicles on power quality in a Smart Grid context. In Proceedings of the 11th International Conference on Electrical Power Quality and Utilisation, Lisbon, Portugal, 17–19 October 2011; pp. 1–6. [CrossRef]
93. Gomez, J.; Morcos, M. Impact of EV battery chargers on the power quality of distribution systems. *IEEE Trans. Power Deliv.* **2003**, *18*, 975–981. [CrossRef]

94. Zhou, C.; Wang, H.; Zhou, W.; Qian, K.; Meng, S. Determination of maximum level of EV penetration with consideration of EV charging load and harmonic currents. *IOP Conf. Ser. Earth Environ. Sci.* **2019**, *342*, 012010. [CrossRef]
95. Hilshey, A.D.; Rezaei, P.; Hines, P.D.H.; Frolik, J. Electric vehicle charging: Transformer impacts and smart, decentralized solutions. In Proceedings of the 2012 IEEE Power Energy Society General Meeting, San Diego, CA, USA, 22–26 July 2012; pp. 1–8. [CrossRef]
96. Papadopoulos, P.; Skarvelis-Kazakos, S.; Grau, I.; Awad, B.; Cipcigan, L.M.; Jenkins, N. Impact of residential charging of electric vehicles on distribution networks, a probabilistic approach. In Proceedings of the 45th International Universities Power Engineering Conference, Cardiff, UK, 31 August–3 September 2010; pp. 1–5.
97. Verzijlbergh, R.A.; Lukszo, Z.; Slootweg, J.G.; Ilic, M.D. The impact of controlled electric vehicle charging on residential low voltage networks. In Proceedings of the 2011 International Conference on Networking, Sensing and Control, Delft, The Netherlands, 11–13 April 2011; pp. 14–19. [CrossRef]
98. Turker, H.; Chatroux, D.; Hably, A.; Bacha, S. Low-Voltage Transformer Loss-of-Life Assessments for a High Penetration of Plug-In Hybrid Electric Vehicles (PHEVs). *IEEE Trans. Power Deliv.* **2012**, *27*, 1323–1331. [CrossRef]
99. Hilshey, A.; Hines, P.; Rezai, P.; Dowds, J. Estimating the impact of electric vehicle smart charging on distribution transformer aging. In Proceedings of the 2013 IEEE Power Energy Society General Meeting, Vancouver, BC, Canada, 21–25 July 2013; p. 1. [CrossRef]
100. Razeghi, G.; Zhang, L.; Brown, T.; Samuelsen, S. Impacts of plug-in hybrid electric vehicles on a residential transformer using stochastic and empirical analysis. *J. Power Sources* **2014**, *252*, 277–285. [CrossRef]
101. Yan, Q.; Kezunovic, M. Impact analysis of Electric Vehicle charging on distribution system. In Proceedings of the 2012 North American Power Symposium (NAPS), Champaign, IL, USA, 9-11 September 2012; pp. 1–6.
102. Akhavan-Rezai, E.; Shaaban, M.F.; El-Saadany, E.F.; Zidan, A. Uncoordinated charging impacts of electric vehicles on electric distribution grids: Normal and fast charging comparison. In Proceedings of the 2012 IEEE Power and Energy Society General Meeting, San Diego, CA, USA, 22–26 July 2012; pp. 1–7.
103. Fernandez, L.P.; GomezSanRoman, T.; Cossent, R.; Domingo, C.M.; Frias, P. Assessment of the Impact of Plug-in Electric Vehicles on Distribution Networks. *IEEE Trans. Power Syst.* **2010**, *26*, 206–213. [CrossRef]
104. Pillai, J.R.; Bak-Jensen, B. Impacts of electric vehicle loads on power distribution systems. In Proceedings of the 2010 IEEE Vehicle Power and Propulsion Conference, Lille, France, 1–3 September 2010; pp. 1–6.
105. Masoum, M.A.; Moses, P.S.; Smedley, K.M. Distribution transformer losses and performance in smart grids with residential Plug-In Electric Vehicles. In Proceedings of the ISGT 2011, Anaheim, CA, USA, 17–19 January 2011.
106. Deb, S.; Tammi, K.; Kalita, K.; Mahanta, P. Impact of Electric Vehicle Charging Station Load on Distribution Network. *Energies* **2018**, *11*, 178. [CrossRef]
107. Chang, M.; Bae, S.; Yoon, G.-G.; Park, S.-H.; Choy, Y. Impact of Electric Vehicle Charging Demand on a Jeju Island Radial Distribution Network. In Proceedings of the 2019 IEEE Power & Energy Society Innovative Smart Grid Technologies Conference (ISGT), Bucharest, Romania, 29 September–2 October 2019; pp. 1–5.
108. Tomić, J.; Kempton, W. Using fleets of electric-drive vehicles for grid support. *J. Power Sources* **2007**, *168*, 459–468. [CrossRef]
109. Pavić, I.; Capuder, T.; Kuzle, I. Value of flexible electric vehicles in providing spinning reserve services. *Appl. Energy* **2015**, *157*, 60–74. [CrossRef]
110. Izadkhast, S.D. Aggregation of Plug-in Electric Vehicles in Power Systems for Primary Frequency Control. *Delft Univ. Technol. Comillas Pontif. Univ. KTH R. Inst.* **2017**, 240. [CrossRef]
111. Yang, Z.; Li, K.; Foley, A.; Guo, Y. Computational scheduling methods for integrating plug-in electric vehicles with power systems: A review. *Renew. Sustain. Energy Rev.* **2015**, *51*, 396–416. [CrossRef]
112. Prabha, K. *Power System Stability and Control*; McGraw-Hill Education: New York, NY, USA, 1994; p. 1176.
113. Babahajiani, P.; Shafiee, Q.; Bevrani, H. Intelligent Demand Response Contribution in Frequency Control of Multi-Area Power Systems. *IEEE Trans. Smart Grid* **2016**, *9*, 1282–1291. [CrossRef]
114. Ulbig, A.; Borsche, T.; Andersson, G. *Impact of Low Rotational Inertia on Power System Stability and Operation*; Elsevier: Amsterdam, The Netherlands, 2014; Volume 47, pp. 7290–7297.
115. Rodrigues, E.M.G.; Osório, G.; Godina, R.; Bizuayehu, A.; Lujano-Rojas, J.; Catalão, J. Grid code reinforcements for deeper renewable generation in insular energy systems. *Renew. Sustain. Energy Rev.* **2016**, *53*, 163–177. [CrossRef]

116. Cabrera-Tobar, A.; Bullich-Massagué, E.; Aragues-Penalba, M.; Gomis-Bellmunt, O. Review of advanced grid requirements for the integration of large scale photovoltaic power plants in the transmission system. *Renew. Sustain. Energy Rev.* **2016**, *62*, 971–987. [CrossRef]
117. Marinelli, M.; Martinenas, S.; Knezović, K.; Andersen, P.B. Validating a centralized approach to primary frequency control with series-produced electric vehicles. *J. Energy Storage* **2016**, *7*, 63–73. [CrossRef]
118. Mu, Y.; Wu, J.; Ekanayake, J.; Jenkins, N.; Jia, H. Primary Frequency Response From Electric Vehicles in the Great Britain Power System. *IEEE Trans. Smart Grid* **2012**, *4*, 1142–1150. [CrossRef]
119. Wang, L.; Chen, B. Dual-level consensus-based frequency regulation using vehicle-to-grid service. *Electr. Power Syst. Res.* **2019**, *167*, 261–276. [CrossRef]
120. Almeida, P.R.; Soares, F.J.; Lopes, J.P. Electric vehicles contribution for frequency control with inertial emulation. *Electr. Power Syst. Res.* **2015**, *127*, 141–150. [CrossRef]
121. Meng, J.; Mu, Y.; Jia, H.; Wu, J.; Yu, X.; Qu, B. Dynamic frequency response from electric vehicles considering travelling behavior in the Great Britain power system. *Appl. Energy* **2016**, *162*, 966–979. [CrossRef]
122. Jia, H.; Li, X.; Mu, Y.; Xu, C.; Jiang, Y.; Yu, X.; Wu, J.; Dong, C. Coordinated control for EV aggregators and power plants in frequency regulation considering time-varying delays. *Appl. Energy* **2018**, *210*, 1363–1376. [CrossRef]
123. Vahedipour-Dahraie, M.; Rashidizadeh-Kermani, H.; Ghamsari-Yazdel, M.; Khaloie, H.; Guerrero, J.M. Coordination of EVs Participation for Load Frequency Control in Isolated Microgrids. *Appl. Sci.* **2017**, *7*, 539. [CrossRef]
124. Debbarma, S.; Dutta, A. Utilizing Electric Vehicles for LFC in Restructured Power Systems Using Fractional Order Controller. *IEEE Trans. Smart Grid* **2016**, *8*, 2554–2564. [CrossRef]
125. Hajforoosh, S.; Masoum, M.A.; Islam, S. Online optimal variable charge-rate coordination of plug-in electric vehicles to maximize customer satisfaction and improve grid performance. *Electr. Power Syst. Res.* **2016**, *141*, 407–420. [CrossRef]
126. Nour, M.; Said, S.M.; Ramadan, H.; Ali, A.; Farkas, C. Control of Electric Vehicles Charging Without Communication Infrastructure. In Proceedings of the 2018 Twentieth International Middle East Power Systems Conference (MEPCON), Cairo, Egypt, 18–20 December 2018; pp. 773–778.
127. Ramadan, H.; Ali, A.; Nour, M.; Farkas, C. Smart Charging and Discharging of Plug-in Electric Vehicles for Peak Shaving and Valley Filling of the Grid Power. In Proceedings of the 2018 Twentieth International Middle East Power Systems Conference (MEPCON), Cairo, Egypt, 18–20 December 2018; pp. 735–739.
128. Marra, F.; Yang, G.Y.; Fawzy, Y.T.; Traeholt, C.; Larsen, E.; Garcia-Valle, R.; Jensen, M.M. Improvement of Local Voltage in Feeders with Photovoltaic Using Electric Vehicles. *IEEE Trans. Power Syst.* **2013**, *28*, 3515–3516. [CrossRef]
129. Kisacikoglu, M.C.; Ozpineci, B.; Tolbert, L.M. Examination of a PHEV bidirectional charger system for V2G reactive power compensation. In Proceedings of the 2010 Twenty-Fifth Annual IEEE Applied Power Electronics Conference and Exposition (APEC) 2010, Palm Springs, CA, USA, 21–25 February 2010; pp. 458–465. [CrossRef]
130. Yong, J.Y.; Ramachandaramurthy, V.K.; Tan, K.M.; Mithulananthan, N. Bi-directional electric vehicle fast charging station with novel reactive power compensation for voltage regulation. *Int. J. Electr. Power Energy Syst.* **2015**, *64*, 300–310. [CrossRef]
131. Azzouz, M.A.; Shaaban, M.F.; El-Saadany, E.F. Real-Time Optimal Voltage Regulation for Distribution Networks Incorporating High Penetration of PEVs. *IEEE Trans. Power Syst.* **2015**, *30*, 1–12. [CrossRef]
132. Nour, M.; Ali, A.; Farkas, C. Evaluation of Electric Vehicles Charging Impacts on A Real Low Voltage Grid. *Int. J. Power Eng. Energy* **2018**, 1–6.
133. Zhang, K.; Xu, L.; Ouyang, M.; Wang, H.; Lu, L.; Li, J.; Li, Z. Optimal decentralized valley-filling charging strategy for electric vehicles. *Energy Convers. Manag.* **2014**, *78*, 537–550. [CrossRef]
134. Su, J.; Lie, T.; Zamora, R. Modelling of large-scale electric vehicles charging demand: A New Zealand case study. *Electr. Power Syst. Res.* **2019**, *167*, 171–182. [CrossRef]
135. Staudt, P.; Schmidt, M.; Gärttner, J.; Weinhardt, C. A decentralized approach towards resolving transmission grid congestion in Germany using vehicle-to-grid technology. *Appl. Energy* **2018**, *230*, 1435–1446. [CrossRef]
136. Wang, Z. Grid Power Peak Shaving and Valley Filling Using Vehicle-to-Grid Systems. *IEEE Trans. Power Deliv.* **2013**, *28*, 1822–1829. [CrossRef]

137. Colmenar-Santos, A.; Linares-Mena, A.-R.; Borge-Diez, D.; Quinto-Alemany, C.-D. Impact assessment of electric vehicles on islands grids: A case study for Tenerife (Spain). *Energy* **2017**, *120*, 385–396. [CrossRef]
138. Dagdougui, H.; Ouammi, A.; Dessaint, L.A. Peak Load Reduction in a Smart Building Integrating Microgrid and V2B-Based Demand Response Scheme. *IEEE Syst. J.* **2019**, *13*, 3274–3282. [CrossRef]
139. Martinenas, S.; Knezović, K.; Marinelli, M. Management of Power Quality Issues in Low Voltage Networks using Electric Vehicles: Experimental Validation. *IEEE Trans. Power Deliv.* **2016**, *32*, 1. [CrossRef]
140. Akhavan-Rezai, E.; Shaaban, M.F.; El-Saadany, E.F.; Karray, F. Managing Demand for Plug-in Electric Vehicles in Unbalanced LV Systems with Photovoltaics. *IEEE Trans. Ind. Inform.* **2017**, *13*, 1057–1067. [CrossRef]
141. Farahani, H.F. Improving voltage unbalance of low-voltage distribution networks using plug-in electric vehicles. *J. Clean. Prod.* **2017**, *148*, 336–346. [CrossRef]
142. Jiménez, A.; García, N. Voltage unbalance analysis of distribution systems using a three-phase power flow ans a Genetic Algorithm for PEV fleets scheduling. In Proceedings of the 2012 IEEE Power Energy Society General Meeting 2012, San Diego, CA, USA, 22–26 July 2012; pp. 1–8. [CrossRef]
143. Jabalameli, N.; Su, X.; Ghosh, A. Online Centralized Charging Coordination of PEVs With Decentralized Var Discharging for Mitigation of Voltage Unbalance. *IEEE Power Energy Technol. Syst. J.* **2019**, *6*, 152–161. [CrossRef]
144. Gallardo-Lozano, J.; Milanés-Montero, M.I.; Guerrero-Martínez, M.A.; Romero-Cadaval, E. Electric vehicle battery charger for smart grids. *Electr. Power Syst. Res.* **2012**, *90*, 18–29. [CrossRef]
145. Rauchfuss, L.; Foulquier, J.; Werner, R. Charging station as an active filter for harmonics compensation of smart grid. In Proceedings of the 2014 16th International Conference on Harmonics and Quality of Power (ICHQP), Bucharest, Romania, 5–28 May 2014; Institute of Electrical and Electronics Engineers (IEEE); pp. 181–184.
146. Ali, A.; Raisz, D.; Mahmoud, K. Voltage fluctuation smoothing in distribution systems with RES considering degradation and charging plan of EV batteries. *Electr. Power Syst. Res.* **2019**, *176*, 105933. [CrossRef]
147. García-Villalobos, J.; Zamora, I.; Knezović, K.; Marinelli, M. Multi-objective optimization control of plug-in electric vehicles in low voltage distribution networks. *Appl. Energy* **2016**, *180*, 155–168. [CrossRef]
148. Brinkel, N.B.G.; Gerritsma, M.K.; AlSkaif, T.A.; Lampropoulos, I.; van Voorden, A.M.; Fidder, H.A.; van Sark, W.G.J.H.M. Impact of rapid PV fluctuations on power quality in the low-voltage grid and mitigation strategies using electric vehicles. *Int. J. Electr. Power Energy Syst.* **2020**, *118*, 105741. [CrossRef]
149. Magdy, G.; Nour, M.; Shabib, G.; Elbaset, A.A.; Mitani, Y. Supplementary Frequency Control in a High-penetration Real Power System by Renewables Using SMES Application. *J. Electr. Syst.* **2019**, *15*, 526–538.
150. Liu, L.; Kong, F.; Liu, X.; Peng, Y.; Wang, Q. A review on electric vehicles interacting with renewable energy in smart grid. *Renew. Sustain. Energy Rev.* **2015**, *51*, 648–661. [CrossRef]
151. Teng, F.; Aunedi, M.; Strbac, G. Benefits of flexibility from smart electrified transportation and heating in the future UK electricity system. *Appl. Energy* **2016**, *167*, 420–431. [CrossRef]
152. Dallinger, D.; Wietschel, M. Grid integration of intermittent renewable energy sources using price-responsive plug-in electric vehicles. *Renew. Sustain. Energy Rev.* **2012**, *16*, 3370–3382. [CrossRef]
153. Dallinger, D.; Gerda, S.; Wietschel, M. Integration of intermittent renewable power supply using grid-connected vehicles—A 2030 case study for California and Germany. *Appl. Energy* **2013**, *104*, 666–682. [CrossRef]
154. Lund, H.; Kempton, W. Integration of renewable energy into the transport and electricity sectors through V2G. *Energy Policy* **2008**, *36*, 3578–3587. [CrossRef]
155. Li, Y.; Li, K. Incorporating Demand Response of Electric Vehicles in Scheduling of Isolated Microgrids With Renewables Using a Bi-Level Programming Approach. *IEEE Access* **2019**, *7*, 116256–116266. [CrossRef]
156. Carrión, M.; Zárate-Miñano, R. Operation of renewable-dominated power systems with a significant penetration of plug-in electric vehicles. *Energy* **2015**, *90*, 827–835. [CrossRef]
157. Saber, A.Y.; Venayagamoorthy, G.K.; Member, S. Plug-in Vehicles and Renewable Energy Sources for Cost and Emission Reductions Plug-in Vehicles and Renewable Energy Sources for Cost and Emission Reductions. *IEEE Trans. Ind. Electron.* **2016**, *58*, 1229–1238. [CrossRef]
158. Kordkheili, R.A.; Pourmousavi, S.A.; Savaghebi, M.; Guerrero, J.M.; Nehrir, M.H. Assessing the Potential of Plug-in Electric Vehicles in Active Distribution Networks. *Energies* **2016**, *9*, 34. [CrossRef]

159. Ahmadian, A.; Sedghi, M.; Mohammadi-Ivatloo, B.; Elkamel, A.; Golkar, M.A.; Fowler, M. Cost-Benefit Analysis of V2G Implementation in Distribution Networks Considering PEVs Battery Degradation. *IEEE Trans. Sustain. Energy* **2017**, *9*, 961–970. [CrossRef]
160. Arias, N.B.; Hashemi, S.; Andersen, P.B.; Traeholt, C.; Romero, R. Distribution System Services Provided by Electric Vehicles: Recent Status, Challenges, and Future Prospects. *IEEE Trans. Intell. Transp. Syst.* **2019**, *20*, 4277–4296. [CrossRef]
161. Zecchino, A.; Knezović, K.; Marinelli, M. Identification of conflicts between transmission and distribution system operators when acquiring ancillary services from electric vehicles. In Proceedings of the 2017 IEEE PES Innovative Smart Grid Technologies Conference Europe (ISGT-Europe), Torino, Italy, 26–29 September 2017.
162. Gerard, H.; Puente, E.I.R.; Six, D. Coordination between transmission and distribution system operators in the electricity sector: A conceptual framework. *Util. Policy* **2018**, *50*, 40–48. [CrossRef]
163. Zeb, M.Z.; Imran, K.; Khattak, A.; Janjua, A.K.; Pal, A.; Nadeem, M.; Zhang, J.; Khan, S. Optimal Placement of Electric Vehicle Charging Stations in the Active Distribution Network. *IEEE Access* **2020**, *8*, 68124–68134. [CrossRef]
164. Zhang, H.; Moura, S.J.; Hu, Z.; Song, Y. PEV Fast-Charging Station Siting and Sizing on Coupled Transportation and Power Networks. *IEEE Trans. Smart Grid* **2018**, *9*, 2595–2605. [CrossRef]

MDPI
St. Alban-Anlage 66
4052 Basel
Switzerland
Tel. +41 61 683 77 34
Fax +41 61 302 89 18
www.mdpi.com

Energies Editorial Office
E-mail: energies@mdpi.com
www.mdpi.com/journal/energies

www.ingramcontent.com/pod-product-compliance
Lightning Source LLC
LaVergne TN
LVHW071258170726
843515LV00006BA/1435

* 9 7 8 3 0 3 9 4 3 4 8 7 9 *